T0221718

Ricci Solitons in Low Dimensions

GRADUATE STUDIES
IN MATHEMATICS **235**

Ricci Solitons in Low Dimensions

Bennett Chow

AMERICAN
MATHEMATICAL
SOCIETY
Providence, Rhode Island

EDITORIAL COMMITTEE

Matthew Baker
Marco Gualtieri
Gigliola Staffilani (Chair)
Jeff A. Viaclovsky
Rachel Ward

2020 *Mathematics Subject Classification.* Primary 53E20, 53E10, 53E30, 58J05, 58J35, 58J60, 57K30, 57M50, 30F20, 30F45.

For additional information and updates on this book, visit
www.ams.org/bookpages/gsm-235

Library of Congress Cataloging-in-Publication Data

Cataloging-in-Publication Data has been applied for by the AMS.
See http://www.loc.gov/publish/cip/.
DOI: https://doi.org/10.1090/gsm/235

Copying and reprinting. Individual readers of this publication, and nonprofit libraries acting for them, are permitted to make fair use of the material, such as to copy select pages for use in teaching or research. Permission is granted to quote brief passages from this publication in reviews, provided the customary acknowledgment of the source is given.

Republication, systematic copying, or multiple reproduction of any material in this publication is permitted only under license from the American Mathematical Society. Requests for permission to reuse portions of AMS publication content are handled by the Copyright Clearance Center. For more information, please visit www.ams.org/publications/pubpermissions.

Send requests for translation rights and licensed reprints to reprint-permission@ams.org.

© 2023 by the author. All rights reserved.
Printed in the United States of America.

∞ The paper used in this book is acid-free and falls within the guidelines
established to ensure permanence and durability.
Visit the AMS home page at https://www.ams.org/

10 9 8 7 6 5 4 3 2 1   28 27 26 25 24 23

Contents

Preface

It can't be that cold, the ground is still warm to touch
– From "Red Rain" by Peter Gabriel

The Ricci flow was invented by Hamilton. Initially he used it to classify
closed 3-manifolds with positive Ricci curvature [**173**]. Then he considered mainly dimensions at most 4 and developed a detailed approach to
the 3-dimensional Thurston geometrization and Poincaré (subsumed) conjectures. This involved a plethora of geometric estimates, and he established
the framework for singularity analysis and to a large extent carried out his
program, including understanding fundamental aspects of singularity formation, an approach to surgery, and the long-time behavior of the Ricci
flow in dimension 3 [**179, 181, 182**]. It turned out that several deep issues
were unresolved in his program, and Perelman finally completed Hamilton's
program, and their combined works yielded a proof of both the Poincaré
conjecture and Thurston's geometrization conjecture [**250–252**]. Hamilton's
and Perelman's works went far beyond such a proof and laid the ground-
work for further understanding of the Ricci flow in all dimensions. There
are a number of expositions of the Hamilton–Perelman theory that con-
tain original work, including [**35, 70, 201, 227, 228, 300**]. With more recent
works of Brendle and Bamler, the Ricci flow in dimension 3 is well understood. In particular, Brendle resolved conjectures of Perelman regarding
3-dimensional Ricci solitons and ancient solutions [**42, 46**]. Bamler has resolved Perelman's conjecture about the finiteness of the number of surgeries
[**17–21**]. Bamler and Kleiner understood important aspects of the linearized
Ricci flow and applied this to resolve the generalized Smale conjecture on
diffeomorphism groups [**26, 27, 29, 30**].

ix

The understanding of the Ricci flow in dimensions at least 4 is much more wide open. Notable successes are the sphere theorems of Böhm and Wilking [**37**] and of Brendle and Schoen [**52, 53**], where the latter pair of authors resolved a more general form of the 1/4-pinching spherical space form conjecture. A recent fundamental development is the singularity theory of Bamler [**14–16**]. Bamler's work has revived the hope that Ricci flow may be used to understand aspects of 4-dimensional smooth topology, a subject whose present understanding is largely based on Donaldson's gauge theory [**140**] and Seiberg–Witten theory [**152, 226, 229**] in combination with Freedman's classification of simply connected closed topological 4-manifolds [**150**]. See Bamler [**22**] for a survey.

In any case, Ricci solitons are the prototypical singularity models for the Ricci flow. As such, one is particularly motivated to understand their geometry and topology. This book is an introduction to the study of Ricci solitons, with an emphasis on the basic theory and dimensions at most 3. In particular, this book focuses on some aspects of the theory that are the most well established. Currently, the theory in higher dimensions is developing rapidly. For example, there are works of Munteanu and Wang [**239, 241–243**] on the geometry at infinity of Ricci solitons, some of which are strongest in dimension 4. There are recent exciting developments in the classification of Kähler–Ricci solitons including the papers [**25, 122**].

A word about notation: When dealing with tensors expressed in local coordinates, we use the **Einstein summation convention** where pairs of repeated indices consisting of an upper index and a lower index are summed.

Acknowledgments

This book originated from lecture notes begun around the year 2009 related to talks at East China Normal University on the topic of Ricci solitons. I would like to thank Yu Zheng for invitations to visit ECNU. I am grateful to Brett Kotschwar and Yongjia Zhang who have ghostwritten parts of the book and to Peng Lu who has made a number of improvements to the book. This book in fact follows Brett's vision for the contents and structure of a first volume on Ricci solitons.

I would like to thank Ben Andrews, Richard Bamler, Simon Brendle, Xiaodong Cao, Pak-Yeung Chan, Eric Chen, Tobias Colding, Ronan Conlon, Xianzhe Dai, Panagiota Daskalopoulos, Yuxing Deng, Alix Deruelle, Michael Freedman, Laiyuan Gao, Huabin Ge, David Glickenstein, Hongxin Guo, Max Hallgren, Richard Hamilton, Tom Ilmanen, Wenshuai Jiang, Brett Kotschwar, Yi Lai, Mat Langford, Peng Lu, Feng Luo, Zilu Ma, Rafe Mazzeo, William Minicozzi, Ovidiu Munteanu, Lei Ni, Gongping Niu, Nataša Šešum, Weimin Sheng, Yuguang Shi, Henry Shin, Luca Spolaor, Jeffrey Streets, Gang Tian, Peter Topping, Fadi Twainy, Guofang Wei, Jiayong Wu, William Wylie, Bo Yang, Deane Yang, Yu Yuan, Qi Zhang, Shijin Zhang, Yongjia Zhang, Zhenlei Zhang, Zhou Zhang, Yu Zheng, and Xiaohua Zhu for interest, help, and encouragement for this book project and the topics that it covers. We would also like to thank Zilu Ma for his assistance in producing some of the MATLAB figures. We would like to thank Pak-Yeung Chan and Alix Deruelle for reading through parts of the book and their helpful suggestions which have improved the book.

Special thanks to AMS publisher Sergei Gelfand and AMS book acquisitions editor Ina Mette. We are grateful to Ina for all of her splendid work and help as our book editor. We would like to thank the editors of the

AMS Graduate Studies in Mathematics series and the anonymous reviewers for their careful reading and their plethora of help and suggestions, which have vastly improved the book. In particular, two of the reviewers have made remarkably exhaustive, detailed, and fantastic suggestions that have transformed the quality of the book. We owe them a debt of gratitude. We would like to thank associate editor Christine Thivierge for her help with the book. Special thanks to Arlene O'Sean for her magnificent copyediting. During the finishing of this book, we learned of the passing of editorial assistant Marcia Almeida. We are greatly indebted to her for all of the years of enthusiastic and expert help with the books we have co-authored.

Ben would like to thank his wife, Jingwei, his daughters, Michelle, Isabelle, and Gloriana, his brother, Peter, and his parents for their encouragement. He is especially grateful to his wife for her support and patience through the writing process. This book is dedicated, with deep admiration and respect, to Richard Hamilton.

Bennett Chow
University of California San Diego

Notation and Symbols

Here we list some of the notation and symbols used throughout the book. Regarding tensor calculations, we use the Einstein summation convention and we do not bother to raise indices.

0^n	the origin in $\mathbb{R}^n$
$\vec{0}$	the origin in a tangent space
II	second fundamental form of a hypersurface
$\measuredangle$	angle
$\int$	integral
$\oint$	contour integral
∇	covariant derivative
∇^2	Hessian
$.^{\top}$	tangential component or projection
$.^{\perp}$	normal component or projection
∂_t	$\frac{\partial}{\partial t}$ partial time-derivative
$\alpha^\natural$	dual vector field to the 1-form α
$\alpha(g)$	aperture of a metric g
$\overset{\circ}{\alpha}$	trace-free part of a 2-tensor α
$A \setminus B$	the set difference between sets A and B
Area	area of a surface or volume of a hypersurface
ACR	asymptotic (Riemann) curvature ratio
ASCR	asymptotic scalar curvature ratio
AVR	asymptotic volume ratio

$B_r(p)$	open ball of radius r centered at p
$B_{ijk\ell}$	the curvature quadratic $-\sum_{p,q=1}^{n} R_{ipj}^{q} R_{\ell qk}^{p}$
χ	Euler characterisic
$\mathrm{CA}(g)$	cone angle at infinity of g
$C_0\Sigma$	cone on Σ
$\mathrm{Cut}(p)$	cut locus of $p \in \mathcal{M}^n$
Δ	Laplacian
Δ_f	f-Laplacian
Δ_X	X-Laplacian $= \Delta - X \cdot \nabla$
Δ_Σ	Laplacian of a hypersurface Σ
Δ_L	Lichnerowicz Laplacian
$\Delta_{L,f}$	f-Lichnerowicz Laplacian
$\frac{d_-}{dt}(t)$	lim inf of backward difference quotients
$d(x,y)$	Riemannian distance between x and y
dA	area form (volume form in dimension 2)
$d\mu$	Riemannian volume form or measure
$d\sigma$	volume form or measure on a hypersurface
D_p°	injectivity domain of $\exp_p$
$\mathrm{diam}(\mathcal{M}^n)$	diameter of $\mathcal{M}^n$
$\mathrm{Diff}(\mathcal{M}^n)$	diffeomorphism group of $\mathcal{M}^n$
div	divergence
div_f	f-divergence
$\{e_i\}_{i=1}^n$	orthonormal frame (field)
$\exp_p$	exponential map at $p \in \mathcal{M}^n$
f_{Gau}	potential function for Gaussian soliton
γ_V	geodesic with initial velocity V
Γ_{ij}^k	Christoffel symbols
g	Riemannian metric
g_∞	limit Riemannian metric
$g(t)$	1-parameter family of metrics, often a solution to Ricci flow
g_{Bry}	metric of Bryant soliton
g_{Euc}	Euclidean metric on $\mathbb{R}^n$
$g_{\mathbb{S}^n}$	standard metric on $\mathbb{S}^n$
g_Σ	cigar soliton metric
GRS	gradient Ricci soliton(s)
H_f	f-mean curvature of a hypersurface

id_X	identity map of a set X
inj	injectivity radius
$\mathrm{Isom}(\mathcal{M}^n, g)$	isometry group of $(\mathcal{M}^n, g)$
J	almost complex structure
J	Jacobian of the exponential map
J_f	f-Jacobian of the exponential map
K	Gauss curvature of a surface
K_{orb}	orbital sectional curvature of a rotationally symmetric metric
K_{rad}	radial sectional curvature of a rotationally symmetric metric
ker	kernel
$\Lambda^k T^* \mathcal{M}$	vector bundle of k-forms on $\mathcal{M}^n$
L	length
$\mathcal{L}$	Lie derivative
$\mathcal{M}^n$	manifold
$\tilde{\mathcal{M}}^n$	(usually universal) cover of $\mathcal{M}^n$
$\mathcal{M}^n_\infty$	limit manifold
$M^-(\mathbf{z}_0)$	ODE unstable manifold at $\mathbf{z}_0$
M_{ij}, P_{ijk}	components of Hamilton's matrix Harnack quadratic form
$\mathrm{Met}(\mathcal{M}^n)$	space of Riemannian metrics on $\mathcal{M}^n$
$N(\mathcal{M}^n)$	number of ends of $\mathcal{M}^n$
$n\omega_n$	volume of the $(n-1)$-dimensional Euclidean unit sphere
ω_n	volume of the n-dimensional Euclidean unit ball
ω_i^j	connection 1-form
$\{\omega^i\}_{i=1}^n$	orthonormal coframe (field)
Ω_i^j	curvature 2-form
$\mathcal{O}$	orbifold
$Q(\mathrm{Rm})$	quadratic of Rm arising in Hamilton's evolution equation
$r_R(x)$	scalar curvature scale of x
R	scalar curvature
R_f	f-scalar curvature
$\mathbb{R}_+$	set of strictly positive real numbers
$\mathbb{R}^n$	n-dimensional Euclidean space
Ric	Ricci tensor
Rm	Riemann curvature operator
Ric_f	f-Ricci tensor
Σ_c	level set of a function

$\mathfrak{so}(n)$	special orthogonal group
$\mathbb{S}^n$	n-dimensional unit sphere
$\mathbb{S}^n(r)$	n-dimensional sphere of radius r
$S_r(p)$	geodesic sphere of radius r centered at p
$\mathcal{S}_p^{n-1}$	the unit sphere in $T_p\mathcal{M}$
$S^2(T^*\mathcal{M})$	bundle of symmetric 2-tensors on $\mathcal{M}^n$
sect	sectional curvature
Sym	symmetization (of a 2-tensor)
$T_p\mathcal{M}$	tangent space at $p \in \mathcal{M}^n$
tr	trace
Vol	volume
Vol_f	f-volume
W	Lambert W function
$W^-(\mathbf{z}_0)$	ODE unstable set at $\mathbf{z}_0$
$X^\flat$	dual 1-form to the vector field X

Chapter 1

Ricci Flow Singularity Formation

Hamilton's Ricci flow consists of the evolution of Riemannian metrics on a given manifold as determined by the nonlinear system of heat-type partial differential equations (1.1) given below. Although the metric "smooths out" under the Ricci flow, *singularities*, where the curvature tends to infinity, typically form in finite time due in part to the nonlinearity of the Ricci flow equation as well as for geometric and topological reasons. To define a Ricci flow that continues past singularity times, ideally one would like to understand and classify the singularities which form. Short of a complete classification, a good understanding of singularity formation should be necessary to study flows past singularity times such as *Ricci flow with surgery*.

Hamilton developed Ricci flow singularity analysis and surgery theory to approach the Poincaré and Thurston geometrization conjectures through a series of seminal and highly original works [**173**–**179**, **181**, **182**]. In dimension 3, Perelman's spectacular work [**250**–**252**] on Hamilton's program proves Thurston's geometrization conjecture (and, in particular, the Poincaré conjecture) via such a theory. The work of Bamler [**14**–**16**] lays the foundation for singularity analysis, and hence potentially Ricci flow surgery theory, in higher dimensions. The significance of *gradient Ricci solitons*, the topic of this book, is that they are, at the very least, prototypical singularity models for the Ricci flow.

In this book, our focus is on Ricci solitons in low dimensions, that is, dimensions 2 and 3. Along the way, we also develop some of the general dimensional theory.

In this chapter, whose purpose is to motivate the study of Ricci solitons, we survey what is known about Ricci flow singularity formation in low dimensions. The reader is referred to the ensuing chapters for some of the notions regarding gradient Ricci solitons mentioned in this chapter.

1.1. The Ricci flow and Ricci soliton equations

Hamilton's **Ricci flow** equation on an n-dimensional smooth manifold without boundary $\mathcal{M}^n$ is the equation

$$(1.1) \qquad\qquad\qquad \partial_t g = -2\mathrm{Ric}$$

for a 1-parameter family of smooth metrics $g(t)$ on $\mathcal{M}^n$ defined for t in some interval $I \subset \mathbb{R}$. Here, $\partial_t = \frac{\partial}{\partial t}$ denotes the time derivative and $\mathrm{Ric} = \mathrm{Ric}_{g(t)}$ denotes the Ricci tensor of $g(t)$.

By a theorem of Hamilton [**173**, Theorem 4.2], if $\mathcal{M}^n$ is compact and if g_0 is a smooth metric on $\mathcal{M}^n$, then there exists a unique solution $g(t)$ to the Ricci flow on a short time interval $[0, \varepsilon)$ with $g(0) = g_0$, where $\varepsilon > 0$ depends only on g_0 (his *short-time existence* theorem has been simplified by DeTurck [**137**, **138**]). By another theorem of Hamilton [**173**, Theorem 14.1], this solution exists on a maximum time interval $[0, T)$. In his theorem, if $T < \infty$, then the space-time supremum of the norm of the Riemann curvature operator is infinite. Šešum [**281**] improved this to show that the space-time supremum of the norm of the Ricci curvature tensor is infinite. Subsequently, there have been various improvements. In this case, where $T < \infty$, we say that the Ricci flow $g(t)$ **forms a singularity** at time T; we also say that $g(t)$, $t \in [0, T)$, is a **singular solution**.

Gradient Ricci solitons are, loosely speaking, those solutions to the Ricci flow whose shapes do not change under the flow. Recall that the motivation for studying gradient Ricci solitons is that they model singularity formation for the Ricci flow. For now, we give a quick definition which we will revisit later:

A **gradient Ricci soliton** is a Riemannian manifold $(\mathcal{M}^n, g)$, a function $f : \mathcal{M}^n \to \mathbb{R}$, and $\lambda = 1, 0$, or -1 satisfying the tensor equation

$$(1.2) \qquad\qquad\qquad \mathrm{Ric} + \nabla^2 f = \frac{\lambda}{2} g,$$

where $\nabla^2 f$ denotes the Hessian of f. We say that the gradient Ricci soliton is **shrinking**, **steady**, or **expanding** corresponding to whether $\lambda = 1, 0$, or -1, respectively.

1.2. Compactness, noncollapsing, and singularity models

A key tool in the understanding of the Ricci flow and its singularities is the Cheeger–Gromov compactness theorem, which we describe in this section.

Given a Riemannian manifold $(\mathcal{M}^n, g)$, let $\mathrm{Rm} = \mathrm{Rm}_g$ and $R = R_g$ denote its Riemann curvature operator and scalar curvature, respectively. Given $p \in \mathcal{M}^n$ and $r > 0$, let $B_r(p) = B_r^g(p) = \{x \in \mathcal{M}^n : d(x,p) < r\}$ denote the open ball of radius r centered at p, where d is the distance function. Let $\mathrm{Vol} = \mathrm{Vol}_g$ denote the volume.

1.2.1. The Cheeger–Gromov compactness theorem.

A cornerstone of Riemannian geometry is the Cheeger–Gromov compactness theorem (cf. Cheeger [83], Greene and Wu [160], Gromov [162], and Peters [253]).

Firstly, we have the notion of pointed C^∞ Cheeger–Gromov convergence, which is defined as follows.

Definition 1.1 (C^∞-convergence of manifolds modulo pointed diffeomorphisms). A sequence $\{(\mathcal{M}_i^n, g_i, p_i)\}_{i\in\mathbb{N}}$ of complete pointed Riemannian manifolds **converges in the pointed** C^∞ **Cheeger–Gromov sense** to a complete pointed Riemannian manifold $(\mathcal{M}_\infty^n, g_\infty, p_\infty)$ provided there exist:

(1) an exhaustion $\{U_i\}_{i\in\mathbb{N}}$ of $\mathcal{M}_\infty^n$ by open sets with $p_\infty \in U_i$ and

(2) a sequence of diffeomorphisms $\Phi_i : U_i \to V_i := \Phi_i(U_i) \subset \mathcal{M}_i^n$ with $\Phi_i(p_\infty) = p_i$

such that $\left(U_i, \Phi_i^*(g_i|_{V_i})\right)$ converges in C^k for all $k \in \mathbb{N}$ to $(\mathcal{M}_\infty^n, g_\infty)$ uniformly on compact sets in $\mathcal{M}_\infty^n$.

The **pointed** C^∞ **Cheeger–Gromov compactness theorem** for Riemannian manifolds is the following.

Theorem 1.2 (Cheeger and Gromov). *Let* $\{(\mathcal{M}_i^n, g_i, p_i)\}_{i\in\mathbb{N}}$ *be a sequence of pointed complete Riemannian manifolds satisfying the uniform bounds*

$$|\nabla_{g_i}^k \mathrm{Rm}_{g_i}| \le C_k \quad and \quad \mathrm{Vol}_{g_i} B_1^{g_i}(p_i) \ge c,$$

where $c > 0$ *and* $C_k < \infty$, $k \ge 0$, *are constants independent of* i. *Then there exists a subsequence* $\{(\mathcal{M}_i^n, g_i, p_i)\}$ *which converges in the pointed* C^∞ *Cheeger–Gromov sense to a pointed complete Riemannian manifold* $(\mathcal{M}_\infty^n, g_\infty, p_\infty)$ *with the same bounds on its curvature and derivatives of curvature.*

Note that there are two key hypotheses in this compactness theorem: a uniform curvature bound and a lower bound for the injectivity radius at the basepoints.

As an example of Cheeger–Gromov convergence, we observe that for any Riemannian manifold $(\mathcal{M}^n, g)$, $p \in \mathcal{M}^n$, and $\lambda_i \to \infty$, we have that $(\mathcal{M}^n, \lambda_i g, p)$ converges in the pointed C^∞ Cheeger–Gromov sense to the tangent space $(T_p\mathcal{M}, g_p, \vec{0})$, which is isometric to Euclidean space $\mathbb{R}^n$; see Figure 1.1.

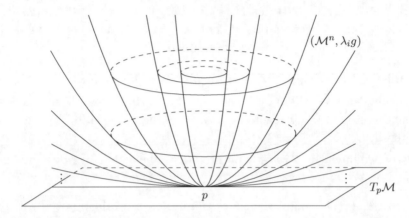

Figure 1.1. Scalings $(\mathcal{M}^n, \lambda_i g, p)$ converging to $(T_p\mathcal{M}, g_p, \vec{0})$ as $\lambda_i \to \infty$.

Remark 1.3. In the presence of a curvature bound, such as assumed in the theorem, the condition $\mathrm{Vol}_{g_i} B_1^{g_i}(p_i) \geq c$ above is qualitatively equivalent to the condition $\mathrm{inj}_{g_i}(p_i) \geq c$, where inj denotes the injectivity radius.

Example 1.4. As a more concrete example of the Cheeger–Gromov convergence to a tangent space that we observed above, consider the 1-parameter family of metrics on $\mathbb{R}^n$ defined by

$$(1.3) \qquad g_\lambda(\mathbf{x}) = \left(\frac{\lambda^2}{1 + |\mathbf{x}|^2} \right) g_{\mathrm{Euc}}, \quad \lambda \in (0, \infty).$$

Then $(\mathbb{R}^n, g_\lambda, 0^n)$ converges in the pointed C^∞ Cheeger–Gromov sense, as $\lambda \to \infty$, to $(\mathbb{R}^n, g_{\mathrm{Euc}}, 0^n)$. To see this, observe that in Euclidean coordinates $\mathbf{x} = (x^1, \ldots, x^n)$ we have

$$(1.4) \qquad g_\lambda(\mathbf{x}) = \frac{\lambda^2}{1 + |\mathbf{x}|^2} \sum_{i=1}^n dx^i \otimes dx^i.$$

So, with the change of variables $\mathbf{y} = \lambda \mathbf{x}$, so that $y^i = \lambda x^i$, we have

$$(1.5) \qquad g_\lambda(\mathbf{x}) = \frac{1}{1 + \lambda^{-2}|\mathbf{y}|^2} \sum_{i=1}^n dy^i \otimes dy^i.$$

In other words, defining the diffeomorphisms $\varphi_\lambda : \mathbb{R}^n \to \mathbb{R}^n$ by $\varphi_\lambda(\mathbf{y}) = \frac{1}{\lambda}\mathbf{y}$, which satisfy $\varphi_\lambda(0^n) = 0^n$, we have

$$(\varphi_\lambda^* g_\lambda)(\mathbf{y}) = \frac{1}{1 + \lambda^{-2}|\mathbf{y}|^2} g_{\text{Euc}}.$$

We see that $\varphi_\lambda^* g_\lambda$ converges to g_{Euc} as $\lambda \to \infty$ in each C^k norm and on each ball $B_k(0^n)$, where $k \geq 1$. This is true even though each metric g_λ is asymptotically cylindrical.

Remark 1.5. The metric g_2 on $\mathbb{R}^2$, defined by (1.3) with $n = \lambda = 2$, is called the **cigar soliton**. We will discuss the cigar soliton, which plays an important role in the study of the Ricci flow, in more detail in Chapter 3.

1.2.2. Local derivative of curvature estimates.

In view of the requisite derivative of curvature bounds in the hypothesis of the Cheeger–Gromov compactness theorem (see Theorem 1.2), we now recall the local derivative of curvature bounds for the Ricci flow.

Let ∇ denote the Riemannian covariant derivative acting on tensors and let $\nabla^m = \underbrace{\nabla \circ \cdots \circ \nabla}_{m \text{ times}}$. Shi's [**264**, §7] (cf. Hamilton [**179**, §13] or the expository [**102**, Chapter 14]) Bernstein-type *local* covariant derivative estimates for the curvature tensor say the following. (A global version of this result on closed manifolds was proved earlier by Bando [**31**].)

In the following, $\mathcal{U}^n$ denotes an n-dimensional manifold. We use this notation instead of the usual $\mathcal{M}^n$ to emphasize that we do not assume the metrics on it are complete. In practice, $\mathcal{U}^n$ is an open subset of the underlying manifold of a complete solution to the Ricci flow.

Theorem 1.6 (Shi). *For any positive integers n and m and for any positive constant c, there exists a constant $C < \infty$ such that if for some K, $(\mathcal{U}^n, g(t), p)$, $t \in [0, T]$, $0 < T \leq cK^{-1}$, is a (not necessarily complete) solution to the Ricci flow such that the closed ball $\bar{B}_{cK^{-1/2}}^{g(0)}(p)$ is compact and satisfies*

$$(1.6) \qquad |\text{Rm}|(x, t) \leq K \quad \text{in } B_{cK^{-1/2}}^{g(0)}(p) \times [0, T],$$

then

$$(1.7) \qquad |\nabla^m \text{Rm}|(x, t) \leq \frac{CK}{t^{m/2}} \quad \text{in } B_{\frac{1}{2}cK^{-1/2}}^{g(0)}(p) \times (0, T].$$

In particular,

$$(1.8) \qquad |\nabla^m \text{Rm}|(p, cK^{-1}) \leq Cc^{-m/2}K^{1+\frac{m}{2}}.$$

Remark 1.7. In the statement of the theorem above, since the metrics are uniformly equivalent on the time interval $[0, T]$, we may replace $B^{g(0)}$ by $B^{g(T)}$.

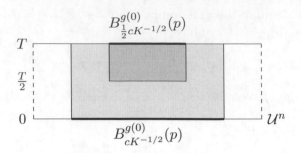

Figure 1.2. Assuming the local curvature bound (1.6), we have the derivative bound $|\nabla^m \mathrm{Rm}|(x,t) \leq \frac{CK}{(T/2)^{m/2}}$ in $B_{\frac{1}{2}cK^{-1/2}}^{g(0)}(p) \times [T/2, T]$.

See Figure 1.2. The genesis of the idea of Bernstein-type estimates can be seen for the heat equation $(\partial_t - \Delta)u = 0$ on a closed manifold $(\mathcal{M}^n, g)$ with nonnegative Ricci curvature. Using the Bochner formula

$$(1.9) \qquad \frac{1}{2}\Delta|\nabla u|^2 = \langle \nabla u, \nabla \Delta u \rangle + |\nabla^2 u|^2 + \mathrm{Ric}(\nabla u, \nabla u)$$

(see Exercise 1.3 for its proof), we compute that

$$(\partial_t - \Delta)\left(t|\nabla u|^2 + u^2/2\right) = -2t\left(|\nabla^2 u|^2 + \mathrm{Ric}(\nabla u, \nabla u)\right) \leq 0.$$

Hence, if $|u| \leq C$, then by the parabolic maximum principle (Lemma 1.21 below) we obtain Bernstein's estimate

$$|\nabla u|(x,t) \leq \frac{C}{\sqrt{t}} \quad \text{for } x \in \mathcal{M}^n, \, t > 0.$$

This estimate can be localized by using a cutoff function; see [**102**, Chapter 14].

1.2.3. Hamilton's Cheeger–Gromov compactness theorem for the Ricci flow.

For sequences of solutions to the Ricci flow, the definition of pointed C^∞ Cheeger–Gromov convergence is given as follows.

Definition 1.8. We say that a sequence $\{(\mathcal{M}_i^n, g_i(t), p_i)\}_{i \in \mathbb{N}}$, $t \in (\alpha, \omega)$, of complete pointed solutions to the Ricci flow **converges in the pointed** C^∞ **Cheeger–Gromov sense** to a complete pointed solution to the Ricci flow $(\mathcal{M}_\infty^n, g_\infty(t), p_\infty)$, $t \in (\alpha, \omega)$, provided there exist:

(1) an exhaustion $\{U_i\}_{i \in \mathbb{N}}$ of $\mathcal{M}_\infty^n$ by open sets with $p_\infty \in U_i$ and

(2) a sequence of diffeomorphisms $\Phi_i : U_i \to V_i := \Phi_i(U_i) \subset \mathcal{M}_i^n$ with
$$\Phi_i(p_\infty) = p_i$$

such that $\left(U_i, \Phi_i^*(g_i(t)|_{V_k})\right)$ converges in C^∞ to $(\mathcal{M}_\infty^n, g_\infty(t))$ uniformly on compact sets in $\mathcal{M}_\infty^n \times (\alpha, \omega)$.

With the help of Shi's local derivative of curvature estimates discussed in the previous subsection, Hamilton [**178**] proved the following important version of the Cheeger–Gromov compactness theorem for solutions to the Ricci flow (see [**101**, Chapters 3 and 4] for an exposition).

Theorem 1.9 (Compactness theorem for Ricci flow solutions). *Suppose that* $\{(\mathcal{M}_i^n, g_i(t), p_i)\}_{i \in \mathbb{N}}$, $t \in (\alpha, \omega)$, *where* $-\infty \leq \alpha < 0 < \omega \leq \infty$, *is a sequence of complete pointed solutions to the Ricci flow satisfying the following properties:*

(1) (**uniformly bounded curvatures**)

$$|\mathrm{Rm}\, g_i|_{g_i} \leq C_0 \quad on \ \mathcal{M}_i^n \times (\alpha, \omega)$$

for some constant $C_0 < \infty$ *independent of* i *and*

(2) (**injectivity radius estimate at** $t = 0$)

$$\mathrm{inj}_{g_i(0)}(p_i) \geq \iota_0$$

for some constant $\iota_0 > 0$.

Then there exists a subsequence $\{j_i\}_{i \in \mathbb{N}}$ *such that* $\left\{\left(\mathcal{M}_{j_i}^n, g_{j_i}(t), p_{j_i}\right)\right\}_{i \subset \mathbb{N}}$ *converges in the pointed* C^∞ *Cheeger–Gromov sense to a complete pointed solution to the Ricci flow* $(\mathcal{M}_\infty^n, g_\infty(t), p_\infty)$, $t \in (\alpha, \omega)$, *as* $i \to \infty$.

We remark that since we are studying Ricci solitons, which are the fixed points of the Ricci flow (see §2.2 below), we will mostly be able to appeal to just the Riemannian version of the compactness theorem.

Definition 1.10 (Hamilton [**179**, §16]). Let $(\mathcal{M}^n, g(t))$, $t \in [0, T)$, be a singular solution to the Ricci flow. Suppose that (x_i, t_i) is a sequence of spacetime points with $K_i := |\mathrm{Rm}|(x_i, t_i) \to \infty$ and satisfying the following property: The associated pointed sequence of rescaled solutions $(\mathcal{M}^n, g_i(t), x_i)$, defined by

$$(1.10) \qquad\qquad g_i(t) := K_i g(t_i + K_i^{-1} t),$$

converges to a complete solution to the Ricci flow $(\mathcal{M}_\infty^n, g_\infty(t), x_\infty)$ in the pointed C^∞ Cheeger–Gromov sense. If this property is true, then we call $(\mathcal{M}_\infty^n, g_\infty(t))$ a **singularity model** of the singular solution $g(t)$.

Since $K_i \to \infty$, a singularity model $(\mathcal{M}_\infty^n, g_\infty(t))$ is defined for t in the infinite time interval $(-\infty, 0]$. The rescaling in (1.10) is called **parabolic rescaling**. Heuristically, time scales like distance squared for heat-type equations such as the Ricci flow. Since the metric also scales like distance squared, we have the particular rescaling in (1.10).

Definition 1.11. Whenever we have a solution to the Ricci flow that exists on some time interval of the form $(-\infty, \omega)$, where $\omega \in (-\infty, \infty]$, we say that the solution is an **ancient solution**. If $\omega = \infty$, we say that the solution is an **eternal solution**.

For example, if a family of Riemannian metrics on a topological sphere shrinks to a point under the Ricci flow while becoming asymptotically round, then the associated singularity model is a round (constant curvature) shrinking sphere, which is an ancient solution; see Figures 1.3 and 1.4. Hamilton proved that a condition sufficient to ensure this is that the initial metric on a closed 3-manifold has positive Ricci curvature, in which case the associated singularity model must be a spherical space form.

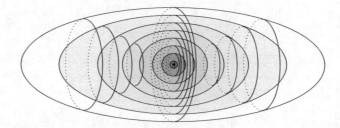

Figure 1.3. A positively curved solution to the Ricci flow on a sphere shrinking to a round point.

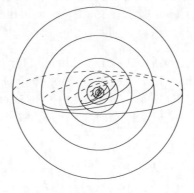

Figure 1.4. The singularity model associated to the singular solution in Figure 1.3 is the shrinking round sphere. This is because the asymptotic shape is round.

1.2.4. Perelman's no local collapsing theorem.

1.2.4.1. *The definition of noncollapsing.* Recall the following notion [**250**, Definition 4.2], which is important in view of the Cheeger–Gromov compactness theorem (see Theorem 1.2).

Definition 1.12. Let $\kappa, \rho > 0$. We say that a Riemannian manifold $(\mathcal{M}^n, g)$ is κ-**noncollapsed on the scale** ρ if for any $p \in \mathcal{M}^n$ and $r \in (0, \rho)$ satisfying $R \leq r^{-2}$ in $B_r(p)$, we have

$$(1.11) \qquad \operatorname{Vol} B_r(p) \geq \kappa r^n.$$

If $(\mathcal{M}^n, g)$ is κ-noncollapsed on the scale ρ for all $\rho \in (0, \infty)$, then we say that $(\mathcal{M}^n, g)$ is κ-**noncollapsed on all scales**.

We can define the **scalar curvature scale** $r_R(x)$ of a point x in a Riemannian manifold $(\mathcal{M}^n, g)$ as the supremum of all $r > 0$ such that $R \leq r^{-2}$ in $B_r(p)$. The manifold $(\mathcal{M}^n, g)$ being noncollapsed below the scale ρ means that for all $x \in \mathcal{M}^n$ and for all $r \leq \min\{r_R(x), \rho\}$, we have the volume κ-noncollapsing condition $\operatorname{Vol} B_r(p) \geq \kappa r^n$.

1.2.4.2. *Statement of Perelman's no local collapsing theorem.* The following is Perelman's celebrated **no local collapsing theorem** for Ricci flow [**250**, §4 and §8]. This answers a conjecture of Hamilton, who formulated it as his "little loop conjecture" (see [**179**, §15]). Hamilton proved it in dimension 2 (see [**100**, §5]) and for 3-dimensional "Type I" singular solutions under certain assumptions (see [**179**, §23]) and for certain limits with nonnegative curvature (see [**179**, §25]).[1]

Theorem 1.13 (Perelman's no local collapsing theorem). *For any closed Riemannian manifold* $(\mathcal{M}^n, g_0)$, $\rho > 0$, *and* $T \in (0, \infty)$ *there exists* $\kappa > 0$ *depending only on* g_0, ρ, *and* T *such that for any solution* $(\mathcal{M}^n, g(t))$, $t \in [0, T)$, *to the Ricci flow with* $g(0) = g_0$, *each metric* $g(t)$ *is* κ-*noncollapsed on the scale* ρ.

Remark 1.14. A typical way of defining noncollapsing is to assume instead that $|\mathrm{Rm}| \leq r^{-2}$ in $B_r(p)$ implies $\operatorname{Vol} B_r(p) \geq \kappa r^{-n}$. We chose to use the stronger Definition 1.12 of noncollapsing because of Theorem 1.13.

Example 1.15. The shrinking sphere solution to the Ricci flow, $\mathbb{S}^n$ with $g(t) = (r_0^2 - 2(n-1)t)g_{\mathbb{S}^n}$, $t \in \left(-\infty, \frac{r_0^2}{2(n-1)}\right)$, is noncollapsing. Indeed, letting $r(t) = \sqrt{r_0^2 - 2(n-1)t}$, we see that $\operatorname{Vol} B_{\pi r(t)}^{g(t)}(p) = \operatorname{Vol}(g(t)) = (n+1)\omega_{n+1} r(t)^n$ for any $p \in \mathbb{S}^n$, and $R_{g(t)} \equiv n(n-1)r(t)^{-2}$. Here, ω_{n+1} is the volume of the $(n+1)$-dimensional unit Euclidean ball, so that $(n+1)\omega_{n+1}$ is the surface area of the n-dimensional unit sphere. A more general way to see this is to observe that noncollapsing is a scale-invariant condition and the round sphere evolves only by scaling under the Ricci flow.

---

[1] See §B.2 for the categorization of types of singular solutions.

1.2.4.3. *Existence of singularity models.* Perelman's no local collapsing theorem has numerous consequences. In this subsection we discuss the existence of singularity models.

Firstly, in the presence of a curvature bound, noncollapsing is equivalent to an injectivity radius lower bound. In particular, we have the following consequence of Perelman's no local collapsing theorem; see e.g. [**111**, Corollary 5.41].

Corollary 1.16 (Injectivity radius estimate for Ricci flow). *Let* $(\mathcal{M}^n, g(t))$, $t \in [0, T)$, $T < \infty$, *be a solution to the Ricci flow on a closed manifold. For every constant* $C < \infty$, *there exists a constant* $\iota > 0$ *depending only on* C, $g(0)$, *and* T *such that if* $(x, t) \in [0, T)$ *satisfies*

$$|\operatorname{Rm}_{g(t)}|_{g(t)} \leq CK$$

in $B_{g(t)}(x, 1/(\sqrt{CK}))$, *where* $K := |\operatorname{Rm}_{g(t)}|_{g(t)}$, *then*

$$(1.12) \qquad\qquad \operatorname{inj}_{g(t)}(x) \geq \frac{\iota}{\sqrt{K}}.$$

The existence of singularity models, following from Perelman's no local collapsing theorem, is given by the following.

Corollary 1.17 (Existence of singularity models). *Let* $(\mathcal{M}^n, g(t))$, $t \in [0, T)$, *where* $T \in (0, \infty)$, *be a singular solution to the Ricci flow on a closed manifold. Suppose that* (x_i, t_i) *is a sequence of points with the property that*

$$|\operatorname{Rm}|(x, t) \leq CK_i := C|\operatorname{Rm}|(x_i, t_i) \to \infty$$

for all $(x, t) \in \mathcal{M}^n \times [0, t_i]$. *Then, in the sense of Definition* 1.10, *the corresponding sequence of rescaled solutions* $(\mathcal{M}^n, g_i(t), x_i)$ *defined by* (1.10) *converges in the pointed* C^∞ *Cheeger–Gromov sense to a singularity model with bounded curvature.*

1.2.4.4. *Ruling out the cigar soliton as a singularity model.* Not only does Perelman's no local collapsing theorem imply the existence of singularity models, it also gives information about the geometry of singularity models. In particular, it rules out certain types of singularity models. For example, since the flat Riemannian product manifold $\mathbb{S}^1 \times \mathbb{R}^{n-1}$ is collapsed, namely $\operatorname{Vol} B_r(p) \sim c_n r^{n-1}$ for r large, it is not a singularity model of the Ricci flow; see Figure 1.5 and the related Figure 1.6. Perelman's no local collapsing theorem rules out, as Hamilton conjectured, the formation of the *cigar soliton* (defined in Remark 1.5) as a Cartesian factor of a singularity model.

Figure 1.5. The flat manifold $\mathbb{S}^1 \times \mathbb{R}^{n-1}$. It collapses as the scales, visualized as the radii of geodesic balls centered at a fixed point (shaded in the picture), tend to infinity.

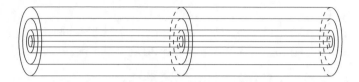

Figure 1.6. A sequence of collapsing flat finite 2-dimensional cylinders $\mathbb{S}^1(\varepsilon_i) \times [-\pi, \pi]$, where $\varepsilon_i \to 0$. If we identify the ends, then we obtain a sequence of 2-tori collapsing to a circle.

1.2.4.5. Idea of the proof of Perelman's no local collapsing theorem. Perelman's first proof of noncollapsing for the Ricci flow uses his entropy monotonicity formula. Perelman's $\mathcal{W}$**-entropy** of $(\mathcal{M}^n, g, f, \tau)$, where f is a function on $\mathcal{M}^n$ and $\tau \in \mathbb{R}_+$, is

$$(1.13) \qquad \mathcal{W}(g, f, \tau) = \int_{\mathcal{M}} \left(\tau \left(R + |\nabla f|^2 \right) + f - n \right) \tau^{-\frac{n}{2}} \mathrm{e}^{-f} d\mu,$$

where e is Euler's number. The entropy monotonicity formula is that under the coupled Ricci flow equations

$$(1.14\mathrm{a}) \qquad\qquad \partial_t g = -2\mathrm{Ric},$$

$$(1.14\mathrm{b}) \qquad\qquad \partial_t f = -R - \Delta f + |\nabla f|^2 + \frac{n}{2\tau},$$

$$(1.14\mathrm{c}) \qquad\qquad \partial_t \tau = -1,$$

we have

$$(1.15) \quad \frac{d}{dt} \mathcal{W}\big(g(t), f(t), \tau(t)\big) = 2\tau \int_{\mathcal{M}} \left| \mathrm{Ric} + \nabla^2 f - \frac{1}{2\tau} g \right|^2 \tau^{-\frac{n}{2}} \mathrm{e}^{-f} d\mu \geq 0.$$

Moreover, $\frac{d}{dt} \int_{\mathcal{M}} \tau(t)^{-\frac{n}{2}} \mathrm{e}^{-f(t)} d\mu_{g(t)} = 0$.

We remark that equations (1.14b) and (1.14c) are motivated by the consideration of *shrinking gradient Ricci solitons*. In particular, the 2-tensor $\mathrm{Ric} + \nabla^2 f - \frac{1}{2\tau} g$ on the right-hand side of the consequent equation (1.15) vanishes on gradient Ricci solitons as defined in (1.2).

Intuitively, we can see the link between entropy and noncollapsing as follows. Ignoring the fact that a characteristic function is not continuous, choose for $\mathcal{W}(g, f, r^2)$ the test function $\mathrm{e}^{-f} = \rho \chi_{B_r(p)}$, where $\rho := \frac{r^n}{\mathrm{Vol}\, B_r(p)}$ so that $\int_{\mathcal{M}} r^{-n} \mathrm{e}^{-f} d\mu = 1$. A cavalier computation of the entropy of this test function yields

$$\mathcal{W}(g, f, r^2) \leq r^2 \sup_{B_r(p)} R - \ln \rho.$$

Assuming that $r^2 \sup_{B_r(p)} R \leq C$ for some constant C, we obtain

$$\frac{\operatorname{Vol} B_r(p)}{r^n} = \frac{1}{\rho} \geq e^{\mathcal{W}(g, f, r^2) - C}.$$

Since the entropy is nondecreasing under the Ricci flow, assuming the scalar curvature upper bound on $B_r(p)$, we should get the noncollapsing of $B_r(p)$.

1.2.5. Manifolds with nonnegative sectional curvature.

Recall the following result.

Theorem 1.18 (Gromoll and Meyer [**161**]). *If $(\mathcal{M}^n, g)$ is a complete noncompact Riemannian manifold with positive sectional curvature, then $\mathcal{M}^n$ is diffeomorphic to $\mathbb{R}^n$. Moreover, if the sectional curvature is bounded above by a constant K, then*

$$(1.16) \qquad \operatorname{inj}(g) := \inf_{p \in \mathcal{M}} \operatorname{inj}_g(p) \geq \pi / \sqrt{K}.$$

A failed attempt at producing a counterexample to Theorem 1.18 is given by Figure 1.7.

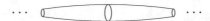

Figure 1.7. A Riemannian manifold can locally have positive sectional curvature bounded above by K and injectivity radius less than $\pi/\sqrt{K}$, but such a manifold cannot be isometric to a subset of a complete noncompact manifold with positive sectional curvature bounded by K.

Recall the following result about the geometry at infinity of manifolds with nonnegative curvature. For the proof, see Appendix A.

Theorem 1.19. *Let $(\mathcal{M}^n, g)$ be a complete Riemannian manifold with nonnegative sectional curvature and let $o \in \mathcal{M}^n$. Suppose there exist sequences $x_i \in \mathcal{M}^n$ and $r_i > 0$ with $r_i^{-1} d(o, x_i) \to \infty$ such that $(\mathcal{M}^n, g_i, x_i)$, where $g_i = r_i^{-2} g$, converges in the pointed C^∞ Cheeger–Gromov sense to a complete limit $(\mathcal{M}_\infty^n, g_\infty, x_\infty)$. Then $(\mathcal{M}_\infty^n, g_\infty)$ is the product of a line with a C^∞ Riemannian manifold with nonnegative sectional curvature.*

1.3. 2-Dimensional singularity formation

In this section we outline the work of Hamilton [**175**] on the normalized Ricci flow on closed surfaces. His work is the first instance of gradient Ricci solitons being used in the Ricci flow. Hamilton's ideas carry over to the Ricci flow on both compact 2-dimensional orbifolds and noncompact surfaces.

1.3.1. The Ricci flow on closed surfaces.

Let $\mathcal{M}^2$ be a closed orientable surface. Recall that for dimension $n = 2$ we have that the Ricci tensor is given by $R_{ij} = \frac{R}{2}g_{ij}$, where the scalar curvature R is twice the Gauss curvature K.

The **Ricci flow** equation on surfaces reduces to

$$(1.17) \qquad \partial_t g_{ij} = -R g_{ij}.$$

Figures 1.8 and 1.9 illustrate the fact that neckpinches do not occur in dimension 2.

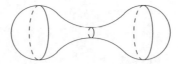

Figure 1.8. In dimension 2, neck regions have negative curvature and the Ricci flow expands the metric in those regions. Pictured is a dumbbell-shaped metric on $\mathbb{S}^2$, which can serve as an initial metric for the Ricci flow.

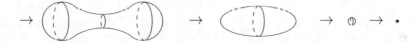

Figure 1.9. Some time later, under the Ricci flow on a surface, the positively curved caps have shrunk and the negatively curved neck has expanded. Eventually, the metric becomes positively curved and then shrinks to a round point.

1.3.1.1. *The normalized Ricci flow on surfaces.*

Definition 1.20. The **normalized Ricci flow** equation on surfaces is

$$(1.18) \qquad \partial_t g_{ij} = (r - R)g_{ij},$$

where $r = \int_{\mathcal{M}} R d\mu / \int_{\mathcal{M}} d\mu$ is the average scalar curvature and $d\mu$ is the area (volume) form of $(\mathcal{M}^2, g(t))$.

Note that the $r g_{ij}$ term in (1.18) makes the area of a solution $g(t)$ stay constant (the "normalization"):

$$(1.19) \qquad \text{Area}(g(t)) \equiv \text{Area}(g(0)).$$

1.3.1.2. *Global existence of the normalized flow.* Hamilton first proved that for any C^∞ Riemannian metric g_0 on $\mathcal{M}^2$ there exists a unique "global" solution $g(t)$, $t \in [0, \infty)$, to (1.18) with $g(0) = g_0$. We now briefly survey his estimates for proving the convergence of $g(t)$ as $t \to \infty$ to a constant curvature metric on $\mathcal{M}^2$. For the details of the proof, see Hamilton [175] and the expository [108, Chapter 5]. The main idea is to find suitable monotone quantities, a.k.a. monotonicity formulas. Throughout this subsection, we will assume that $g(t)$ is a solution to the normalized Ricci flow on a closed surface and that the geometric quantities discussed are those associated to $g(t)$.

Let $\Delta := g^{ij}\nabla_i\nabla_j$ denote the **Laplacian** acting on functions. When acting on tensors, this is called the rough Laplacian. From the formula (see Exercise 3.15)

$$(1.20) \qquad R_{e^u g} = e^{-u}(R_g - \Delta_g u)$$

on surfaces, we obtain the following.

Heat-type equation for the scalar curvature: Under the normalized Ricci flow on a closed surface,

$$(1.21) \qquad \partial_t R = \Delta R + R^2 - rR.$$

The quadratic term R^2 on the right-hand side poses a problem for proving global existence of the flow, in particular for obtaining an upper bound for the curvature. This difficulty is solved as follows.

Since $\int_{\mathcal{M}}(r - R)d\mu = 0$, there exists by Hodge theory a C^∞ function f on $\mathcal{M}^2$ (labeled the potential function), well-defined up to an additive constant, such that

$$(1.22) \qquad \Delta f = r - R.$$

Equation (1.21) is equivalent to

$$(1.23) \qquad \Delta(\partial_t f) = \Delta(\Delta f + rf).$$

Hence, for each t we may adjust the additive constant so that f is smooth in both the space and time variables and satisfies the following.

Heat-type equation for the potential function:

$$(1.24) \qquad \partial_t f = \Delta f + rf,$$

where the Laplacian is with respect to the evolving metric $g(t)$ on $\mathcal{M}^2$.

Solutions to heat-type equations and inequalities can be estimated using the first and second derivative tests from calculus.

Lemma 1.21 (Parabolic (weak) maximum principle). *Let $g(t)$, $t \in [0, T)$, be a 1-parameter family of Riemannian metrics on a closed manifold $\mathcal{M}^n$. If $u : \mathcal{M}^n \times [0, T) \to \mathbb{R}$ satisfies $\partial_t u \leq \Delta_{g(t)} u$, then*

$$(1.25) \qquad u(x, t) \leq \max_{y \in \mathcal{M}} u(y, 0) \quad \text{for } (x, t) \in \mathcal{M}^n \times [0, T).$$

Similarly, if $\partial_t v \geq \Delta_{g(t)} v$, then

$$(1.26) \qquad v \geq \min_{y \in \mathcal{M}} v(y, 0) \quad \text{on } \mathcal{M}^n \times [0, T).$$

Remark 1.22. We will usually just call this "weak" (versus "strong") parabolic maximum principle the parabolic maximum principle.

By applying the parabolic maximum principle to (1.24), we obtain the estimate

$$(1.27) \qquad |f| \leq Ce^{rt}.$$

The first key estimate in the study of the Ricci flow on surfaces is obtained by defining the function

$$(1.28) \qquad h := R + |\nabla f|^2 - r.$$

The motivation for considering h is that, paradoxically, sometimes it is easier to estimate from above a larger quantity (h) than a smaller quantity ($R - r$); the reason for this is that the nonnegative quantity $|\nabla f|^2$ has nonpositive terms on the right-hand side of its evolution equation which we can use to good effect.

Define the symmetric 2-tensor

$$(1.29) \qquad \begin{aligned} H_{ij} &:= \nabla_i \nabla_j f - \frac{1}{2} \Delta f g_{ij} \\ &= R_{ij} + \nabla_i \nabla_j f - \frac{r}{2} g_{ij}. \end{aligned}$$

Here, $\nabla_i \nabla_j f = (\nabla \nabla f)_{ij} = (\nabla^2 f)_{ij}$ are the components of the **Hessian** of f. Observe that the Laplacian is the trace of the Hessian. Since $n = 2$, we have that H is traceless: $\text{tr}(H) = 0$.

Remark 1.23. As we shall see in the next chapter, by definition the 2-tensor H vanishes on gradient Ricci solitons (where $r > 0, = 0, < 0$ correspond to shrinking, steady, expanding gradient Ricci solitons, respectively). As we shall also see there, the scalar quantity h is likewise also motivated by gradient Ricci solitons.

Using (1.21) and (1.24), one may compute that under the normalized Ricci flow on surfaces,

$$(1.30) \qquad \partial_t h = \Delta h - 2|H|^2 + rh,$$

where the norm of H is defined by $|H|^2 = g^{ik}g^{j\ell}H_{ij}H_{k\ell}$. One may also compute that

$$(1.31) \qquad \partial_t |H|^2 = \Delta |H|^2 - 2|\nabla H|^2 - 2R|H|^2.$$

In view of the sharpness of the two displayed equations above, we consider them as *parabolic Bochner formulas* or *pointwise monotonicity formulas*.

Let $A = \text{Area}(g(t))$, which is independent of t. By the Gauss–Bonnet formula, $r = 4\pi\chi(\mathcal{M}^2)/A$ is constant in time, where $\chi(\mathcal{M}^2)$ is the Euler characteristic of $\mathcal{M}^2$. Hence the sign of r is the same as the sign of $\chi(\mathcal{M}^2)$.

Equation (1.21) may be rewritten as

$$(1.32) \qquad \partial_t(R-r) = \Delta(R-r) + (R-r)^2 + r(R-r)$$
$$\geq \Delta(R-r) + r(R-r),$$

as long as the solution exists. By applying the parabolic maximum principle (Lemma 1.21) to the second line of (1.32), we have that

$$(1.33) \qquad R - r \geq -Ce^{rt}.$$

On the other hand, by applying the parabolic maximum principle to (1.21) and (1.30), it follows that exists a constant C such that

$$(1.34) \qquad R - r \leq R - r + |\nabla f|^2 \leq Ce^{rt}.$$

Summarizing, we have:

Estimate for the scalar curvature: Under the normalized Ricci flow on a closed surface $\mathcal{M}^2$,

$$(1.35) \qquad -Ce^{rt} \leq R - r \leq Ce^{rt}.$$

One can infer from this time-dependent curvature bound and from the curvature derivative estimates of Theorem 1.6 that for any initial data g_0 the solution to the normalized Ricci flow on $\mathcal{M}^2$ exists for all time $t \in [0, \infty)$.

1.3.1.3. *The case* $\chi(\mathcal{M}^2) < 0$. Suppose $\chi(\mathcal{M}^2) < 0$, so that $r < 0$ and g ≥ 2, where g is the genus of $\mathcal{M}^2$ ($\chi(\mathcal{M}^2) = 2 - 2$g). This is the easiest case. Since $r < 0$, (1.33) and (1.34) imply that $R - r \to 0$ exponentially as $t \to \infty$; see Figure 1.10. From this and the consequent higher derivative estimates, one shows that $g(t)$ converges exponentially fast in each C^k norm to a C^∞ metric g_∞ with constant curvature in the same conformal class as g_0.

We remark that in this case, where $\chi(\mathcal{M}^2) < 0$, we have that the metric g_∞ is the unique metric in the conformal class of g_0 with constant curvature equal to r. We can see the uniqueness of the constant curvature metric in a conformal class as follows. Suppose that Riemannian metrics g and $e^u g$ on $\mathcal{M}^2$ satisfy $R_g = R_{e^u g} \equiv r < 0$. Then, by (1.20), we have that

$$(1.36) \qquad r = e^{-u}(r - \Delta_g u),$$

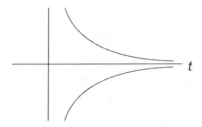

Figure 1.10. When $\chi(\mathcal{M}^2) < 0$, $R(x,t)$ lies between the two curves.

which implies that

$$r(e^u - 1) = -\Delta_g u.$$

Therefore,

$$r \int_{\mathcal{M}} (e^u - 1)^2 d\mu_g = -\int_{\mathcal{M}} (e^u - 1)\Delta_g u \, d\mu_g = \int_{\mathcal{M}} e^u |\nabla u|^2 d\mu_g.$$

Since $r < 0$, we conclude that $u \equiv 0$, so that $e^u g = g$.

1.3.1.4. *The case* $\chi(\mathcal{M}^2) = 0$. With a little more work, one can show exponential convergence to a flat metric when $\chi(\mathcal{M}^2) = 0$ (where $r = 0$ and $g = 1$). In fact, by only using the parabolic maximum principle, one can show that for each nonnegative integer k there exists a constant C_k depending only on $g(0)$ and k such that[2]

$$(1.37) \qquad |\nabla^k R|(x,t) \leq \frac{C_k}{(1+t)^{\frac{k}{2}+1}} \quad \text{on } \mathcal{M}^2 \times [0, \infty).$$

From $\partial_t g = (\Delta f)g = (\partial_t f)g$ (here we used $r = 0$) and from f being uniformly bounded, we have that the metrics $g(t)$, $t \in [0, \infty)$, are uniformly equivalent, and hence the diameters are uniformly bounded. From this and integrating the higher derivative estimates along minimal geodesics, it is not difficult to show that for all $k, \ell \geq 0$ there exist constants $C_{k,\ell}$ such that

$$(1.38) \qquad |\nabla^k R|(x,t) \leq \frac{C_{k,\ell}}{(1+t)^\ell} \quad \text{on } \mathcal{M}^2 \times [0, \infty).$$

That is, for derivatives of curvature of each order, we obtain polynomial decay in time of arbitrary degree. This implies smooth convergence to a flat metric as $t \to \infty$. Using integral estimates instead, one can prove smooth convergence at an exponential rate (see Hamilton [**175**, §5]).

---

[2] The reader familiar with PDEs may recognize this as a Bernstein-type estimate.

1.3.1.5. *The case* $\chi(\mathcal{M}^2) > 0$. Suppose now that $\chi(\mathcal{M}^2) > 0$, so that $r > 0$ and $\mathcal{M}^2$ is diffeomorphic to the 2-sphere. First suppose $R_{g_0} > 0$. Then $R_{g(t)} > 0$ for $t \geq 0$. In this case we have:

Definition 1.24. For a closed Riemannian surface $(\mathcal{M}^2, g)$ with positive curvature, **Hamilton's surface entropy** is

$$(1.39) \qquad\qquad N(g) = \int_{\mathcal{M}} R \ln R \, d\mu.$$

One may compute the **surface entropy monotonicity formula**:

(1.40)

$$d_t N(g(t)) = -2 \int_{\mathcal{M}} \left| \mathrm{Ric} + \nabla^2 f - \frac{r}{2} g \right|^2 d\mu - \int_{\mathcal{M}} \frac{|\nabla R - 2 \mathrm{Ric}(\nabla f)|^2}{R} d\mu$$

$$= -2 \int_{\mathcal{M}} |H|^2 d\mu - 4 \int_{\mathcal{M}} \frac{|\mathrm{div}(H)|^2}{R} d\mu$$

$$\leq 0.$$

In particular, by Remark 1.23 we have that $d_t N = 0$ if and only if $g(t)$ is a shrinking gradient Ricci soliton.

Following the Li–Yau method for obtaining differential Harnack estimates [**218**], we have the following [**175**, §6].

Theorem 1.25 (Hamilton's trace Harnack estimate). *Under the normalized Ricci flow on a closed surface* $\mathcal{M}^2$ *with positive curvature, the quantity*

$$(1.41) \qquad\qquad Q := \Delta \ln R + R - r = \partial_t \ln R - |\nabla \ln R|^2$$

satisfies the estimate

$$(1.42) \qquad\qquad Q(x, t) \geq \frac{Cre^{rt}}{1 - Ce^{rt}} =: q(t),$$

where C *is a constant depending only on* g_0 *with* $C > 1$ *or* $C \leq 0$.

Let β be an m-tensor. Its **divergence** is the $(m-1)$-tensor defined by

$$(1.43) \qquad\qquad \mathrm{div}(\beta)_{k_1 \cdots k_{m-1}} := g^{ij} \nabla_i \beta_{j k_1 \cdots k_{m-1}}.$$

Define the double divergence of a 2-tensor β by

$$\mathrm{div}^2(\beta) := \mathrm{div}(\mathrm{div}(\beta)) = g^{il} g^{jk} \nabla_i \nabla_j \beta_{kl},$$

which is a scalar. We observe that the symmetric 2-tensor H defined by (1.29) satisfies

$$\mathrm{div}^2(H) = \Delta R - \langle \nabla R, \nabla f \rangle + R(R - r) = RQ - \frac{2}{R} \langle \nabla R, \mathrm{div}(H) \rangle.$$

So Q vanishes on any gradient Ricci soliton (recall Remark 1.23).

The proof of (1.42) is simply to derive the following heat-type inequality:

$$(1.44) \qquad \partial_t Q \geq \Delta Q + 2 \langle \nabla \ln R, \nabla Q \rangle + Q^2 + rQ.$$

Let $q_0 := \min_{\mathcal{M}^2} Q(\cdot, 0) \leq 0$, where the nonpositivity is because the integral of Q over $\mathcal{M}^2$ is equal to 0 by Stokes's theorem. If $q_0 \neq -r$, then the function $q(t) = \frac{Cre^{rt}}{1 - Ce^{rt}}$, with $C := \frac{q_0}{q_0 + r}$, satisfies the ODE $\dot{q} = q^2 + rq$ with $q(0) = q_0$. This proves (1.42) when $q_0 \neq -r$; note that $C > 1$ or $C \leq 0$. On the other hand, if $q_0 = -r$, then the parabolic maximum principle implies that $Q \geq -r$, and so (1.42) is true for any $C > 1$ in this case.

Now, applying the parabolic maximum principle to (1.44) yields the Harnack estimate.

Using Hamilton's entropy monotonicity and trace Harnack estimate, one can prove that there exists a constant $c > 0$ such that $R \geq c$ for all $t \geq 0$. With this, one can simply apply the parabolic maximum principle to (1.31) to conclude that

$$(1.45) \qquad |H|(x, t) \leq C e^{-ct}.$$

Definition 1.26. We say that $g(t)$ is a solution to the **modified Ricci flow** if

$$(1.46) \qquad \partial_t g_{ij} = (r - R) g_{ij} - (\mathcal{L}_{\nabla f} g)_{ij} = -2 H_{ij},$$

where $\mathcal{L}$ denotes the Lie derivative.

Let $\tilde{g}(t)$ be the solution to the modified Ricci flow with $\tilde{g}(0) = g_0$. Then $\tilde{g}(t) = \varphi_t^* g(t)$ for suitable diffeomorphisms φ_t of $\mathcal{M}^2$, where $g(t)$ is the solution to the (unmodified) normalized Ricci flow with $g(0) = g_0$. In particular, the estimate (1.45), which is for a diffeomorphism-invariant quantity (because $H_{\varphi^* g} = \varphi^* H_g$ for any diffeomorphism φ), still holds under the modified Ricci flow (1.46).

From higher derivative estimates for the tensor $H_{\tilde{g}(t)}$ one concludes that $\tilde{g}(t)$ and the corresponding $\tilde{f}(t)$ converge to a C^∞ metric $\tilde{g}_\infty$ and function $\tilde{f}_\infty$ on $\mathcal{M}^2$, respectively, exponentially fast in each C^k norm, satisfying the **shrinking gradient Ricci soliton equation** (in the next chapter we begin to discuss the properties of solutions to this important equation):

$$(1.47) \qquad \mathrm{Ric}_{\tilde{g}_\infty} + \nabla^2_{\tilde{g}_\infty} \tilde{f}_\infty = \frac{r}{2} \tilde{g}_\infty.$$

Using the uniformization theorem in complex analysis, Hamilton showed that any solution to this equation on the 2-sphere must satisfy $R_{\tilde{g}_\infty} = r$ (and $\tilde{f}_\infty$ is constant). Later, Chen, Lu, and Tian [**93**] proved this without assuming the uniformization theorem (we present their proof in Theorem 3.8 below). From all of this, one concludes that the original normalized Ricci flow converges to a constant curvature metric exponentially fast in each C^k norm.

In [**100**] it was shown that for solutions on the 2-sphere $\mathbb{S}^2$ one can remove the condition that $R_{g_0} > 0$. The idea is that both Hamilton's entropy and Harnack estimates, with slight modifications, generalize to the curvature changing sign case. Note that for $\chi > 0$ in the variable sign curvature case, by applying the parabolic maximum principle to the first line of (1.32), we have

$$(1.48) \qquad\qquad\qquad R \geq -Ce^{-rt}.$$

Following the methods of Hamilton, one can show that the curvature becomes positive in finite time and so reduces to the previous case; see Figure 1.11.

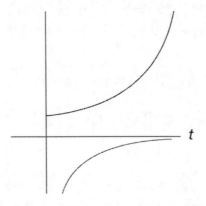

Figure 1.11. When $\chi > 0$, the parabolic maximum principle tells us that $R(x, t)$ is bounded between the two curves. Hamilton's methods yield improved estimates.

1.3.1.6. *Convergence of the normalized Ricci flow on surfaces.* To summarize, by Hamilton's work and a slight extension, we have the following.

Theorem 1.27. *For any Riemannian metric g_0 on a closed orientable surface $\mathcal{M}^2$ there exists a unique solution $g(t)$, $t \in [0, \infty)$, to the normalized Ricci flow (1.18) with $g(0) = g_0$ that converges as $t \to \infty$ to a C^∞ constant curvature metric in the same conformal class.*

In terms of singularity formation, we have the following.

Corollary 1.28. *The only finite-time singularity models associated to the Ricci flow on closed surfaces are the round shrinking 2-sphere and its $\mathbb{Z}_2$-quotient (a shrinking constant curvature real projective plane).*

All other "singularities" form at infinite time and their associated models have constant nonpositive curvature. Historically, a key takeaway is that the *cigar soliton* cannot occur as a singularity model of the Ricci flow on closed surfaces.

1.3.2. The Ricci flow on closed 2-orbifolds.

The result above can be generalized to 2-dimensional closed oriented **orbifolds** $\mathcal{O}^2$ with isolated cone points. In this context, $\mathcal{O}^2$ is locally modelled on $\mathbb{R}^2/\mathbb{Z}_q$, where $q \in \mathbb{N}$ and the action of $\mathbb{Z}_q$ on $\mathbb{R}^2$ is generated by a rotation of angle $2\pi/q$. Here, the origin in $\mathbb{R}^2$ corresponds to a **cone point** in $\mathcal{O}^2$ of order q with cone angle $2\pi/q$.[3] The underlying topological space of $\mathcal{O}^2$ is a closed topological surface of genus g.

Suppose that $\mathcal{O}^2$ has a finite number of cone points of orders $p_1, \ldots, p_k$. Then the (**orbifold**) **Euler characteristic** is defined by

$$(1.49) \qquad \chi(\mathcal{O}^2) = 2 - 2g + \sum_{i=1}^{k} \left(\frac{1}{p_i} - 1 \right).$$

When $\chi(\mathcal{O}^2) \leq 0$, the orbifold $\mathcal{O}^2$ is **good** (i.e., covered by a surface) and Hamilton observed that his work carries over to show that $g(t)$ converges to a constant curvature metric for any initial metric g_0. In the case where $\chi(\mathcal{O}^2) > 0$, the orbifold $\mathcal{O}^2$ can be **bad** (i.e., not good). The bad closed oriented 2-orbifolds have been classified by Thurston as either a **teardrop**, which is a topological $\mathbb{S}^2$ with one cone point, or a **football**, which is a topological $\mathbb{S}^2$ with two cone points of different orders; see Figures 1.12 and 1.13.

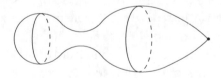

Figure 1.12. A bad teardrop orbifold.

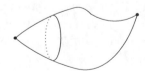

Figure 1.13. A bad football orbifold.

---

[3]For the general definition of orbifold, see Thurston [**270**].

In spite of the presence of cone points, smooth structures can be defined on orbifolds. For example, we say that a function $f : \mathcal{O}^2 \to \mathbb{R}$ is smooth if in each local coordinate chart, when lifted to (an open subset of) $\mathbb{R}^2$, it is smooth; see Figure 1.14. Similarly, we can define the smoothness of tensors on $\mathcal{O}^2$. An important consequence of this is that we may consider smooth Riemannian metrics on $\mathcal{O}^2$. (In a local coordinate chart about a cone point of order q, the metric lifts to a $\mathbb{Z}_q$-equivariant smooth metric on an open subset of $\mathbb{R}^2$.) For such metrics, their scalar curvatures are smooth functions. In particular, the curvature at a cone point is defined and finite. The Gauss–Bonnet theorem holds for a Riemannian 2-orbifold $(\mathcal{O}^2, g)$ and is given by the usual formula

$$(1.50) \qquad \int_{\mathcal{O}^2} R \, d\mu = 4\pi \chi(\mathcal{O}^2).$$

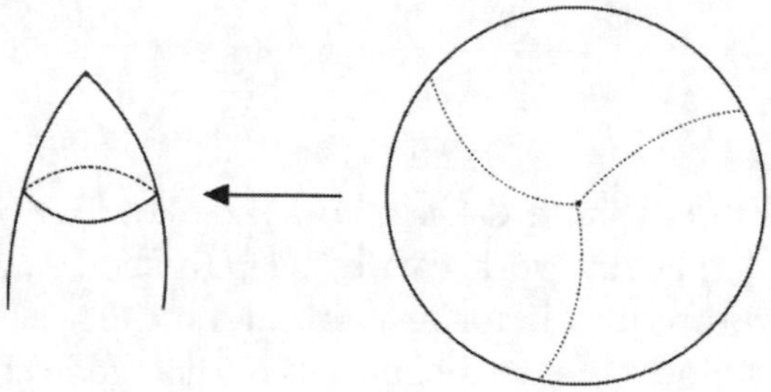

Figure 1.14. A function is smooth near a conical singularity if its local lift (pictured) is smooth.

We may consider the Ricci flow on 2-orbifolds with isolated cone points. When the initial metric has positive curvature $R_{g_0} > 0$, L.-F. Wu [**290**] proved that under the modified Ricci flow (1.46), $g(t)$ converges to a shrinking gradient Ricci soliton metric exponentially fast in each C^k norm. In [**114**] the $R_{g_0} > 0$ condition was removed. In conclusion, we have:

Theorem 1.29 (Uniformization of closed 2-orbifolds). *Let g_0 be any Riemannian metric on a 2-dimensional closed orientable orbifold $\mathcal{O}^2$.*

(1) *(Good orbifolds) If $\mathcal{O}^2$ is good, then there exists a unique solution $g(t)$, $t \in [0, \infty)$, to the normalized Ricci flow (1.18) with $g(0) = g_0$ that converges as $t \to \infty$ to a C^∞ constant curvature metric g_∞ in the same conformal class as g_0; i.e., $R_{g_\infty} \equiv r$.*

(2) *(Bad orbifolds) If $\mathcal{O}^2$ is bad, then there exists a unique solution $g(t)$, $t \in [0, \infty)$, to the modified Ricci flow (1.46) with $g(0) = g_0$ that converges as $t \to \infty$ to a C^∞ shrinking gradient Ricci soliton metric g_∞ with nonconstant positive curvature.*

In terms of the solution $g(t)$ to the normalized Ricci flow (1.18) and the normalized dynamical shrinking gradient Ricci soliton $\tilde{g}_\infty(t)$ (which evolves purely by the pullback by diffeomorphisms), part (2) of the theorem says that $\|g(t) - \tilde{g}_\infty(t)\|_{C^k(\tilde{g}_\infty(t))}$ decays exponentially for each $k \geq 0$.

The main ideas of the proof are the same as those of the proof by Hamilton for the convergence of the Ricci flow on $\mathbb{S}^2$. A key new ingredient is the noncollapsing estimate of Wu [290], which was extended to the variable sign curvature case. One difference with [100] in the variable sign curvature case is that the relative energy is not bounded from below on bad 2-orbifolds, and so the entropy estimate needs to be proved another way, via a delicate modification of Hamilton's original proof. With all of this, one can obtain the exponential decay estimate (1.45) for $|H|$ in the same way as in Hamilton's original proof.

Hamilton has classified the shrinking gradient Ricci soliton metrics on bad 2-orbifolds as rotationally symmetric and unique up to isometry.

Corollary 1.30. *For the Ricci flow on closed 2-orbifolds, the singularity models are the rotationally symmetric shrinking gradient Ricci soliton with cone points of order $q \geq p \geq 1$, where this gradient Ricci soliton has constant curvature if and only if $p = q$.*

For further work on the Ricci flow on surfaces and its variants, see the notes and commentary at the end of this chapter.

1.4. 3-Dimensional singularity formation

1.4.1. Overview.

In Hamilton's original paper [173] he both introduced the Ricci flow and used it to uniformize closed 3-manifolds with positive Ricci curvature as topological spherical space forms. The singularity theory for Ricci flow, especially in dimension 3, was presented by Hamilton in [179], which applied his earlier works [176–178]. The surgery theory was formulated in [181] in dimension 4 for the purpose of the potential implementation in dimension 3. The vision for the long-time behavior of the normalized Ricci flow in dimension 3 was clarified by showing that bounded curvature solutions existing on the "immortal" time interval $[0, \infty)$ are geometrizable [182]. Thus, the foundations for understanding 3-dimensional singularity formation and surgery for Ricci flow to approach Thurston's geometrization conjecture were laid down by Hamilton in these papers.

In a revolutionary series of papers Perelman [250–252] used Hamilton's program to prove Thurston's geometrization conjecture and the subsumed Poincaré conjecture.

For expositions of various aspects of the Perelman–Hamilton proof, see Kleiner and Lott [**201**], Morgan and Tian [**227, 228**], Bessières, Besson, Maillot, Boileau, and Porti [**35**], and Cao and Zhu [**70**]. Colding and Minicozzi [**119**] gave an alternate proof of Perelman's finite extinction time. Q. Zhang [**300**] presents alternative, heat kernel and entropy based, approaches to parts of Perelman's work. Bamler's fundamental works [**14–16**] provide new proofs of aspects of the Perelman–Hamilton theorem.

1.4.2. Neckpinches and degenerate neckpinches.

Heuristically, finite-time singularity formation is governed by neckpinches and degenerate neckpinches; see Hamilton [**179**, §3].

Definition 1.31. We say that a singular solution to the Ricci flow $(\mathcal{M}^n, g(t))$ develops a **neckpinch** if there exists a sequence of space-time points with an associated singularity model being a round shrinking cylinder $\mathbb{S}^{n-1} \times \mathbb{R}$; see Figures 1.15 and 1.16.

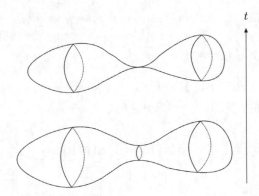

Figure 1.15. Formation of a neckpinch singularity.

Figure 1.16. Rescalings of the forming neckpinch limit to the cylinder. Time t increases from left to right.

A degenerate neckpinch is more difficult to define rigorously. We choose the following.

Definition 1.32. We say that a singular solution to the Ricci flow $(\mathcal{M}^n, g(t))$ develops a **degenerate neckpinch** if there exists a sequence of space-time points with an associated singularity model being the unique rotationally symmetric steady gradient Ricci soliton, called the **Bryant soliton**; see Figure 1.17. The existence and properties of the Bryant soliton are discussed in Chapter 6 of this book.

We remark that degenerate neckpinches are a different class of singularities than neckpinches.

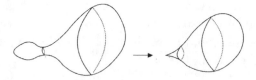

Figure 1.17. Formation of a degenerate neckpinch singularity. Time t increases from left to right.

For now, we just mention that the Bryant soliton is a rotationally symmetric steady gradient Ricci soliton which has positive curvature operator and an underlying smooth manifold diffeomorphic to $\mathbb{R}^n$. It opens up like a paraboloid at infinity and has the norm of the curvature decaying inverse linearly. In addition, any rescaled pointed limit, based at space-time points on the Bryant soliton at a fixed time and tending to spatial infinity, is the round shrinking cylinder; see Figure 1.18. Hence, if a singular solution develops a degenerate neckpinch, then it also develops a neckpinch. We remark that the Bryant soliton is κ-noncollapsed in the sense of Definition 1.12, as is the case for all singularity models by Perelman's no local collapsing theorem.

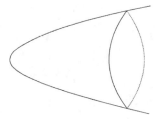

Figure 1.18. The Bryant soliton: asymptotically paraboloidal.

1.4.3. Classification of 3-dimensional singularity models.

Following Perelman's proof of the geometrization theorem, there remained some fundamental open questions regarding 3-dimensional singularity models. Brendle [**42**] proved Perelman's assertion that the Bryant steady gradient Ricci soliton is the unique 3-dimensional complete κ-noncollapsed steady gradient Ricci soliton with positive sectional curvature. Chapter 8 of this book is devoted to an exposition of Brendle's result. Brendle [**46**] further proved Perelman's conjecture that the Bryant steady gradient Ricci soliton is the unique 3-dimensional complete noncompact κ-noncollapsed ancient solution with positive bounded sectional curvature (see also [**45**] for the compact case). An alternative proof was later given by Bamler and Kleiner [**28**], which differed in part from Brendle's proof. In view of the aforementioned works, the understanding of the finite-time behavior of 3-dimensional Ricci flow is relatively complete.

A ramification of the results of Hamilton, Perelman, and Brendle is that 3-dimensional singularity models have been classified. Namely, any 3-dimensional singularity model must be either a *shrinking* or *steady gradient Ricci soliton* and moreover these gradient Ricci solitons are classified. In particular, the only possibilities are: a spherical space form shrinking gradient Ricci soliton, the cylinder shrinking gradient Ricci soliton and its $\mathbb{Z}_2$-quotient, and the *Bryant steady gradient Ricci soliton*. See Chapter 5 of this book for an exposition of the classification of 3-dimensional shrinking gradient Ricci solitons.

We say that a singular solution to the Ricci flow $(\mathcal{M}^n, g(t))$ develops a $\mathbb{Z}_2$-**neckpinch** if there exists a sequence of space-time points with the associated singularity model being the $\mathbb{Z}_2$-quotient of a round shrinking cylinder $\mathbb{S}^{n-1} \times \mathbb{R}$, where the $\mathbb{Z}_2$-action is generated by the map $(x, y) \mapsto (-x, -y)$ for $x \in \mathbb{S}^{n-1}$ and $y \in \mathbb{R}$.

Summarizing, we have the following.

Theorem 1.33 (Classification of 3-dimensional singularity models). *Suppose that $(\mathcal{M}^3, g(t))$ is a singular solution on a closed, orientable 3-manifold. Then any singularity model is either a spherical space form, a round shrinking cylinder $\mathbb{S}^2 \times \mathbb{R}$ (develops a neckpinch), the $\mathbb{Z}_2$-quotient of $\mathbb{S}^2 \times \mathbb{R}$ (develops a $\mathbb{Z}_2$-neckpinch), or the Bryant soliton (develops a degenerate neckpinch).*

1.5. Notes and commentary

§1.1. The Ricci flow was introduced by Hamilton [**173**] in 1982, who used it to prove that any closed 3-manifold with positive Ricci curvature is diffeomorphic to a spherical space form. He developed the theory of singularities in

[**179**]. One of the cornerstones for this theory is Hamilton's matrix Harnack estimate [**177**]. He initially developed surgery theory, for a 4-dimensional test case, in [**181**]. Hamilton's Ricci flow program for the geometrization conjecture crystallized in [**182**] on the infinite-time behavior of the Ricci flow. In revolutionary works, Perelman completed Hamilton's program and proved the Poincaré and geometrization conjectures in [**250–252**]. References for detailed expositions of Perelman's and Hamilton's work are given earlier in this chapter.

§1.2. Hamilton conjectured no local collapsing in dimension 3; cf. [**179**, §15]. Perelman gave two proofs of no local collapsing in all dimensions; see [**250**, §4 and §8].

The local curvature derivative estimates for solutions to Ricci flow were proved by Shi [**264**, §7], following the Bernstein method for obtaining *a priori* derivative estimates for solutions to partial differential equations. Earlier, global derivative estimates on compact manifolds were given by Bando [**31**].

The fundamental compactness theorem for Riemannian manifolds is due to Cheeger [**83**] and Gromov [**162**]. Some further development is by Greene and Wu [**160**], Peters [**253**], and Hamilton [**178**], the latter of whom proved a compactness theorem for solutions to the Ricci flow. A vastly more general compactness theorem for Ricci flow, related to the study of the conjugate heat kernel coupled to the Ricci flow, has been proved by Bamler [**14**].

The topological classification as Euclidean space of complete noncompact Riemannian manifolds with positive sectional curvature is due to Gromoll and Meyer [**161**].

§1.3. For Lemma 1.21 on the parabolic maximum principle, see pp. 101–102 of Hamilton [**172**].

Besides Thurston's book [**270**], for further background on the theory of orbifolds, see e.g. Cooper, Hodgson, and Kerckhoff [**123**].

New proofs of the convergence of the Ricci flow on compact surfaces are given by Bartz, Struwe, and Ye [**32**], Hamilton [**180**], Struwe [**267**], and Andrews and Bryan [**4**].

The Ricci flow on compact surfaces with lower regularity initial data has been studied by Chen and Ding [**92**], Topping and Yin [**279**], and Richard [**261**]. The Ricci flow on compact surfaces with conical singularities has been considered by Yin [**296**], Mazzeo, Rubinstein, and Šešum [**225**], and Ramos [**260**]. For the Ricci flow on surfaces with reverse cusps, see Topping [**277**].

The Ricci flow on noncompact surfaces was initiated by Wu [**291**]; see also Yokota [**297**]. For the Ricci flow on surfaces with cusps, see Daskalopoulos and Hamilton [**125**], Ji, Mazzeo, and Šešum [**198**], Yin [**295**], and Albin, Aldana, and Rochon [**1**]. For incomplete surfaces, see Giesen and Topping [**155, 156**]. The Ricci flow on asymptotically conical surfaces is discussed by Isenberg, Mazzeo, and Šešum [**191**].

For ancient solutions on surfaces, see Hamilton [**179**] and, in the physics literature, King [**200**] and Rosenau [**262**]. For eternal solutions (i.e., existing on the time interval $(-\infty, \infty)$) and "Type II" ancient solutions, see Daskalopoulos and Šešum [**127**] and Chu [**117**].[4] For ancient compact solutions, see Daskalopoulos, Hamilton, and Šešum [**126**].

For flows on surfaces related to the Ricci flow, see Streets [**266**] on the Ricci–Yang–Mills flow and Buzano and Rupflin [**57**] for the harmonic Ricci flow.

§1.4. Bamler [**17–21**] proved Perelman's conjecture that Ricci flow with surgery may be formulated so that only a finite number of surgeries occur. Furthermore, he proved that immortal solutions must be "Type III", i.e., satisfy $|\operatorname{Rm}|(t) \leq C/(1 + t)$.

The **generalized Smale conjecture** says that for any spherical space form $(S^3/\Gamma, g)$, the inclusion map $i : \operatorname{Isom}(S^3/\Gamma, g) \to \operatorname{Diff}(S^3/\Gamma)$ of the isometry group into the diffeomorphism group is a homotopy equivalence. When Γ is the trivial group, this is known as the **Smale conjecture** and was proved by Hatcher [**186**]. A number of cases of the generalized Smale conjecture were proved by Hong, Kalliongis, McCullough, and Rubinstein [**189**]. Finally, Bamler and Kleiner [**27, 29, 30**] gave a complete and remarkable proof of the generalized Smale conjecture by using Ricci flow with surgery. One of the ingredients of their proof is a version of stability for Ricci flow with surgery. The Smale conjecture for hyperbolic 3-manifolds was proved by Gabai [**153**]; there are earlier works by Hatcher and by Ivanov.

In general dimensions, differentiable spherical space form theorems were proved via Ricci flow convergence by Böhm and Wilking [**37**] assuming 2-positivity of the curvature operator and by Brendle and Schoen [**40, 52, 53**] assuming positive isotropic curvature on $\mathcal{M}^n \times \mathbb{R}^2$, a condition which is weaker than 1/4-sectional curvature pinching.

Perelman's techniques [**250, 251**] complete the work of Hamilton on 4-manifolds with positive isotropic curvature. This was carried out in detail by Chen and Zhu [**90**] and by Chen, Tang, and Zhu [**87**], where a technical condition in the former paper was removed. Brendle [**44**] extended Ricci flow

---

[4]See §B.2 for the definitions of Types I, II, and III solutions.

with surgery to higher dimensions assuming certain positivity of curvature conditions on the initial data, which are preserved under the Ricci flow.

1.6. Exercises

1.6.1. A fundamental Bochner-type formula.

In the first three exercises you will derive a formula for the Laplacian of the gradient squared of a function. This Bochner-type formula is fundamental for many basic geometric calculations. This is one of the starting points for the role of Ricci curvature in geometric analysis.

Exercise 1.1. Prove that for any function u on a Riemannian manifold,

$$(1.51) \qquad \Delta |\nabla u|^2 = 2\langle \Delta \nabla u, \nabla u \rangle + 2|\nabla^2 u|^2.$$

HINT: Can you justify calculating as follows? Using the Einstein summation convention and computing at a point where $g_{ij} = \delta_{ij}$,

$$\nabla_j \nabla_j (\nabla_i u \nabla_i u) = 2\nabla_j \nabla_j \nabla_i u \nabla_i u + 2\nabla_j \nabla_i u \nabla_j \nabla_i u.$$

Note, for example, that we denote $\nabla_j \nabla_j \nabla_i u := (\nabla^3 u)(\partial_j, \partial_j, \partial_i)$.

Exercise 1.2 (Commutator formula for Δ and d). Prove the standard Laplacian and covariant derivative commutator formula on a Riemannian manifold:[5]

$$(1.52) \qquad \Delta(du) = d(\Delta u) + \text{Ric}(\nabla u).$$

HINT: Trace the standard covariant derivative formula

$$(1.53) \qquad \nabla_i \nabla_j \nabla_k u = \nabla_j \nabla_i \nabla_k u - R_{ijk\ell} \nabla_\ell u.$$

Note that this last equation exposes our sign convention for Rm. Invariantly, we have

$$(1.54) \qquad \nabla^2_{X,Y}(du) = \nabla^2_{Y,X}(du) - (\text{Rm}(X,Y)\nabla u)^\flat,$$

where for a vector field Z, $Z^\flat$ denotes its dual 1-form.

Exercise 1.3 (Bochner-type formula). Using (1.52), prove the fundamental Bochner-type formula: For any C^2 function u on a Riemannian manifold $(\mathcal{M}^n, g)$,

$$(1.55) \qquad \frac{1}{2}\Delta |\nabla u|^2 = \langle \nabla u, \nabla \Delta u \rangle + |\nabla^2 u|^2 + \text{Ric}(\nabla u, \nabla u).$$

---

[5]This is also written as $\Delta \nabla u = \nabla \Delta u + \text{Ric}(\nabla u)$.

1.6.2. The divergence theorem and integration by parts.

Integration by parts is a basic tool used to study integral quantities in geometric analysis.

Exercise 1.4. Using Stokes's theorem, prove the divergence theorem: For any smooth 1-form α on a compact Riemannian manifold $(\mathcal{M}^n, g)$ with boundary $\partial\mathcal{M}$, we have

$$(1.56) \qquad \int_{\mathcal{M}} \operatorname{div}(\alpha)\, d\mu = \int_{\partial\mathcal{M}} \alpha(\nu)\, d\sigma,$$

where $\operatorname{div}(\alpha) = g^{ij}\nabla_i \alpha_j$ denotes the divergence of α, ν denotes the outward unit normal, and $d\sigma$ denotes the volume form of the boundary. In particular, for any vector field X,

$$(1.57) \qquad \int_{\mathcal{M}} \operatorname{div}(X)\, d\mu = \int_{\partial\mathcal{M}} \langle X, \nu \rangle\, d\sigma,$$

where $\operatorname{div}(X) := \sum_i \nabla_i X^i$. As a further special case, for any smooth function f on a closed Riemannian manifold, we have

$$(1.58) \qquad \int_{\mathcal{M}} \Delta f = 0.$$

Exercise 1.5. Prove Green's identity: If f and h are smooth functions on a compact Riemannian manifold $(\mathcal{M}^n, g)$ with boundary $\partial\mathcal{M}$, then

$$(1.59) \qquad \int_{\mathcal{M}} (f\Delta h - h\Delta f)\, d\mu = \int_{\partial\mathcal{M}} \big(f\nu(h) - h\nu(f)\big)\, d\sigma,$$

where $\nu(h)$ denotes the directional derivative of h in the direction ν.

1.6.3. Exercises on the Ricci flow on surfaces.

In this set of exercises we will explore various steps of Hamilton's proof of the convergence to constant curvature of the Ricci flow on surfaces as outlined in §1.3. Basic ideas include: the parabolic maximum principle, gradient Ricci solitons, the surface entropy monotonicity, and the Harnack estimate. For the full proof, see Hamilton's original research paper [**175**] or the expository book chapter [**108**, Chapter 5]. Recall that the contracted second Bianchi identity says that on any Riemannian manifold $(\mathcal{M}^n, g)$,

$$(1.60) \qquad\qquad \operatorname{div}(\operatorname{Ric}) = \frac{1}{2}dR.$$

Exercise 1.6 (Scalar curvature evolution). Using formula (1.20) for the scalar curvature of a conformal change of a Riemannian metric on a surface, prove formula (1.21) for the evolution of the scalar curvature R under the normalized Ricci flow on surfaces:

$$(1.61) \qquad\qquad \partial_t R = \Delta R + R^2 - rR.$$

Exercise 1.7 (First variation of the Christoffel symbols). Let $g(s)$ be a 1-parameter family of Riemannian metrics on a manifold $\mathcal{M}^n$. Prove that if $v_{ij} = \partial_s g_{ij}$, then

$$(1.62) \qquad \partial_s \Gamma_{ij}^k = \frac{1}{2} g^{k\ell} \left(\nabla_i v_{j\ell} + \nabla_j v_{i\ell} - \nabla_\ell v_{ij} \right).$$

HINT: Compute at a point p and in coordinates $\{x^i\}$ such that $\Gamma_{ij}^k(p) = 0$. Then use the fact that both sides of the equation are the components of tensors to convert partial derivatives into covariant derivatives. This yields a tensorial equation that holds in any coordinates.

Exercise 1.8. Conclude from the previous exercise that, under the same hypotheses,

$$(1.63) \qquad g^{ij} \partial_s \Gamma_{ij}^k = \operatorname{div}(v)^k - \frac{1}{2} \nabla^k \operatorname{tr}(v).$$

In particular, show that under either the Ricci flow or the normalized Ricci flow in all dimensions,

$$(1.64) \qquad g^{ij} \partial_t \Gamma_{ij}^k = 0.$$

Show that if $n = 2$ and $v = ug$, where u is a function, then

$$g^{ij} \partial_s \Gamma_{ij}^k = 0.$$

Exercise 1.9. Assume that $n = 2$. Prove that if a 1-parameter family of metrics $g(s)$ on $\mathcal{M}^2$ satisfies $\partial_s g(s) = u(s) g(s)$, then

$$(1.65) \qquad \partial_s (\Delta_{g(s)} v(s)) = -u(s) \Delta_{g(s)} v(s) + \Delta_{g(s)} (\partial_s v(s)).$$

Conclude that under the normalized Ricci flow on surfaces, if $v(t)$ is a 1-parameter family of functions on $\mathcal{M}^2$, then

$$(1.66) \qquad \partial_t (\Delta v) = (R - r) \Delta v + \Delta(\partial_t v),$$

where all quantities besides v are with respect to $g(t)$.

Exercise 1.10. Prove (1.23): Under the normalized Ricci flow on surfaces, the potential function f satisfies

$$\Delta(\partial_t f) = \Delta(\Delta f + rf).$$

Then derive (1.24) for the evolution of the potential function:

$$\partial_t f = \Delta f + rf.$$

Exercise 1.11 (Gradient of potential function evolution). Using (1.24), show that under the normalized Ricci flow on surfaces, we have

$$(1.67) \qquad \partial_t |\nabla f|^2 = \Delta |\nabla f|^2 - 2|\nabla^2 f|^2 + r|\nabla f|^2.$$

HINT: By the commutator formula (1.52) for Δ and ∇, since $n = 2$ we have $\Delta \nabla f = \nabla \Delta f + \frac{1}{2} R \nabla f$. This will help you compute $(\partial_t - \Delta) \nabla f$. Note also that $|\nabla f|^2 = g^{ij} \nabla_i f \nabla_j f$ and that $\partial_t g^{ij} = (R - r) g^{ij}$.

Exercise 1.12. Use the previous exercise to derive the evolution equation (1.30):

$$\partial_t h = \Delta h - 2|H|^2 + rh$$

for $h = R + |\nabla f|^2 - r$ under the normalized Ricci flow on surfaces.

Exercise 1.13 (Difference from being a Ricci soliton evolution). Using the fact that $\partial_t H = \Delta H + (r - 2R)H$ (see the original [**175**, §9.1] or the expository [**108**, (5.12)]), prove the evolution equation (1.31):

$$\partial_t |H|^2 = \Delta |H|^2 - 2|\nabla H|^2 - 2R|H|^2$$

under the normalized Ricci flow on surfaces.

Exercise 1.14 (Hamilton's surface entropy evolution). Using the fact that under the normalized Ricci flow on surfaces, Hamilton's entropy satisfies

$$(1.68) \qquad d_t N = -\int_{\mathcal{M}} \frac{|\nabla R|^2}{R} d\mu + \int_{\mathcal{M}} (R - r)^2 d\mu$$

(cf. [**175**, §7]), derive the surface entropy formula (1.40) via integration by parts.

HINT: $|H|^2 = |\nabla^2 f|^2 - \frac{1}{2}(\Delta f)^2$, and $\nabla f \cdot \nabla \Delta f = \nabla f \cdot \Delta \nabla f - \frac{1}{2}R|\nabla f|^2$.

Exercise 1.15 (Log curvature evolution). Show that under the normalized Ricci flow on surfaces (1.18), if $R > 0$, then

$$\partial_t \ln R = \Delta \ln R + |\nabla \ln R|^2 + R - r.$$

Use this, the fact that $\partial_t \Delta = (R - r)\Delta$, and the formula for commuting Δ and ∇ to prove the trace Harnack formula (1.44) on surfaces.

Exercise 1.16 (Entropy integrand evolution). Show that under the normalized Ricci flow on surfaces with $R > 0$,

$$(1.69)$$

$$(\partial_t - \Delta)(R \ln R \, d\mu) = \left(-\frac{|\nabla R - R\nabla f|^2}{R} - 2|H|^2 + \Delta\left(|\nabla f|^2 - rf\right) \right) d\mu,$$

where H is defined by (1.29). Give another derivation of (1.40) using this formula.

Exercise 1.17 (Harnack quantity evolution). Prove that under (1.18) with $R > 0$, Hamilton's Harnack quantity Q defined in (1.41) satisfies

$$(1.70) \qquad \left(\frac{\partial}{\partial t} - \Delta - 2\nabla \ln R \cdot \nabla \right) Q = 2 \left| \mathrm{Ric} + \nabla^2 \ln R - \frac{r}{2} g \right|^2 + rQ.$$

Deduce (1.44) and (1.42). This gives a proof of Hamilton's trace Harnack theorem (see Theorem 1.25).

Exercise 1.18 (Gradient Ricci solitons on $\mathbb{S}^2$ have constant curvature). Let g be a Riemannian metric on a surface $\mathcal{M}^2$ diffeomorphic to $\mathbb{S}^2$ and with vanishing 2-tensor $H := \mathrm{Ric} + \nabla^2 f - \frac{r}{2} = 0$, as defined in (1.29). Firstly, observe that $X := \nabla f$ is a **conformal vector field**; that is,

$$(1.71) \qquad \mathcal{L}_X g = \frac{1}{n} \mathrm{tr}(\mathcal{L}_X g)\, g,$$

where $n = 2$ in our case. Secondly, use the Kazdan–Warner identity (Theorem A.1 in Appendix A of this book) and integration by parts to show that $R \equiv r$ on $\mathcal{M}^2$.

1.6.4. Exercises on general Ricci flow properties.

In this set of exercises we consider various Ricci flow properties that are relevant to gradient Ricci solitons.

Exercise 1.19 (Product solutions). Let $(\mathcal{M}^{n_1}, g_1(t))$ and $(\mathcal{M}^{n_2}, g_2(t))$ be solutions to the Ricci flow. Show that $(\mathcal{M}_1^{n_1} \times \mathcal{M}_2^{n_2}, g_1(t) + g_2(t))$ is a solution to the Ricci flow.

Exercise 1.20 (Preservation of isometries). Let $(\mathcal{M}^n, g(t))$, $t \in I$, be a solution to the Ricci flow on a closed manifold. Let $t_1 < t_2$ be two times in I. Prove that if $\psi : \mathcal{M}^n \to \mathcal{M}^n$ is an isometry of $g(t_1)$, then ϕ is an isometry of $g(t_2)$.
HINT: Use the uniqueness of solutions of the Ricci flow with a given initial metric.

Exercise 1.21 (Covering Ricci flows). Let $(\mathcal{M}^n, g(t))$ be a solution to the Ricci flow and let $\pi : \tilde{\mathcal{M}}^n \to \mathcal{M}^n$ be a covering space of $\mathcal{M}^n$. Prove that $(\tilde{\mathcal{M}}^n, \pi^* g(t))$ is a solution to the Ricci flow.

Exercise 1.22 (Einstein solutions). Show that if $(\mathcal{M}^n, g)$ is an Einstein manifold satisfying $\mathrm{Ric} = \frac{\lambda}{2} g$, where $\lambda \in \mathbb{R}$, then $g(t) := (1 - \lambda t) g$, defined for all $t \in \mathbb{R}$ satisfying $1 - \lambda t > 0$, is a solution to the Ricci flow on $\mathcal{M}^n$.

Exercise 1.23 (Positive scalar curvature Einstein solutions are κ-noncollapsed). Let $g(t) = (1 - \lambda t) g$ be as in the previous exercise, where we now assume that $\lambda > 0$. Show that $g(t)$ is a singular solution. Show that there exists $\kappa > 0$ such that for each $t \in \left(-\infty, \frac{1}{\lambda} \right)$, $g(t)$ is κ-noncollapsed on all scales.

Exercise 1.24 (Noncollapsing and products). Suppose that $(\mathcal{M}^n, g)$ is κ-noncollapsed on the scale ρ and that $m \in \mathbb{Z}_+$. Prove that there exists $\kappa' > 0$ such that $(\mathcal{M}^n, g) \times \mathbb{R}^m$ is κ'-noncollapsed on the scale ρ.

Exercise 1.25 (κ-Noncollapsedness of 2-dimensional cylinders). Consider the noncompact surface $\mathcal{M}^2 = S^1 \times \mathbb{R}$ with the standard metric. Suppose that $\mathcal{M}^2$ is $\kappa(\rho)$-noncollapsed on the scale ρ. Prove that $\kappa(\rho) \to 0$ as $\rho \to \infty$. Can you generalize this example?

Exercise 1.26 (Scalar curvature lower bound). Under the Ricci flow of a metric $g(t)$ on a compact manifold $\mathcal{M}^n$, the scalar curvature satisfies the heat-type equation $\partial_t R = \Delta R + 2|\mathrm{Ric}|^2$. Use this and the parabolic maximum principle to prove that for any solution on a time interval $[\alpha, \omega)$, we have the lower bound

$$(1.72) \qquad R(x,t) \geq -\frac{n}{2(t-\alpha)} \quad \text{for all } x \in \mathcal{M}^n,\ t \in (\alpha, \omega).$$

HINT: Use the inequality $|\mathrm{Ric}|^2 \geq \frac{1}{n}R^2$.

1.6.5. Further exercises in Riemannian geometry and geometric analysis.

Some basic tools we will use to study Ricci solitons are the Cheeger–Gromov compactness theorem, injectivity radius estimates, and local derivatives of curvature estimates. In this set of exercises, we get more familiar with these tools.

Exercise 1.27 (Model solutions to the heat equation for derivative estimates). Define a sequence of solutions to the heat equation on the unit circle defined by

$$u_k(x,t) = e^{-k^2 t}\sin(kx), \quad x \in \mathbb{S}^1 = \mathbb{R}/2\pi\mathbb{Z},\ t \geq 0.$$

Prove that for each $m \in \mathbb{Z}_+$ there exists a constant C_m such that

$$|\partial_x^m u_k(x,t)| \leq \frac{C_m}{t^{m/2}} \quad \text{for all } x \in \mathbb{S}^1,\ t > 0,\ k \in \mathbb{Z}_+.$$

Exercise 1.28 (Scaling up compact manifolds). Let $(\mathcal{M}^n, g)$ be a compact Riemannian manifold, let $p_i \in \mathcal{M}^n$, and let $0 < r_i \to \infty$. Prove that any pointed C^∞ Cheeger–Gromov limit of $(\mathcal{M}^n, r_i^2 g, p_i)$ is Euclidean n-space (recall Definition 1.8).

Exercise 1.29 (Scaling up noncompact manifolds). Give an example of a complete noncompact Riemannian manifold $(\mathcal{M}^n, g)$, a sequence of points p_i in $\mathcal{M}^n$, and $0 < r_i \to \infty$ such that the pointed Cheeger–Gromov limit of $(\mathcal{M}^n, r_i^2 g, p_i)$ is not Euclidean n-space.

Exercise 1.30 (Collapsing sequence of spherical space forms). Prove that there exists a sequence of compact 3-manifolds with constant sectional curvature 1 with injectivity radius tending to zero.

Exercise 1.31 (Berger spheres). Prove that there exists a sequence of Riemannian metrics on $\mathbb{S}^3$ with positive sectional curvature bounded above by 1 and with injectivity radii tending to zero.

Exercise 1.32. Show that there exists a sequence of 4-dimensional complete noncompact Riemannian manifolds with positive Ricci curvature bounded above by 1 and with injectivity radii tending to zero.

1.6.6. Exercises on calculations in geometric analysis.

Formulas are at the heart of geometric analysis. Calculations appear throughout the book. In this set of exercises, we consider a few calculations.

Exercise 1.33 (First variation of the determinant). Let $A(s)$ be a 1-parameter family of invertible square matrices. Show that

$$(1.73) \qquad \frac{d}{ds} \ln \det A = \left(A^{-1}\right)^{ij} \frac{d}{ds} A_{ij}.$$

Exercise 1.34 (Laplacian in local coordinates). Show that the Laplacian acting on functions on a Riemannian manifold $(\mathcal{M}^n, g)$ may be expressed as

$$(1.74) \qquad \Delta f = \frac{1}{\sqrt{|g|}} \frac{\partial}{\partial x^i} \left(\sqrt{|g|} g^{ij} \frac{\partial f}{\partial x^j} \right),$$

where $|g| := \det(g_{ij})$.
HINT: Use

$$(1.75) \qquad \Delta u = \sum_{i,j=1}^{n} g^{ij} \nabla_i \nabla_j u = \sum_{i,j=1}^{n} g^{ij} \left(\frac{\partial^2 u}{\partial x^i \partial x^j} - \Gamma_{ij}^{k} \frac{\partial u}{\partial x^k} \right).$$

Exercise 1.35 (Laplacian of the norm squared). Prove that if α is a 1-form on a Riemannian manifold $(\mathcal{M}^n, g)$, then

$$(1.76) \qquad \frac{1}{2} \Delta |\alpha|^2 = \langle \Delta \alpha, \alpha \rangle + |\nabla \alpha|^2.$$

HINT: Justify the local coordinate calculation

$$\frac{1}{2} \nabla_j \nabla_j \left((\alpha_i)^2 \right) = \nabla_j (\alpha_i \nabla_j \alpha_i) = \nabla_j \alpha_i \nabla_j \alpha_i + \alpha_i \nabla_j \nabla_j \alpha_i.$$

Exercise 1.36 (Identity applicable to Killing vector fields). Show that for any vector field U on a Riemannian manifold $(\mathcal{M}^n, g)$,

$$(1.77) \qquad \Delta U + \operatorname{Ric}(U) = \operatorname{div}(\mathcal{L}_U g) - \frac{1}{2} \nabla \operatorname{tr}(\mathcal{L}_U g).$$

In particular, if V is a Killing vector field, i.e., $\mathcal{L}_V g = 0$, then

$$\Delta V + \operatorname{Ric}(V) = 0.$$

Chapter 2

The Ricci Soliton Equation

In this chapter we familiarize ourselves with the Ricci soliton equation. In particular, we see how Ricci solitons are dynamically self-similar solutions to the Ricci flow and we consider special examples. We consider the special case of gradient Ricci solitons, which are the main objects of study in this book. By differentiating the Ricci soliton equation, we derive fundamental and useful identities. Regarding the qualitative study of Ricci solitons, we discuss the lower bound for the scalar curvature, completeness of the Ricci soliton vector field, and the uniqueness theorem for compact Ricci solitons.

A **Ricci soliton structure** is a quadruple $(\mathcal{M}^n, g, X, \lambda)$ consisting of a smooth manifold $\mathcal{M}^n$, a Riemannian metric g, a smooth vector field X, and a real constant λ, which together satisfy the equation

$$(2.1) \qquad \operatorname{Ric} + \frac{1}{2}\mathcal{L}_X g = \frac{\lambda}{2}g$$

on $\mathcal{M}^n$, where Ric denotes the Ricci tensor of g and where $\mathcal{L}$ denotes the Lie derivative. We include the factor of one half in order to slightly simplify certain fundamental equations which follow.

Tracing (2.1), we have

$$(2.2) \qquad R + \operatorname{div} X = \frac{n\lambda}{2},$$

where R is the scalar curvature of g and $\operatorname{div} X = \operatorname{tr}(\nabla X) = \sum_{i=1}^{n} \nabla_i X^i$ denotes the divergence of X; cf. (1.43). Here, ∇ is the Riemannian covariant derivative.

Note that when we write ∇f, where f is a function, this could mean either (1) the covariant derivative, which is equal to the exterior derivative,

$\nabla f = df$, or (2) the gradient ∇f, which is the vector field metrically dual to the 1-form df. In local coordinates,

$$\nabla_i f := (df)_i = \frac{\partial f}{\partial x^i} \quad \text{and} \quad \nabla^i f := (\nabla f)^i = g^{ij} \nabla_j f.$$

The most important class of Ricci solitons, and the primary focus of this book, is those for which $X = \nabla f$ for some smooth function f on $\mathcal{M}^n$. For these so-called **gradient Ricci solitons**, equation (2.1) simplifies to

$$(2.3) \qquad\qquad\qquad \mathrm{Ric} + \nabla^2 f = \frac{\lambda}{2} g,$$

since $\mathcal{L}_{\nabla f} g = 2\nabla^2 f$ (see (2.28) below if you have not seen this formula). Here, ∇^2 denotes the Hessian, i.e., the second covariant derivative. This acts on tensors, and when acting on a function f, $\nabla^2 f = \nabla df$. We will often use the abbreviation **GRS** for gradient Ricci soliton.

We will use the notation $(\mathcal{M}^n, g, f, \lambda)$ to denote a gradient Ricci soliton structure. When the **expansion constant** (or **scale**) λ is fixed and the **potential function** f is known or can be determined from the context at hand, we will often simply refer to the underlying manifold $(\mathcal{M}^n, g)$ as *the* Ricci soliton.[1]

2.1. Riemannian symmetries and notions of equivalence

The groups $\mathbb{R}_+$ of positive real numbers and $\mathrm{Diff}(\mathcal{M}^n)$ of diffeomorphisms act naturally by dilation via $\alpha \cdot g = \alpha g$ and pull back via $\phi \cdot g = \phi^* g$, respectively, on the space $\mathrm{Met}(\mathcal{M}^n)$ of Riemannian metrics on $\mathcal{M}^n$. Via the scaling and diffeomorphism invariances

$$(2.4) \qquad\qquad \mathrm{Ric}(\alpha g) = \mathrm{Ric}(g), \quad \mathrm{Ric}(\phi^* g) = \phi^* \mathrm{Ric}(g)$$

of the Ricci tensor, they act on Ricci solitons $(\mathcal{M}^n, g, X, \lambda)$ as follows:

(1) (Metric scaling) If $\alpha \in \mathbb{R}_+$, then $(\mathcal{M}^n, \alpha g, \alpha^{-1} X, \alpha^{-1} \lambda)$ is a Ricci soliton.

(2) (Diffeomorphism invariance) If $\varphi : \mathcal{N}^n \to \mathcal{M}^n$ is a diffeomorphism, then $(\mathcal{N}^n, \varphi^* g, \varphi^* X, \lambda)$ is a Ricci soliton.

Observe also that if K is a Killing vector field, then $(\mathcal{M}^n, g, X + K, \lambda)$ is a Ricci soliton. We leave it as an exercise to check these properties (see Exercise 2.6). Only the *sign* of the expansion constant λ is of material significance, since, according to property (1), we can adjust the magnitude of a nonzero λ arbitrarily by multiplying g and X by appropriate positive

---

[1] In the case where $\mathcal{M}^n$ is closed, the function f is the same as the potential function defined by (1.22) since by tracing (2.3) we have that $\Delta f = \frac{n\lambda}{2} - R$ and $\frac{n\lambda}{2}$ must be equal to the average scalar curvature.

factors. We will see shortly that each Ricci soliton gives rise at least to a lo-cally defined *self-similar* solution to the Ricci flow, with the scaling behavior determined by whether λ is positive, negative, or zero. This characteristic scaling behavior motivates the following terminology.

Definition 2.1 (Types of Ricci solitons). A Ricci soliton $(\mathcal{M}^n, g, X, \lambda)$ is said to be **shrinking** if $\lambda > 0$, **expanding** if $\lambda < 0$, and **steady** if $\lambda = 0$.

For brevity, we will often simply refer to such Ricci solitons as **shrinkers**, **expanders**, or **steadies**. When working within one of these classes of Ricci solitons, we will usually normalize the structure so that λ is 1, -1, or 0 and suppress further mention of it.[2] For example, the shrinking GRS equation is

$$(2.5) \qquad \mathrm{Ric} + \nabla^2 f = \frac{1}{2}g.$$

In §2.2 we will see, via the equivalent dynamical version of Ricci solitons, the reasons for the terminologies shrinking, expanding, and steady.

We will say that two Ricci soliton structures $(\mathcal{M}_i^n, g_i, X_i, \lambda_i)$, $i = 1, 2$, are **equivalent** if $\lambda_1 = \lambda_2$ and the underlying Riemannian manifolds $(\mathcal{M}_i^n, g_i)$ are isometric. An isometry $\phi : (\mathcal{M}_1^n, g_1) \to (\mathcal{M}_2^n, g_2)$ need not pull back X_2 to X_1, however, since

$$(2.6) \qquad \mathrm{Ric}(g_1) - \frac{\lambda_1}{2}g_1 = \phi^*\left(\mathrm{Ric}(g_2) - \frac{\lambda_2}{2}g_2\right),$$

and we have (see Exercise 2.3)

$$\mathcal{L}_{X_1}g_1 = \phi^*(\mathcal{L}_{X_2}g_2) = \mathcal{L}_{\phi^*X_2}\phi^*g_2 = \mathcal{L}_{\phi^*X_2}g_1,$$

so

$$(2.7) \qquad \mathcal{L}_{(\phi^*X_2-X_1)}g_1 = 0;$$

i.e., the difference $\phi^*X_2 - X_1$ will at least be a Killing vector field on $(\mathcal{M}_1^n, g_1)$. In particular, it is not difficult to see that $(\mathcal{M}_1^n, g, X_1, \lambda)$ and $(\mathcal{M}^n, g, X_2, \lambda)$ are equivalent if and only if $X_2 - X_1$ is a Killing vector field.

2.2. Ricci solitons and Ricci flow self-similarity

The scaling and diffeomorphism invariances of the Ricci tensor (2.4) manifest themselves in symmetries of the Ricci flow equation. If $g(t)$ is a solution to the Ricci flow on $\mathcal{M}^n \times [c, d]$, then, for any fixed $\alpha > 0$ and $\phi \in \mathrm{Diff}(\mathcal{M}^n)$,

$$\tilde{g}(t) := \alpha(\phi^*g)(t/\alpha)$$

is a solution on $\mathcal{M}^n \times [\alpha c, \alpha d]$. From a geometric perspective, these solutions are essentially the same: For each t, $g(t/\alpha)$ and $\tilde{g}(t)$ are isometric but for a

---

[2]Strictly speaking, no normalization is required if $\lambda = 0$.

homothetical constant. A solution to the Ricci flow which moves exclusively under these symmetries, that is, which has the form

$$(2.8) \qquad\qquad g(t) = c(t)\phi_t^* \bar{g}$$

for some fixed metric $\bar{g}$ and positive smooth function $c(t)$ and smooth family of diffeomorphisms ϕ_t, is therefore essentially stationary from a geometric perspective. To wit, Ricci solitons are the fixed points of the Ricci flow in the space of metrics modulo scalings and diffeomorphisms. Such solutions are said to be **self-similar**.

The following proposition demonstrates that Ricci solitons and self-similar solutions are two sides of the same coin: A self-similar solution defines a Ricci soliton structure on each time-slice, and conversely a Ricci soliton structure gives rise to an (at least locally defined) self-similar solution.[3] The interplay between the two perspectives, one static and one dynamic, is fundamental to the analysis of Ricci solitons. The following is our first formulation; we reformulate it slightly later.

Proposition 2.2 (Canonical form, I). *Let $(\mathcal{M}^n, g_0)$ be a Riemannian manifold.*

(a) *Suppose that $g(t) = c(t)\phi_t^* g_0$ satisfies the Ricci flow on $\mathcal{M}^n \times (\alpha, \omega)$ for some positive smooth function $c : (\alpha, \omega) \to \mathbb{R}$ and smooth family of diffeomorphisms $\{\phi_t\}_{t \in (\alpha, \omega)}$. Then, for each $t \in (\alpha, \omega)$, there is a vector field $X(t)$ and a scalar $\lambda(t)$ such that $(\mathcal{M}^n, g(t), X(t), \lambda(t))$ satisfies the Ricci soliton equation (2.1).*

(b) *Suppose that $(\mathcal{M}^n, g_0, X, \lambda)$ satisfies the Ricci soliton equation (2.1) for some smooth vector field X and constant λ. Then, for each $x_0 \in \mathcal{M}^n$, there is a neighborhood U of x_0, an interval (α, ω) containing 0, a smooth family $\phi_t : U \to \mathcal{M}^n$ of injective local diffeomorphisms, and a smooth positive function $c : (\alpha, \omega) \to \mathbb{R}$ such that $g(t) = c(t)\phi_t^* g_0$ solves the Ricci flow on $U \times (\alpha, \omega)$ with $g(0) = g_0$.*

Proof. Suppose first that $g(t) = c(t)\phi_t^* g_0$ solves the Ricci flow on $\mathcal{M}^n \times (\alpha, \omega)$. Fix $a \in (\alpha, \omega)$. Differentiating $g(t)$ at a yields

$$\left.\frac{\partial}{\partial t}\right|_{t=a} g(t) = c'(a)\phi_a^* g_0 + c(a) \left.\frac{\partial}{\partial t}\right|_{t=a} \phi_t^* g_0.$$

Now,

$$\left.\frac{\partial}{\partial t}\right|_{t=a} \phi_t^* g_0 = \left.\frac{\partial}{\partial t}\right|_{t=0} (\phi_a^{-1} \circ \phi_{a+t})^* \phi_a^* g_0 = \mathcal{L}_{X(a)}\phi_a^* g_0,$$

---

[3] If g is complete, then one obtains a globally defined self-similar solution; see Theorem 2.27 below.

where $X(a)$ is the generator of the family $\phi_a^{-1} \circ \phi_{a+t}$, so, taking $\lambda(a) = -c'(a)/c(a)$ and using that $g(t)$ solves the Ricci flow, we obtain a solution $(\mathcal{M}^n, g(a), X(a), \lambda(a))$ to the Ricci soliton equation (2.1).

On the other hand, suppose that $(\mathcal{M}^n, g_0, X, \lambda)$ satisfies (2.1) and that $x_0 \in \mathcal{M}^n$. By the local existence theory for ODEs (see, for example, Theorem 9.12 of [**216**]), there are open neighborhoods U, V of x_0 with $U \subset V$, $\epsilon > 0$, and a smooth family of injective local diffeomorphisms $\psi_s : U \to V$, $s \in (-\epsilon, \epsilon)$, such that $\psi_0(x) = x$ and

$$\frac{\partial}{\partial s}\bigg|_{s=a} \psi_s(x) = X(\psi_a(x))$$

on $U \times (-\epsilon, \epsilon)$.

When $\lambda \neq 0$, define $\omega = \min\{\epsilon, |\lambda|\}$ and $\alpha = -\omega$, and, for $t \in (\alpha, \omega)$, let

$$c(t) = 1 - \lambda t, \quad \phi_t = \psi_{s(t)},$$

where

$$s(t) = -\frac{1}{\lambda} \ln(1 - \lambda t).$$

Then $g(t) = c(t)\phi_t^* g_0$ satisfies $g(0) = g_0$ and

$$\begin{aligned}
\frac{\partial g}{\partial t} &= c'(t)\psi_{s(t)}^* g_0 + c(t)s'(t)\psi_{s(t)}^* \mathcal{L}_X g_0 \\
&= -\lambda \phi_t^* g_0 + \phi_t^*(-2\mathrm{Ric}(g_0) + \lambda g_0) \\
&= -2\mathrm{Ric}(g(t))
\end{aligned}$$

on $U \times (\alpha, \omega)$.

When $\lambda = 0$,

$$\frac{\partial}{\partial t}\psi_t^* g_0 = \psi_t^* \mathcal{L}_X g_0 = -2\psi_t^* \mathrm{Ric}(g_0) = -2\mathrm{Ric}(g(t))$$

on $U \times (-\epsilon, \epsilon)$ so (b) is verified in this case with $c(t) = 1$ and $\phi_t = \psi_t$. $\square$

The interval of existence of the solution in the second half of the above proposition is constrained by the maximum domain of definition of the 1-parameter family of diffeomorphisms generated by the vector field X. However, as we will see in §2.8 below, the vector field X will in most cases of interest generate a flow for all $t \in \mathbb{R}$ (i.e., X is a complete vector field), and in these settings the correspondence between self-similar solutions and Ricci solitons is symmetric.

When the vector field X generates a global flow, the interval of definition for the self-similar solution will be at least as large as that permitted by the Ricci soliton type, namely, $(-\infty, \lambda^{-1})$ for shrinkers, $(-\infty, \infty)$ for steadies, and $(-\lambda^{-1}, \infty)$ for expanders. The lifetime of a self-similar solution may extend beyond these intervals. This phenomenon occurs, for example, in the shrinking and expanding self-similar solutions arising from the Gaussian soliton; see (2.9) immediately below.

2.3. Special and explicitly defined Ricci solitons

In this section we consider some important examples and special classes of Ricci solitons.

2.3.1. The Gaussian soliton.

For $\lambda \in \mathbb{R}$, the structure $(\mathbb{R}^n, g_{\mathrm{Euc}}, f_{\mathrm{Gau}}, \lambda)$, where

$$(2.9) \qquad g_{\mathrm{Euc}} = \sum_{i=1}^{n} dx^i \otimes dx^i \quad \text{and} \quad f_{\mathrm{Gau}}(x) = \frac{\lambda}{4} |x|^2 \,,$$

is called the **Gaussian soliton**. Thus, Euclidean space can be regarded as a Ricci soliton of shrinking, expanding, or steady type. Observe that the choice of potential function $f = f_{\mathrm{Gau}}$ is not unique: Any function of the form $f(x) = \frac{\lambda}{4} |x|^2 + \langle a, x \rangle + b$, where $a \in \mathbb{R}^n$ and $b \in \mathbb{R}$, yields an equivalent Ricci soliton structure.

The self-similar solution to the Ricci flow associated to the Gaussian soliton is static for any choice of λ. It is instructive to carry out the construction in Proposition 2.2 for this simple case explicitly. Integrating the vector field

$$(2.10) \qquad \nabla f = \frac{\lambda x^i}{2} \frac{\partial}{\partial x^i}$$

produces the 1-parameter family of diffeomorphisms $\tilde{\phi}_t(x) = e^{\frac{\lambda t}{2}} x$. Following Proposition 2.2 and taking $\phi_t = \tilde{\phi}_{-\lambda^{-1} \ln(1-\lambda t)}$ when $\lambda \neq 0$ and $\phi_t = \tilde{\phi}_t$ when $\lambda = 0$, we find that

$$(2.11) \qquad \phi_t(x) = (1 - \lambda t)^{-1/2} x$$

and hence that the associated solution $g(t)$ is

$$(2.12) \qquad g(t) = (1 - \lambda t) \phi_t^* g_{\mathrm{Euc}} = g_{\mathrm{Euc}}.$$

When $\lambda \neq 0$, the family of diffeomorphisms ϕ_t — and by extension, the solution provided by Proposition 2.2 — is defined only for $t \in (-\infty, \lambda^{-1})$ or $t \in (\lambda^{-1}, \infty)$ depending on whether λ is positive or negative. However, the solution $g(t)$ is well-defined by the rightmost expression for all $t \in (-\infty, \infty)$.

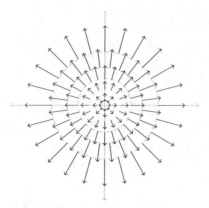

Figure 2.1. The gradient of the potential function $\nabla f = \frac{x^i}{2}\frac{\partial}{\partial x^i}$ for the Gaussian shrinker. Since ∇f points away from the origin, the pullback by ϕ_t expands the metric, which we have to *shrink* to keep the metric static.

2.3.2. Shrinking round spheres.

The metrics of constant positive curvature on the sphere $\mathbb{S}^n$ are naturally shrinking gradient Ricci solitons, when paired with any constant potential function. If $g_{\mathbb{S}^n}$ is the round metric of constant sectional curvature equal to one, the rescaled metric

$$(2.13) \qquad g = 2(n-1)g_{\mathbb{S}^n}$$

will satisfy (2.3) with the canonical choice of constant $\lambda = 1$. For definiteness, we will call $(\mathbb{S}^n, g, n/2)$ the **shrinking round sphere**. (The choice of $f = n/2$ is a convenience that we will explain later.)

The associated self-similar solution is the family $g(t) = (1-t)g$ defined for $t \in (-\infty, 1)$ which simply contracts homothetically as time increases before vanishing identically at $t = 1$. For $t < 1$, the metrics $g(t)$ have radius $r(t) = \sqrt{2(n-1)t}$ and constant sectional curvature $\mathrm{sect}(t) \equiv 1/2(n-1)t$.

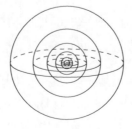

Figure 2.2. A shrinking round sphere.

2.3.3. Einstein manifolds.

The preceding example can be generalized in the following way. To any Einstein manifold $(\mathcal{M}^n, g)$, with

$$(2.14) \qquad\qquad \mathrm{Ric} = \frac{\lambda}{2} g,$$

of constant scalar curvature $n\lambda/2$, we may naturally associate a Ricci soliton structure of the form $(\mathcal{M}^n, g, f, \lambda)$ of (2.3) with $f = \mathrm{const}$. In particular, every manifold of constant sectional curvature admits a Ricci soliton structure.

If a Ricci soliton $(\mathcal{M}^n, g, X, \lambda)$ is Einstein with constant $\lambda/2$, then

$$(2.15) \qquad\qquad \mathcal{L}_X g = \frac{\lambda}{2} g - \mathrm{Ric} = 0;$$

i.e., the vector field X is Killing. Thus it is no loss of generality to assume that such an Einstein soliton is gradient relative to a constant potential f. (However, the example of the Gaussian soliton demonstrates that an Einstein manifold may give rise to Ricci soliton structures of more than one type.)

As with the shrinking spheres, the self-similar solutions corresponding to the Einstein solitons evolve purely by scaling. Depending on the sign of λ, the solution $g(t) = (1 - \lambda t)g$ associated to a metric g satisfying (2.14) will shrink, expand, or remain fixed for all t in a maximal interval determined by λ, that is, for all t such that $1 - \lambda t > 0$.

While non-Einstein (a.k.a. **nontrivial**) Ricci solitons will occupy most of our attention, Einstein solitons are nevertheless of fundamental importance in their own right and as building blocks in the construction of other Ricci solitons.

2.3.4. Product solitons.

If $(\mathcal{M}_1^{n_1}, g_1)$ and $(\mathcal{M}_2^{n_2}, g_2)$ are Riemannian manifolds, then the Ricci tensor of the product manifold $(\mathcal{M}_1^{n_1} \times \mathcal{M}_2^{n_2}, g_1 + g_2)$ is itself a product:

$$(2.16) \qquad\qquad \mathrm{Ric}(g_1 + g_2) = \mathrm{Ric}(g_1) + \mathrm{Ric}(g_2).$$

Here and below, for tensors α_i on $\mathcal{M}_i^{n_i}$, $i = 1, 2$, we will write

$$(2.17) \qquad\qquad \alpha_1 + \alpha_2 := p_1^*(\alpha_1) + p_2^*(\alpha_2),$$

where $p_i : \mathcal{M}_1^{n_1} \times \mathcal{M}_2^{n_2} \to \mathcal{M}_i^{n_i}$ denotes the projection map. It follows that if $(\mathcal{M}_1^{n_1}, g_1, f_1, \lambda)$ and $(\mathcal{M}_2^{n_2}, g_2, , f_2, \lambda)$ are gradient Ricci soliton structures on $\mathcal{M}_1^{n_1}$ and $\mathcal{M}_2^{n_2}$, respectively, then

$$(2.18) \qquad\qquad (\mathcal{M}_1^{n_1} \times \mathcal{M}_2^{n_2}, g_1 + g_2, f_1 + f_2, \lambda)$$

is a gradient Ricci soliton structure on $\mathcal{M}_1^{n_1} \times \mathcal{M}_2^{n_2}$. More generally, given two Ricci soliton structures $(\mathcal{M}_i^{n_i}, g_i, X_i, \lambda)$ on $\mathcal{M}_i^{n_i}$, $i = 1, 2$, we have that $(\mathcal{M}_1^{n_1} \times \mathcal{M}_2^{n_2}, g_1 + g_2, (X_1, X_2), \lambda)$ is a Ricci soliton structure on $\mathcal{M}_1^{n_1} \times \mathcal{M}_2^{n_2}$.

For instance, by taking the product of the Gaussian shrinker with the shrinking round sphere of dimension $k \geq 2$, we obtain the **round-cylindrical shrinkers** $(\mathbb{S}^k \times \mathbb{R}^{n-k}, g_{\text{cyl}}, f_{\text{cyl}}, 1)$, $n \geq 3$, where

$$g_{\text{cyl}} := 2(k-1) g_{\mathbb{S}^k} + g_{\text{Euc}} \quad \text{and} \quad f_{\text{cyl}}(\theta, z) := \frac{|z|^2}{4} + \frac{k}{2}.$$

Here, $|z|^2 = \sum_{i=1}^{n-k}(z^i)^2$, where $z = (z^1, \dots, z^{n-k}) \in \mathbb{R}^{n-k}$ and $\theta \in \mathbb{S}^k$. The shrinking cylindrical solutions that these Ricci solitons define are of paramount importance in the analysis of singularities of the Ricci flow.

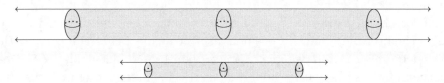

Figure 2.3. *Top*: The shrinker $(\mathbb{S}^{n-1} \times \mathbb{R}^1, g_{\text{cyl}}, f_{\text{cyl}})$. The $\mathbb{S}^{n-1}$ factor is normalized so that its Ricci curvatures are equal to $\frac{1}{2}$.
Bottom: The same shrinker at half the scale.[4] The shading is to indicate the homothetic correspondence. Note however that this is not the correspondence under Ricci flow without diffeomorphism pullback, which shrinks the spheres but not the line.

2.3.5. Quotient solitons.

We will say that a subgroup $\Gamma \subset \text{Isom}(\mathcal{M}^n, g)$ **preserves** the Ricci soliton structure $(\mathcal{M}^n, g, X, \lambda)$ if $\gamma^*(X) = X$ for all $\gamma \in \Gamma$, and preserves the gradient Ricci soliton structure $(\mathcal{M}^n, g, f, \lambda)$ if furthermore $f \circ \gamma = f$ for all $\gamma \in \Gamma$. If Γ is discrete and acts freely and properly discontinuously on $\mathcal{M}^n$, then g and X (respectively, f) descend uniquely to smooth representatives g_{quo} and X_{quo} (respectively, f_{quo}) on the quotient manifold $\mathcal{M}^n / \Gamma$ and define a Ricci soliton structure there.

Example 2.3. The involution $(\theta, r) \mapsto (-\theta, -r)$ on $\mathbb{S}^{n-1} \times \mathbb{R}$ defines a $\mathbb{Z}_2$-quotient of the round-cylindrical shrinker $(\mathbb{S}^{n-1} \times \mathbb{R}, g_{\text{cyl}}, f_{\text{cyl}})$. Here, the underlying manifold is diffeomorphic to a nontrivial real line bundle over $\mathbb{RP}^{n-1}$.

---

[4]That is, the metric of the bottom cylinder is, up to isometry, equal to $\frac{1}{4}$ times the metric of the top cylinder.

The construction in Example 2.3 can be rephrased in the language of covering spaces. Given a covering space $\pi : \tilde{\mathcal{M}}^n \to \mathcal{M}^n$ and a Ricci soliton structure $(\mathcal{M}^n, g, X, \lambda)$ on $\mathcal{M}^n$, defining $\tilde{g} = \pi^* g$ and $\tilde{X} = \pi^* X$ yields a Ricci soliton structure on the cover $\tilde{\mathcal{M}}^n$. If $\pi_1(\tilde{\mathcal{M}}^n) = \{e\}$, we call this structure the **universal covering soliton**.

2.3.6. Nongradient solitons.

The examples we have considered to this point have all been gradient Ricci solitons. They are the most important kind of Ricci soliton from the perspective of singularity analysis, and all examples which have arisen organically thus as a byproduct of this analysis have proven to be gradient. For example, according to [**245**, **251**], any complete shrinking Ricci soliton $(\mathcal{M}^n, g, X, 1)$ of bounded curvature is gradient.

Nevertheless, there are several constructions of nongradient Ricci solitons in the literature and there is no reason to suspect that they are particularly uncommon. Before we give a nontrivial example, let us first describe a superficial means of creating nongradient Ricci solitons from gradient structures. If $(\mathcal{M}^n, g, f, \lambda)$ is a gradient Ricci soliton and $(\mathcal{M}^n, g)$ admits a nontrivial (i.e., not identically zero) Killing vector field K, then adding K to ∇f yields another Ricci soliton structure $(\mathcal{M}^n, g, \nabla f + K, \lambda)$ which will be nongradient provided K is not itself the gradient of a smooth function. Of course this new structure is equivalent to the original one and thus is in a sense "secretly" a gradient Ricci soliton.

The following explicit example of a "true" nongradient Ricci soliton is due to Topping and Yin [**278**].

Example 2.4. The complete Riemannian metric

$$(2.19) \qquad g = \frac{2}{1 + y^2}(dx^2 + dy^2),$$

together with the complete vector field

$$(2.20) \qquad X = -x\frac{\partial}{\partial x} - y\frac{\partial}{\partial y}$$

generated by homotheties, comprises a complete nongradient expanding Ricci soliton structure $(\mathbb{R}^2, g, X, -1)$ on $\mathbb{R}^2$. A short computation shows that the scalar curvature of g is given by (see Figure 2.4)

$$(2.21) \qquad R(x, y) = \frac{1 - y^2}{1 + y^2}.$$

Indeed, this follows from (1.20):

$$(2.22) \qquad R_{e^u g_{\mathbb{E}}} = -e^{-u}\Delta u,$$

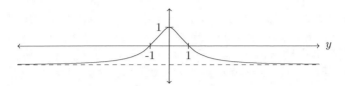

Figure 2.4. The scalar curvature as a function of y: $R(\cdot, y) = \frac{1-y^2}{1+y^2}$.

with $u = \ln\left(\frac{2}{1+y^2}\right)$, and where Δ is the Euclidean Laplacian. We also note that the geometry of $(\mathbb{R}^2, g)$ resembles that of hyperbolic space (with constant sectional curvature $-\frac{1}{2}$) near spatial infinity.

That $(\mathbb{R}^2, g, X, -1)$ is not equivalent to a gradient Ricci soliton structure can be seen by first observing that the Killing vector fields of g are precisely the constant multiples of the vector $\frac{\partial}{\partial x}$.

As we will see below, for any gradient Ricci soliton $(\mathcal{M}^2, g, f, \lambda)$ on an oriented Riemannian surface, the vector $J(\nabla f)$ will be Killing (see Lemma 3.1). Here,

$$(2.23) \qquad J : T\mathcal{M} \to T\mathcal{M}$$

is the almost complex structure defined by the conformal class of g and the orientation on $\mathcal{M}^2$, so J is counterclockwise rotation by 90 degrees and $J^2 = -\operatorname{id}_{T\mathcal{M}}$, but for no $c \in \mathbb{R}$ is $J\left(X + c\frac{\partial}{\partial x}\right)$ a constant multiple of $\frac{\partial}{\partial x}$.[5]

Other nontrivial examples of nongradient expanding Ricci solitons can be found in Lott [**223**] and Baird and Danielo [**12, 13**].

2.4. The gradient Ricci soliton equation

In this section we consider basic properties of gradient Ricci solitons in all dimensions. The basic definitions and derived equations were given by Hamilton in various papers, especially [**175, 176, 179**].

2.4.1. Definitions.

Recall from (2.3) that a *gradient Ricci soliton* is a quadruple $(\mathcal{M}^n, g, f, \lambda)$, where $\lambda \in \mathbb{R}$, satisfying

$$(2.24) \qquad \operatorname{Ric} + \nabla^2 f = \frac{\lambda}{2}g,$$

where, by Definition 2.1, the *expansion constant* $\lambda > 0$, $= 0$, and < 0 (e.g., $\lambda = 1$, 0, and -1) corresponds to being a *shrinking*, *steady*, and *expanding* gradient Ricci soliton, respectively.

---

[5]The counterclockwise rotation by 90 degrees J is characterized by the orthonormal frame $\{U, J(U)\}$ being positively oriented for any unit tangent vector U.

Recall that in all cases, f is called the *potential function*. Evident in the above equations is that there should be some relationships between the geometry of g and the analysis of f. Techniques from Ricci flow also prove to be useful. These themes are prevalent throughout this book.

Recall that the Lie derivative of a k-tensor T on a differentiable manifold $\mathcal{M}^n$ satisfies
(2.25)

$$(\mathcal{L}_X T)(Y_1, \ldots, Y_k) = X(T(Y_1, \ldots, Y_k)) - \sum_{i=1}^k T(Y_1, \ldots, [X, Y_i], \ldots, Y_k),$$

where $X, Y_1, \ldots, Y_k$ are vector fields. In the case where we are on a Riemannian manifold $(\mathcal{M}^n, g)$, we may re-express this formula in terms of the covariant derivative of g as
(2.26)

$$(\mathcal{L}_X T)(Y_1, \ldots, Y_k) = (\nabla_X T)(Y_1, \ldots, Y_k) + \sum_{i=1}^k T(Y_1, \ldots, \nabla_{Y_i} X, \ldots, Y_k).$$

In particular, if T is a 2-tensor, then in local coordinates we have[6]
(2.27) $$(\mathcal{L}_X T)_{ij} = (\nabla_X T)_{ij} + \nabla_i X^k T_{kj} + \nabla_j X^k T_{ik}.$$

Here and throughout the book we use the Einstein summation convention. Notably, (2.25) yields
(2.28) $$\mathcal{L}_{\nabla f} g = 2 \nabla^2 f$$

and we may rewrite the gradient Ricci soliton equation (2.24) in terms of the Lie derivative as
(2.29) $$-2 \operatorname{Ric} = \mathcal{L}_{\nabla f} g - \lambda g.$$

The left-hand side of this equation is the velocity tensor for Hamilton's **Ricci flow**. Equation (2.29) is an **underdetermined system** of PDEs for the pair (g, f)—there are $\frac{n(n+1)}{2}$ equations for $\frac{n(n+1)}{2} + 1$ unknowns. The Lie derivative term represents the infinitesimal action of the diffeomorphism group on the metric by pullback. A consequence of this is the time-dependent Ricci flow form of a gradient Ricci soliton discussed in Proposition 2.2.

As we shall see, the analysis of (2.29) generally uses techniques from elliptic and parabolic partial differential equations, from the comparison geometry of Ricci curvature, and from Ricci flow. Although we cannot decouple the two quantities g and f, it is often useful to consider the gradient Ricci soliton equation from the point of view of one quantity or the other.

---

[6]For the reader unfamiliar with local coordinate calculations, Eisenhart's book [**143**] is an excellent classical reference.

Recall that we have the more general notion of *Ricci soliton* $(\mathcal{M}^n, g, X, \lambda)$, where X is a vector field, satisfying

(2.30) $$2 \operatorname{Ric} + \mathcal{L}_X g = \lambda g.$$

This is also an underdetermined system. In local coordinates,

(2.31) $$2R_{ij} + \nabla_i X_j + \nabla_j X_i = \lambda g_{ij}.$$

Recall that tracing this yields (2.2):

$$R + \operatorname{div} X = \frac{n\lambda}{2}.$$

Observe that if $\mathcal{M}^n$ is closed, then by integrating this and using the divergence theorem, we obtain that the average scalar curvature satisfies

(2.32) $$R_{\text{avg}} := \frac{\int_{\mathcal{M}} R d\mu}{\operatorname{Vol}(g)} = \frac{n\lambda}{2},$$

where $d\mu$ is the volume form of g and $\operatorname{Vol}(g)$ is the volume of $(\mathcal{M}^n, g)$.

2.5. Product and rotationally symmetric solitons

In this section we consider product structures in more detail and the extent of uniqueness of the potential function f of gradient Ricci soliton structures $(\mathcal{M}^n, g, f)$ for the Riemannian metric g fixed. We also state the uniqueness theorem for rotationally symmetric steady gradient Ricci solitons and the nonexistence theorem for rotationally symmetric shrinking gradient Ricci solitons.

2.5.1. Metric products are soliton products.

If a gradient Ricci soliton is a product metrically, then it is a product as a gradient Ricci soliton.

Lemma 2.5. *Suppose that $(\mathcal{M}^n, g, f, \lambda)$ is a gradient Ricci soliton and that $(\mathcal{M}^n, g)$ is isometric to a Riemannian product $(\mathcal{M}_1^{n_1}, g_1) \times (\mathcal{M}_2^{n_2}, g_2)$. Then for any $x_2 \in \mathcal{M}_2^{n_2}$ we have that $(\mathcal{M}_1^{n_1}, g_1, f_1, \lambda)$ is a gradient Ricci soliton, where $f_1 : \mathcal{M}_1^{n_1} \to \mathbb{R}$ is the restriction of f to $\mathcal{M}_1^{n_1} \times \{x_2\} \cong \mathcal{M}_1^{n_1}$. Of course, the same is true for the roles of the indices 1 and 2 switched.*

Proof. Since $g = g_1 + g_2$, we have for $X, Y \in T\mathcal{M}_1 \cong T(\mathcal{M}_1^{n_1} \times \{x_2\}) \subset T\mathcal{M}$,

$$\begin{aligned}
\left(\nabla_g^2 f\right)(X, Y) &= X(Yf) - \left\langle \nabla_X^g Y, \nabla f \right\rangle_g \\
&= X(Yf) - \left\langle \nabla_X^{g_1} Y, \nabla f_1 \right\rangle_{g_1} \\
&= \left(\nabla_{g_1}^2 f_1\right)(X, Y)
\end{aligned}$$

because $\nabla_X^g Y = \nabla_X^{g_1} Y$ is tangential to $\mathcal{M}_1^{n_1} \times \{x_2\}$. Therefore, taking the components of $\mathrm{Ric}_g + \nabla_g^2 f = \frac{\lambda}{2} g$ in the $\mathcal{M}_1^{n_1}$ directions yields

$$\mathrm{Ric}_{g_1} + \nabla_{g_1}^2 f_1 = \frac{\lambda}{2} g_1. \qquad \square$$

2.5.2. Uniqueness and nonuniqueness of the potential function.

Regarding the uniqueness of the potential function of a gradient Ricci soliton with a given metric and a given expansion factor, we have the following.

Proposition 2.6. *Suppose that* $(\mathcal{M}^n, g, \lambda)$*, with either* f_1 *or* f_2 *as its potential function, is a gradient Ricci soliton. Then:*

(1) $f_1 - f_2$ *is a constant or*

(2) $(\mathcal{M}^n, g)$ *is isometric to* $(\mathbb{R}, ds^2) \times (\mathcal{N}^{n-1}, h)$*, where* $(\mathcal{N}^{n-1}, h)$ *is isometric to each level set* $\{f_1 - f_2 = c\}$ *for* $c \in \mathbb{R}$*.*

Moreover, in the second case, $f_1 - f_2$ *is linear on the* $\mathbb{R}$ *factor; that is,*

$$(2.33) \qquad f_2(s, x) = f_1(s, x) + as + b \quad \text{for } s \in \mathbb{R}, \ x \in \mathcal{N}^{n-1},$$

where $a, b \in \mathbb{R}$*.*

Proof. Define $F : \mathcal{M}^n \to \mathbb{R}$ by $F := f_1 - f_2$. Then $\nabla^2 F = 0$; i.e., $\mathcal{L}_{\nabla F} g = 0$. Assume that F is not a constant. Then $|\nabla F| = a$, where a is a positive constant. Let φ_t, $t \in \mathbb{R}$, be the 1-parameter group of isometries of $(\mathcal{M}^n, g)$ generated by ∇F. We have $F \circ \varphi_t = F + a^2 t$. Let

$$(2.34) \qquad \qquad \Sigma_c := \{F = c\},$$

which is a smooth hypersurface with unit normal $\nu = \frac{\nabla F}{|\nabla F|}$ for each $c \in \mathbb{R}$. The **second fundamental form** II of Σ_c vanishes because

$$(2.35) \qquad \mathrm{II}(X, Y) := \langle \nabla_X \nu, Y \rangle = \left\langle \nabla_X \frac{\nabla F}{|\nabla F|}, Y \right\rangle = \frac{\nabla^2 F(X, Y)}{|\nabla F|} = 0$$

for $X, Y \in T\Sigma_c$. Moreover, since $\mathcal{L}_{\nabla F} g = 0$, φ_t maps Σ_c isometrically onto $\Sigma_{c+a^2 t}$. Hence $(\mathcal{M}^n, g)$ is isometric to $(\mathbb{R} \times \mathcal{N}^{n-1}, a^{-2} dF^2 + h)$, where $(\mathcal{N}^{n-1}, h)$ is isometric to each level set $\{F = c\}$. The proposition follows. $\square$

Remark 2.7. To see the nonuniqueness of the potential function in the splitting case, consider the product of an $(n-1)$-dimensional gradient Ricci soliton $(\mathcal{M}^n, g, f, \lambda)$ with $(\mathbb{R}, ds^2, f_a, \lambda)$, where $f_a(s) = \frac{\lambda}{4}(s-a)^2$ and $a \in \mathbb{R}$.

Corollary 2.8. *If* $(\mathcal{M}^n, g, f, \lambda)$ *is a gradient Ricci soliton, where* $(\mathcal{M}^n, g)$ *is equal (isometric) to* $(\mathcal{M}_1^{n_1}, g_1) \times (\mathcal{M}_2^{n_2}, g_2)$*, then there are* $f_i : \mathcal{M}_i^{n_i} \to \mathbb{R}$ *such that the* $(\mathcal{M}_i^{n_i}, g_i, f_i, \lambda)$ *are gradient Ricci solitons, where* $f = f_1 + f_2$*, or* $(\mathcal{M}^n, g)$ *splits off an* $\mathbb{R}$ *factor and* $f - f_1 + f_2$ *is linear on that* $\mathbb{R}$*-factor.*

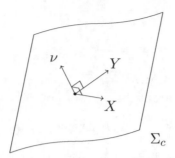

Figure 2.5. A level surface Σ_c of f, a unit normal vector ν to Σ_c, and tangent vectors X, Y to Σ_c.

Proof. Define $f_i : \mathcal{M}_i^{n_i} \to \mathbb{R}$ by Lemma 2.5, so that the $(\mathcal{M}_i^{n_i}, g_i, f_i, \lambda)$ are gradient Ricci solitons. By Proposition 2.6, if $(\mathcal{M}^n, g)$ does not split off an $\mathbb{R}$-factor, then the difference of f and $f_1 + f_2$ is a constant function on $\mathcal{M}^n$, so we may add a constant to, say, f_1 to make them equal. $\qquad\square$

If the expansion constants of the gradient Ricci solitons are different, then we have the following.

Proposition 2.9 (GRS that are metrically the same but have different expansion constants). *Suppose that* $(\mathcal{M}^n, g)$, *with either* (f_1, λ_1) *or* (f_2, λ_2), *is a gradient Ricci soliton, where* $\lambda_1 \neq \lambda_2$. *Then the* $(\mathcal{M}^n, g, f_i, \lambda_i)$, *for* $i = 1, 2$, *are both Gaussian solitons.*

Proof. Without loss of generality, we may assume that $\lambda_1 > \lambda_2$. Define $\psi = f_1 - f_2$. Then

$$(2.36) \qquad\qquad \nabla^2 \psi = cg,$$

where $c := \frac{\lambda_1 - \lambda_2}{2} > 0$. Choose any $p \in \mathcal{M}^n$. Let $\gamma : [0, L] \to \mathcal{M}^n$ be a unit speed geodesic emanating from p and let $\psi(s) := \psi(\gamma(s))$. Then $\psi'(0) \geq -|\nabla \psi|(p)$. Hence $\psi''(s) = c$ implies that

$$\psi(s) \geq \frac{c}{2}s^2 - |\nabla\psi|(p)s + \psi(p) \geq -\frac{1}{2c}|\nabla\psi|^2(p) + \psi(p).$$

This implies that ψ attains its minimum value at some point—call this point $o \in \mathcal{M}^n$—which is unique since ψ is strictly convex. Without loss of generality, we may assume that this minimum value is equal to 0. Hence $\psi > 0$ on $\mathcal{M}^n \setminus \{o\}$.

Now, (2.36) implies that

$$\nabla|\nabla\psi|^2 = 2\nabla^2\psi(\nabla\psi) = 2cg(\nabla\psi) = 2c\nabla\psi.$$

Thus, $|\nabla\psi|^2 = 2c\psi + C$, where C is a constant. Since the minimum of ψ is equal to 0, we have that $C = 0$, so that

$$(2.37) \qquad\qquad\qquad |\nabla\psi|^2 = 2c\psi.$$

Define $\rho := \sqrt{\psi}$. Then

$$(2.38) \qquad\qquad\qquad |\nabla\rho|^2 = \frac{c}{2}$$

on $\mathcal{M}^n \setminus \{o\}$. Moreover, $\nabla(\rho^2) = \nabla\psi$ is a complete vector field which generates a 1-parameter group $\{\varphi_t\}_{t\in\mathbb{R}}$ of homotheties of g. We have that

$$\nabla_{\nabla\rho}(\nabla\rho) = \frac{1}{2}\nabla|\nabla\rho|^2 = 0,$$

where $\nabla\rho$ denotes the gradient of ρ, so that the integral curves to $\nabla\rho$ are geodesics. By Morse theory we have that $\Sigma_t := \rho^{-1}(t)$ is diffeomorphic to $\mathbb{S}^{n-1}$ for all $t \in (0,\infty)$. Since $|\nabla\rho| = 1$, each homothety φ_t of g maps level sets of ρ to level sets of ρ. Hence g may be written as the warped product

$$g = d\rho^2 + \rho^2 \tilde{g}, \quad \text{where } \tilde{g} = g|_{\Sigma_1}.$$

Since g is smooth at o, where $\rho = 0$, we have that $(\Sigma_1, \tilde{g})$ must be isometric to the unit $(n-1)$-sphere. Since $\bigcup_{t\in(0,\infty)} \Sigma_t = \mathcal{M}^n \setminus \{o\}$, we conclude that $(\mathcal{M}^n, g)$ is isometric to Euclidean space. The proposition follows. $\qquad\square$

Remark 2.10. Compare this to Obata's theorem (see [**249**]), which says that if $(\mathcal{M}^n, g)$ is a complete Riemannian manifold with a nonconstant function f satisfying $\nabla^2 f = -fg$, then $(\mathcal{M}^n, g)$ is isometric to the unit n-sphere.

Note that from the equality case of Theorem 2.14 below, we have that a flat shrinking gradient Ricci soliton must be the Gaussian shrinking gradient Ricci soliton.

2.5.3. Uniqueness of rotationally symmetric gradient Ricci solitons.

We have the following uniqueness result, due to Bryant [**54**] in the steady case and Kotschwar [**204**] in the shrinking case.

Theorem 2.11.

(1) *Any complete rotationally symmetric steady gradient Ricci soliton must be flat or the Bryant soliton.*

(2) *Any complete rotationally symmetric shrinking gradient Ricci soliton must be the Gaussian shrinking gradient Ricci soliton on $\mathbb{R}^n$, the round cylinder shrinker on $\mathbb{S}^{n-1}\times\mathbb{R}$, or the round sphere shrinker on $\mathbb{S}^n$.*

Assuming nonflatness, the idea of the proof is to first show that the potential function is rotationally symmetric (see Exercise 6.2 below). The gradient Ricci soliton equation is a nonlinear second-order ODE, which may be then reduced to a first-order system of ODEs. An ODE analysis using the metric's smoothness at any finite end (removable singularity) and completeness at any infinite end yields the classification. A detailed proof of Theorem 2.11(1), with calculations related to the proof of Theorem 2.11(2), will be given in Chapter 6.

Remark 2.12. For an exposition of Bryant's work on rotationally symmetric *expanding* gradient Ricci solitons, see §5 of Chapter 1 in [**101**]. We summarize the results in §7.1.2 of this book.

2.6. Fundamental identities: Differentiating the Ricci soliton equation

In this section we present basic identities satisfied by gradient Ricci solitons. These identities are fundamental to the study of gradient Ricci solitons.

2.6.1. Trace and divergence of the gradient Ricci soliton equation.

Let $(\mathcal{M}^n, g, f, \lambda)$ be a gradient Ricci soliton. By tracing the gradient Ricci soliton equation (2.24), we obtain

$$(2.39) \qquad R + \Delta f = \frac{n\lambda}{2}.$$

On the other hand, taking the divergence of (2.24) while applying the following contracted second Bianchi identity (1.60) yields

$$\frac{1}{2} dR + \Delta (df) = 0.$$

By the commutator formula (1.52), for any function u and by (2.39), we have

$$0 = \frac{1}{2} dR + d(\Delta f) + \operatorname{Ric}(\nabla f) = -\frac{1}{2} dR + \operatorname{Ric}(\nabla f).$$

We write this as the following basic equation:

$$(2.40) \qquad 2 \operatorname{Ric}(\nabla f) = \nabla R.$$

A useful consequence of this is

$$(2.41) \qquad \langle \nabla f, \nabla R \rangle = 2 \operatorname{Ric}(\nabla f, \nabla f).$$

2.6.2. A fundamental identity relating R and f.

Now by (2.24), for any vector field V,

$$V(|df|^2) = 2 \langle \nabla_V df, df \rangle$$
$$= 2 \left\langle -\operatorname{Ric}(V) + \frac{\lambda}{2} g(V), df \right\rangle$$
$$= (-2 \operatorname{Ric}(\nabla f) + \lambda df)(V),$$

so that

$$(2.42) \qquad \nabla |\nabla f|^2 = -2 \operatorname{Ric}(\nabla f) + \lambda \nabla f.$$

Combining this with (2.40) yields

$$(2.43) \qquad \nabla(R + |\nabla f|^2 - \lambda f) = 0.$$

Since $\mathcal{M}^n$ is connected, we conclude that

$$(2.44) \qquad R + |\nabla f|^2 - \lambda f = C,$$

where C is a constant. This equation is used in a fundamental way to understand gradient Ricci solitons. The above equations were obtained by Hamilton.

If $\lambda = \pm 1$ (shrinking or expanding gradient Ricci soliton), then by adding a constant to the potential function f we may assume that $C = 0$, so that

$$(2.45) \qquad R + |\nabla f|^2 = \lambda f.$$

If $\lambda = 0$ (steady gradient Ricci soliton) and g is not Ricci-flat, then by scaling the metric we may take $C = 1$, so that

$$(2.46) \qquad R + |\nabla f|^2 = 1.$$

In other words, we may choose $C = 1 - |\lambda|$. In these cases we say that the gradient Ricci soliton is a **normalized gradient Ricci soliton**. Throughout this book, unless otherwise indicated we shall always assume that we are on a normalized gradient Ricci soliton.

2.6.3. The f-scalar curvature and f-Ricci tensor.

Define the f-**scalar curvature** to be

$$(2.47) \qquad R_f := R + 2\Delta f - |\nabla f|^2.$$

We define the f-**Ricci tensor**, a.k.a. the **Bakry–Emery tensor**, by

$$\operatorname{Ric}_f = \operatorname{Ric} + \nabla^2 f.$$

Then the gradient Ricci soliton equation is

$$(2.48) \qquad \operatorname{Ric}_f = \frac{\lambda}{2} g.$$

Remark 2.13. From (2.39), (2.45), and (2.46), on a (normalized) gradient Ricci soliton we have

$$(2.49) \qquad R_f = -\lambda f + n\lambda - 1 + |\lambda|.$$

2.6.4. f-Laplacian-type equations.

Define the f**-Laplacian** by

(2.50) $$\Delta_f := \Delta - \nabla f \cdot \nabla.$$

This natural elliptic operator is prevalent in computations regarding gradient Ricci solitons. For any functions $A, B : \mathcal{M}^n \to \mathbb{R}$, provided we can integrate by parts (e.g., if A and B have compact support), we have

(2.51) $$\int_{\mathcal{M}} A \Delta_f B \, e^{-f} d\mu = -\int_{\mathcal{M}} \langle \nabla A, \nabla B \rangle e^{-f} d\mu = \int_{\mathcal{M}} B \Delta_f A \, e^{-f} d\mu.$$

That is, the operator Δ_f is formally **self-adjoint** on $L^2(e^{-f} d\mu)$. Moreover, for any $\varphi : \mathcal{M}^n \to \mathbb{R}$ we have that

(2.52) $$\left(\Delta_f - \frac{1}{4} R_f \right) \varphi = e^{f/2} \left(\Delta - \frac{1}{4} R \right) (e^{-f/2} \varphi).$$

By (2.45) and (2.46) and by their differences with (2.39), we obtain the following for each of the three types of normalized gradient Ricci solitons.

(1) For a shrinking gradient Ricci soliton, we have

(2.53) $$R + |\nabla f|^2 = f, \quad \text{so that } R \leq f,$$

and

(2.54) $$\Delta_f f = \frac{n}{2} - f.$$

Hence $f - \frac{n}{2}$ is an eigenfunction of $-\Delta_f$ with eigenvalue 1.

(2) For a non-Ricci-flat steady gradient Ricci soliton, we have

(2.55) $$R + |\nabla f|^2 = 1, \quad \text{so that } R \leq 1,$$

and

(2.56) $$\Delta_f f = -1.$$

(3) For an expanding gradient Ricci soliton, we have

(2.57) $$R + |\nabla f|^2 = -f, \quad \text{so that } R \leq -f,$$

and

(2.58) $$\Delta_f f = f - \frac{n}{2}.$$

By taking the divergence of (2.40) and then applying (1.60) and (2.24), we obtain

(2.59) $$\Delta R = 2 \operatorname{div} (\operatorname{Ric}) (\nabla f) + 2 \left\langle \operatorname{Ric}, \nabla^2 f \right\rangle$$

$$= \langle \nabla R, \nabla f \rangle - 2 \left\langle \operatorname{Ric}, \operatorname{Ric} - \frac{\lambda}{2} g \right\rangle.$$

That is,

(2.60) $\Delta_f R = -2 |\text{Ric}|^2 + \lambda R.$

Thus

(2.61) $\Delta_f R \leq -\frac{2}{n} R^2 + \lambda R.$

It is convenient to define the f-**divergence**

(2.62) $\text{div}_f(T) = \text{div}(T) - \text{tr}^{1,2}(\nabla f \otimes T) = (\text{div} - \iota_{\nabla f})(T) = e^f \text{div}(e^{-f}T)$

acting on tensors, where $\text{tr}^{a,b}$ denotes the trace over the ath and bth components. For example,

$$\Delta_f u = \text{div}_f(du) = \text{div}_f(\nabla u).$$

2.7. Sharp lower bounds for the scalar curvature

2.7.1. Statements and consequences of the lower bounds.

We have seen that every Einstein manifold admits at least one Ricci soliton structure and that these are precisely the Ricci soliton structures of constant scalar curvature. The following theorem shows that the scalar curvature of *any* complete Ricci soliton is bounded from below by a sharp constant. This follows in the gradient case from the work of B.-L. Chen [86] on ancient solutions and from the work of Z.-H. Zhang [303] on GRS. The equality case when $\lambda > 0$ is due to Pigola, Rimoldi, and Setti [258].

Theorem 2.14 (Sharp scalar curvature lower bounds for Ricci solitons). *If $(\mathcal{M}^n, g, X, \lambda)$ is a complete Ricci soliton, then:*

   (a) $R \geq 0$ *if $\lambda \geq 0$.*
   (b) $R \geq \frac{\lambda n}{2}$ *if $\lambda < 0$.*

Moreover, if equality holds at any point of $\mathcal{M}^n$, then $(\mathcal{M}^n, g)$ is Einstein. If $\lambda > 0$ and the shrinker is gradient, that is, $X = \nabla f$ for some function f, with $R = 0$ at some point, then $(\mathcal{M}^n, g, f)$ is a Gaussian shrinker.

Before proving this, we observe that Theorem 2.14 yields a measure of control of the potential function:

Corollary 2.15 (Potential function estimates). *Let $(\mathcal{M}^n, g, f, \lambda)$ be a GRS and let $p \in \mathcal{M}^n$.*

   (1) *On a shrinking GRS $(\lambda = 1)$,*
(2.63)
$$|\nabla f|^2 \leq f, \quad R \leq f, \quad \Delta f \leq \frac{n}{2}, \quad \text{and} \quad \sqrt{f}(x) \leq \sqrt{f}(p) + \frac{1}{2}d(x, p),$$

> *where $d(x,p)$ denotes the Riemannian distance from x to p with respect to the metric g. At a minimum point[7] $o \in \mathcal{M}^n$ of f we have $0 \leq R(o) = f(o) \leq \frac{n}{2}$ and*

$$(2.64) \qquad f(x) \leq \frac{1}{4}\left(d(x,o) + \sqrt{2n}\right)^2.$$

> *(2) On a steady GRS ($\lambda = 0$),*

$$(2.65) \quad |\nabla f|^2 \leq 1, \quad R \leq 1, \quad \Delta f \leq 0, \quad and \quad |f(x) - f(p)| \leq d(x,p).$$

> *(3) On an expanding GRS ($\lambda = -1$),*

(2.66)

$$|\nabla f|^2 \leq \frac{n}{2} - f, \quad \Delta f \leq 0, \quad and \quad \sqrt{\frac{n}{2} - f(x)} \leq \sqrt{\frac{n}{2} - f(p)} + \frac{1}{2}d(x,p).$$

> *In particular, $f \leq \frac{n}{2}$.*

Proof of Corollary 2.15. The upper bounds for Δf follow from (2.39) and Theorem 2.14. The upper bounds for R follow from (2.45) and (2.46). The upper bounds for $|\nabla f|^2$ follow from (2.45), (2.46), and Theorem 2.14. By integrating the bounds for $|\nabla f|$ along minimal geodesics, we obtain the inequalities for f and its square root.

In the case of a shrinking GRS, by (2.54), at a minimum point o of f we have $f(o) - R(o) = |\nabla f|^2(o) = 0$ and

$$(2.67) \qquad 0 \leq \Delta_f f(o) = \frac{n}{2} - f(o).$$

Thus $0 \leq f(o) = R(o) \leq \frac{n}{2}$. Now, integrating the inequality $|\nabla(2\sqrt{f})| \leq 1$ from Theorem 2.14 yields

$$2\sqrt{f(x)} \leq 2\sqrt{f(o)} + d(x,o) \leq \sqrt{2n} + d(x,o),$$

which in turn implies (2.64). $\qquad\square$

2.7.2. Laplacian comparison on Riemannian manifolds.

A basic tool that we will use to prove Theorem 2.14 is the *Laplacian comparison theorem* for the distance function on Riemannian manifolds, which we recall in this subsection.

Let $(\mathcal{M}^n, g)$ be a Riemannian manifold. Recall that the length of a path $\gamma : [a,b] \to \mathcal{M}^n$ is defined by

$$(2.68) \qquad \mathrm{L}(\gamma) := \int_a^b |\gamma'(r)|\, dr.$$

---

[7]We will show in Theorem 4.3 below that the infimum of f over $\mathcal{M}^n$ is attained at some point.

The distance function $d : \mathcal{M}^n \times \mathcal{M}^n \to [0, \infty)$ is defined as an infimum of lengths:

$$(2.69) \qquad d(x, y) = \inf_{\gamma} \mathrm{L}(\gamma),$$

where the infimum is taken over all paths joining x and y.

Let $(\mathcal{M}^n, g)$ be a Riemannian manifold. Let $\gamma_v : [0, L] \to \mathcal{M}^n$ be a 1-parameter family of piecewise smooth paths such that $\gamma := \gamma_0$ (but not necessarily γ_v for $v \neq 0$) is parametrized by arc length. Then the *first variation of arc length formula* says (see Exercise 2.22)

$$(2.70) \quad \frac{d}{dv}\bigg|_{v=0} \mathrm{L}(\gamma_v) = -\int_0^L \langle V(r), \nabla_{\gamma'(r)}\gamma'(r) \rangle \, dr + \langle V(r), \gamma'(r) \rangle \big|_{r=0}^L,$$

where $V(r) := \frac{\partial}{\partial v}\big|_{v=0} \gamma_v(r)$. In particular, by considering the case where both $V(0) = 0$ and $V(L) = 0$, we see that γ is a critical point of the arc length functional L if and only if $\nabla_{\gamma'(r)}\gamma'(r) \equiv 0$; i.e., γ is a geodesic.

The *second variation of arc length formula* tells us the following (see (1.17) in Cheeger and Ebin's book [**84**]); cf. Exercise 2.23.

Proposition 2.16. *Suppose that $p := \gamma_v(0)$ is independent of v and that $\gamma = \gamma_0$ is a unit speed geodesic. Then the second variation of the length L is*

$$(2.71) \quad \frac{d^2}{dv^2}\bigg|_{v=0} \mathrm{L}(\gamma_v) = \int_0^L \left(\left| (\nabla_{\gamma'(r)}V)^{\perp} \right|^2 - \langle \mathrm{Rm}(V, \gamma'(r))\gamma'(r), V \rangle \right) dr$$

$$+ \left\langle \nabla_V \left(\frac{\partial}{\partial v}\gamma_v \right), \gamma'(L) \right\rangle,$$

where $(\nabla_{\gamma'}V)^{\perp} := \nabla_{\gamma'}V - \langle \nabla_{\gamma'}V, \gamma' \rangle \gamma'$ is the projection of $\nabla_{\gamma'}V$ onto the hyperplane $(\gamma')^{\perp} = \{V \in T\mathcal{M} : \langle V, \gamma' \rangle = 0\}$.

We shall also use the notation $\delta_V^2 \mathrm{L}(\gamma) := \frac{\partial^2}{\partial v^2}\big|_{v=0} \mathrm{L}(\gamma_v)$. Since the distance function is only Lipschitz continuous, when considering its Laplacian we shall use the following.

Definition 2.17. Let $\varphi : \mathcal{M}^n \to \mathbb{R}$ be continuous in a neighborhood of a point x. We say that $\Delta\varphi(x) \leq A$ in the **barrier sense** if for any $\varepsilon > 0$ there exists a C^2 function $\psi \geq \varphi$ defined in a neighborhood of x such that $\psi(x) = \varphi(x)$ and $\Delta\psi(x) \leq A + \varepsilon$.

We say that $\Delta\varphi(x) \leq A$ in the **strong barrier sense** if there exists a C^2 function $\psi \geq \varphi$ defined in a neighborhood of x such that $\psi(x) = \varphi(x)$ and $\Delta\psi(x) \leq A$. We have the analogous definitions for the operator Δ_f.

Fix $p \in \mathcal{M}^n$ and denote $r(x) := d(x, p)$. Let $r_x := r(x)$. By applying the second variation of arc length formula, we obtain the following upper bound for the Laplacian of the distance function (cf. Li's book [**217**]).

Proposition 2.18. *Let $x \neq p$, let $\gamma : [0, r_x] \to \mathcal{M}^n$ be a unit speed minimal geodesic joining p to x, and let $\zeta : [0, r_x] \to \mathbb{R}$ be a continuous piecewise C^∞ function satisfying $\zeta(0) = 0$ and $\zeta(r_x) = 1$. Then in the strong barrier sense we have*

$$(2.72) \qquad \Delta r(x) \leq \int_0^{r_x} \left((n-1) \left(\zeta' \right)^2 (r) - \zeta^2(r) \operatorname{Ric} \left(\gamma'(r), \gamma'(r) \right) \right) dr.$$

In particular, the above inequality holds in the classical sense if x is not in the cut locus of p.

Proof. Fix $p \in \mathcal{M}^n$ and let $x \neq p$. Let $\varepsilon \in (0, \operatorname{inj}_g(x))$, where $\operatorname{inj}_g(x)$ denotes the injectivity radius of g at x. We extend γ to an n-parameter family of paths by defining $\gamma^V : [0, r_x] \to \mathcal{M}^n$ for $V \in B_\varepsilon(0) \subset T_x\mathcal{M}$ by

$$\gamma^V(r) := \exp_{\gamma(r)}(\zeta(r) V(r)),$$

where $V(r) \in T_{\gamma(r)}\mathcal{M}$ is the parallel transport of V along γ and where $\zeta : [0, r_x] \to \mathbb{R}$ satisfies $\zeta(0) = 0$ and $\zeta(r_x) = 1$. Note that $V(r_x) = V$.

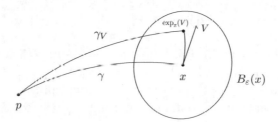

Figure 2.6. A path γ^V, where $V \in B_\varepsilon(0) \subset T_x\mathcal{M}$. Note that γ is a minimal geodesic, but γ_V is not necessarily a geodesic.

The family of paths γ^V have the properties that $\gamma^0(r) = \gamma(r)$, $\gamma^V(0) = p$, $\gamma^V(r_x) = \exp_x(V)$, and

$$\left. \frac{\partial}{\partial t} \right|_{t=0} \gamma^{tV}(r) = \zeta(r) V(r).$$

We have

$$(2.73a) \qquad\qquad \mathrm{L}\left(\gamma^V \right) \geq r(\exp_x(V)),$$

$$(2.73b) \qquad\qquad \mathrm{L}\left(\gamma^0 \right) = r_x.$$

Since $\varepsilon < \operatorname{inj}_g(x)$, $\exp_x : B_\varepsilon(0) \to B_\varepsilon(x)$ is a diffeomorphism. Let $y \in B_\varepsilon(x)$. Note that $\exp_x^{-1}(y) \in B_\varepsilon(0) \subset T_x\mathcal{M}$. So (2.73) implies that the C^∞ function $\varphi : B_\varepsilon(x) \to \mathbb{R}$ defined by

$$\varphi(y) = \mathrm{L}(\gamma^{\exp_x^{-1}(y)})$$

is an *upper barrier* for r at x; that is, $\varphi(y) \geq r(y)$ for $y \in B_\varepsilon(x)$ and $\varphi(x) = r_x$. Thus, in the strong barrier sense of Definition 2.17, we have

$$(2.74) \qquad \Delta r(x) \leq \Delta\varphi(x).$$

Let the vectors $\{e_1, \ldots, e_{n-1}\}$ complete the tangent vector $\gamma'(r_x)$ to an orthonormal basis of $T_x\mathcal{M}$. Then its parallel transport along γ, written as $\{e_1(r), \ldots, e_{n-1}(r), \gamma'(r)\}$, forms an orthonormal basis of $T_{\gamma(r)}\mathcal{M}$ for each $r \in [0, r_x]$. By (2.71), we have

$$\Delta\varphi(x) = \sum_{i=1}^{n-1} \left.\frac{\partial^2}{\partial t^2}\right|_{t=0} \varphi\left(\exp_x(te_i)\right) + \left.\frac{\partial^2}{\partial t^2}\right|_{t=0} \varphi\left(\exp_x(t\gamma'(r_x))\right)$$

$$= \sum_{i=1}^{n-1} \left.\frac{\partial^2}{\partial t^2}\right|_{t=0} \mathrm{L}\left(\gamma^{te_i}\right)$$

$$= \sum_{i=1}^{n-1} \int_0^{r_x} \left((\zeta')^2(r) - \zeta^2(r)\langle \mathrm{Rm}(e_i, \gamma'(r))\gamma'(r), e_i\rangle\right) dr,$$

where we used $\varphi\left(\exp_x(t\gamma'(r_x))\right) = r_x + t$ and $\langle \nabla_{e_i} e_i, \gamma'(r_x)\rangle = 0$ (since $\gamma^{te_i}(r_x) = \exp_x(te_i)$ is a geodesic). The proposition follows. $\square$

The proposition leads to the following question: What are good or optimal choices for $\zeta(r)$ in (2.72)? By taking $\zeta(r) = \frac{r}{r_x}$, a choice which for the case of Euclidean space corresponds to variations comprising straight lines, we obtain the Laplacian comparison theorem:

Corollary 2.19. *If $(\mathcal{M}^n, g)$ is a complete Riemannian manifold with* $\mathrm{Ric} \geq 0$, *then*

$$(2.75) \qquad \Delta r(x) \leq \frac{n-1}{r(x)}$$

in the strong barrier sense.

On the other hand, it is useful to consider a choice of $\zeta(r)$ which corresponds to a frame of parallel unit vector fields except near the ends of the geodesic, where the variations taper down. Now let $x \in \mathcal{M}^n \setminus B_2(p)$ and let $\gamma : [0, r(x)] \to \mathcal{M}^n$ be a unit speed minimal geodesic joining p to x. Define $\zeta : [0, r(x)] \to [0, 1]$ to be the piecewise linear function

$$(2.76) \qquad \zeta(r) = \begin{cases} r & \text{if } 0 \leq r \leq 1, \\ 1 & \text{if } 1 < r \leq r(x) - 1, \\ r(x) - r & \text{if } r(x) - 1 < r \leq r(x). \end{cases}$$

Let $\{e_1, \ldots, e_{n-1}, \gamma'(0)\}$ be an orthonormal basis of $T_p\mathcal{M}$. Define $e_i(r) \in T_{\gamma(r)}\mathcal{M}$ to be the parallel transport of $e_i = e_i(0)$ along γ. Then the frame

$\{e_1(r), \ldots, e_{n-1}(r), \gamma'(r)\}$ forms an orthonormal basis of $T_{\gamma(r)}\mathcal{M}$ for $r \in [0, r(x)]$. Since γ is minimal, by the second variation of arc length formula, we have for each i,

$$0 \le \delta^2_{\zeta e_i} \mathrm{L}(\gamma) = \int_0^{r(x)} \left((\zeta')^2(r) - \zeta^2(r) \left\langle \mathrm{Rm}\left(\gamma'(r), e_i\right) e_i, \gamma'(r) \right\rangle \right) dr.$$

Summing over i, we obtain

(2.77)
$$\int_0^{r(x)} \zeta^2(r) \, \mathrm{Ric}\left(\gamma'(r), \gamma'(r)\right) dr \le 2(n-1).$$

Let

(2.78)
$$\mathrm{S}(x) := \sup_{V \in \mathcal{S}^{n-1}_y, \, y \in B_1(x)} \mathrm{Ric}(V, V)_+,$$

where $\mathcal{S}^{n-1}_y \subset T_y\mathcal{M}$ is the unit $(n-1)$-sphere. We conclude:

Lemma 2.20. *If $x \in \mathcal{M}^n \setminus B_2(p)$ and if $\gamma : [0, r(x)] \to \mathcal{M}^n$ is a unit speed minimal geodesic joining p to x, then*

(2.79)
$$\int_0^{r(x)} \mathrm{Ric}\left(\gamma'(r), \gamma'(r)\right) dr \le 2(n-1) + \frac{2}{3}\left(\mathrm{S}(p) + \mathrm{S}(x)\right).$$

This lemma estimates, in an integral sense, the amount of positive Ricci curvature in the tangential direction that there can be along a minimal geodesic.

We now apply the Laplacian upper bound (2.72) to prove the following differential inequality for the distance function on Ricci solitons in terms of the *X*-**Laplacian** operator:

(2.80)
$$\Delta_X \phi := \Delta \phi - \langle X, \nabla \phi \rangle.$$

Proposition 2.21. *Let $(\mathcal{M}^n, g, X, \lambda)$ be a complete Ricci soliton and let $r = d(p, \cdot)$ be the distance from a fixed $p \in \mathcal{M}^n$. Suppose that $|\mathrm{Ric}| \le K_0$ on $B_p(r_0)$. Then there is a constant $C = C(n)$ such that the inequality*

(2.81)
$$\Delta_X r \le -\frac{\lambda}{2}r + C(n)\left(K_0 r_0 + r_0^{-1}\right) + |X|(p) =: h$$

holds in the barrier sense on $\mathcal{M}^n \setminus B_{r_0}(p)$; that is, for every $x \in \mathcal{M}^n \setminus B_{r_0}(p)$ and $\varepsilon > 0$ there exists a C^2 function $\psi \ge r$ defined in a neighborhood of x such that $\psi(x) = r(x)$ and $\Delta \psi(x) \le h + \varepsilon$.

Proof. Suppose that x is not in the cut locus of p. Since γ is a geodesic, by applying the fundamental theorem of calculus and using the Ricci soliton equation, we obtain

$$(2.82) \quad \langle X, \nabla r \rangle(x) - \langle X(p), \gamma'(0) \rangle = \int_0^{r_x} \frac{d}{dr} \langle X(\gamma(r)), \gamma'(r) \rangle \, dr$$

$$= \int_0^{r_x} (\nabla X)(\gamma'(r), \gamma'(r)) \, dr$$

$$= -\int_0^{r_x} \mathrm{Ric}\left(\gamma'(r), \gamma'(r)\right) dr + \frac{\lambda}{2} r(x).$$

By combining this with (2.72), we obtain

$$(2.83) \quad \Delta_X r(x) \leq \int_0^{r_x} \left((n-1)(\zeta')^2(r) + (1 - \zeta^2(r)) \, \mathrm{Ric}\left(\gamma'(r), \gamma'(r)\right) \right) dr$$

$$- \frac{\lambda}{2} r(x) + \langle X(p), \gamma'(0) \rangle.$$

Let $\zeta(r) = \frac{r}{r_0}$ for $0 \leq r \leq r_0$ and $\zeta(r) = 1$ for $r_0 < r \leq r_x$. We then conclude from (2.83) that

$$\Delta_X r(x) \leq \frac{n-1}{r_0} + \frac{2}{3} r_0 \, \mathrm{S}(p) - \frac{\lambda}{2} r(x) + |X(p)|,$$

where $\mathrm{S}(p)$ is defined by (2.78). The proposition follows. $\square$

2.7.3. Proof of the scalar curvature lower bound.

We are now ready to prove Theorem 2.14. The argument given in [**303**] for gradient Ricci solitons extends essentially verbatim to the nongradient case; we tweak it slightly to obtain a sharp constant in the expanding case.

The proof will also make use of the following specialized *cutoff function*.

Proposition 2.22. *For each $0 < \delta < 1/10$, there exists a smooth function $\varphi = \varphi_\delta : \mathbb{R} \to [0, 1]$ such that*

$$(2.84) \quad \varphi(x) = \begin{cases} 1 & \text{if } x \leq \delta, \\ 0 & \text{if } x \geq 2, \end{cases} \quad -(1+\theta)\sqrt{\varphi} \leq \varphi' \leq 0, \quad |\varphi''| \leq C_0,$$

and

$$(2.85) \quad 1 - \varphi(x) + \frac{x}{2} \varphi'(x) \geq -\varepsilon,$$

where $\theta = \theta(\delta)$ and $\varepsilon = \varepsilon(\delta)$ are positive and tend to 0 as $\delta \to 0$.

Proof of Proposition 2.22. Fix any $0 < \delta < 1/10$. We start with a smooth function $\eta = \eta_\delta$ satisfying

$$\eta(x) = \begin{cases} 1 & \text{if } x \in (-\infty, \delta], \\ \frac{2-\delta-x}{2-3\delta} & \text{if } x \in [3\delta, 2-2\delta], \\ 0 & \text{if } x \in [2, \infty) \end{cases}$$

and

$$-\frac{1}{2}(1+\theta) \leq \eta' \leq 0, \quad |\eta''| \leq C_1,$$

where $C_1 = C_1(\delta) > 0$ and $\theta = \theta(\delta) > 0$ tends to 0 as $\delta \to 0$. Thus η is a smooth approximation to the piecewise linear function that is equal to 1 for $x \leq 2\delta$, decreases linearly to 0 over the interval $[2\delta, 2-\delta]$, and is equal to 0 for $x \geq 2-\delta$. Then $\varphi := \eta^2$ satisfies

$$-(1+\theta)\sqrt{\varphi} \leq \varphi' \leq 0 \quad \text{and} \quad |\varphi''| \leq C_0 := 2C_1.$$

To verify (2.85), we only need to consider $x \in [\delta, 2]$. We consider three cases. First, for $x \in [\delta, 3\delta]$, we have

$$1 - \varphi + \frac{x}{2}\varphi' \geq -3\delta|\varphi'| \geq -3\delta(1+\theta).$$

Next, for $x \in [3\delta, 2-2\delta]$,

$$1 - \varphi(x) + \frac{x}{2}\varphi'(x) = 1 - \eta(x)(\eta(x) - x\eta'(x))$$
$$= 1 - \frac{(2-\delta-x)(2-\delta)}{(2-3\delta)^2}$$
$$= \frac{(2-\delta)x - 8\delta + 8\delta^2}{(2-3\delta)^2}$$
$$\geq -2\delta.$$

Finally, for $x \in [2-2\delta, 2]$, since φ is decreasing, we have that $\varphi(x) \leq \delta^2/(2-3\delta)^2 \leq \delta^2$ and therefore

$$1 - \varphi + \frac{x}{2}\varphi' \geq 1 - \delta^2 - (1+\theta)\delta \geq -\theta\delta.$$

Thus φ satisfies (2.85). $\qquad\square$

Proof of Theorem 2.14. For the case where $\mathcal{M}^n$ is compact, which is quite easy, see Exercise 2.11.

Let $p \in \mathcal{M}^n$ and define $r(x) = d(x, p)$. Choose $0 \leq r_0 < 1$ such that $|X(p)| \leq r_0^{-1}$ and $|\text{Ric}| \leq r_0^{-2}$ on $B_{r_0}(p)$. For each $0 < \delta < 1/10$ and $a > 1/\delta$, let $\varphi = \varphi_\delta$ be as in Proposition 2.22 and define $\phi = \phi_{\delta,a} : \mathcal{M}^n \to [0,1]$ by

$$\phi(x) = \varphi(r(x)/(ar_0)).$$

Let x_0 be a point at which the compactly supported function

$$(2.86) \qquad\qquad F := F_{\delta,a} := \phi_{\delta,a} R : \mathcal{M}^n \to \mathbb{R}$$

achieves its minimum value. We claim that

$$(2.87) \qquad\qquad F(x_0) \geq \begin{cases} -C_1/a & \text{if } \lambda \geq 0, \\ (1+\varepsilon)\frac{n\lambda}{2} - \frac{C_1}{a} & \text{if } \lambda < 0, \end{cases}$$

where $C_1 = C_1(n, \delta, \lambda, r_0)$ is a positive constant independent of a and $\varepsilon = \varepsilon(\delta)$ is positive and tends to 0 as $\delta \to 0$.

To see this, first consider the case that $x_0 \in B_{\delta a r_0}(p)$. Then $F \equiv R$ in a neighborhood of x_0 and

$$(2.88)$$

$$0 \leq \Delta_X F = \Delta_X R = -2|\mathrm{Ric}|^2 + \lambda R = -2\left|\mathrm{Ric} - \frac{R}{n}g\right|^2 - \frac{2}{n}R\left(R - \frac{n\lambda}{2}\right)$$

at x_0, where the second equality is by Exercise 2.30. Since the first term is nonpositive, the second term must be nonnegative. So $F(x_0) = R(x_0) \geq 0$ if $\lambda \geq 0$ and $F(x_0) = R(x_0) \geq n\lambda/2$ if $\lambda < 0$. Either way, (2.87) holds in this situation.

Now suppose that $x_0 \notin B_{\delta a r_0}(p)$. If $F(x_0) \geq 0$, then (2.87) holds and there is nothing to prove, so we may assume that $F(x_0) < 0$. In particular, $x_0 \in B_{2ar_0}(p)$ and $\phi(x_0) > 0$. By Calabi's trick[8], we may assume r is smooth at x_0 and compute that

$$(2.89) \qquad 0 \leq \Delta_X F$$

$$= \phi \Delta_X R + 2\langle \nabla R, \nabla \phi \rangle + R \Delta_X \phi$$

$$\leq -\frac{2F}{n}\left(R - \frac{n\lambda}{2}\right) - 2R\frac{|\nabla \phi|^2}{\phi} + R \Delta_X \phi.$$

Here, we have used that $\nabla R = -R\nabla\phi/\phi$ at x_0, since $\nabla F(x_0) = 0$. By Proposition 2.21 and our choice of r_0, we have

$$(2.90) \qquad\qquad \Delta_X r \leq \begin{cases} C(n)/r_0 & \text{if } \lambda \geq 0, \\ C(n)/r_0 - \frac{\lambda}{2}r & \text{if } \lambda < 0, \end{cases}$$

and hence

$$(2.91) \qquad \Delta_X \phi = \frac{\varphi'}{ar_0}\Delta_X r + \frac{\varphi''}{a^2 r_0^2} \geq \begin{cases} -\frac{C_2}{a} & \text{if } \lambda \geq 0, \\ \frac{\lambda r \varphi'}{2ar_0} - \frac{C_2}{a} & \text{if } \lambda < 0, \end{cases}$$

for some constant $C_2 = C_2(n, \delta)$.

---

[8]For, if x_0 is in the cut locus of p, we may fix $\epsilon > 0$ and replace $F(x)$ by $F_\epsilon(x) = \phi(r_\epsilon(x)/(ar_0))R(x)$ where $r_\epsilon(x) = d(x, \gamma(\epsilon)) + \epsilon$ and γ is a minimal geodesic from p to x_0. We may then apply the elliptic maximum principle to F_ϵ and send $\epsilon \to 0$. See, e.g., §1.2 of Chapter 10 in [**111**] for a more detailed exposition of Calabi's trick.

Consider first the case that $\lambda \geq 0$ (shrinkers and steadies). Using (2.89) and (2.91), we see that

$$0 \leq \frac{2|F|}{n\phi}\left(F - \frac{n\lambda\phi}{2} + \frac{n(1+\theta)^2}{a^2 r_0^2} + \frac{nC_2}{2a}\right) \leq \frac{2|F|}{n\phi}\left(F + \frac{C_3}{a}\right),$$

for an appropriate constant C_3 depending on n, δ, and r_0. So $F(x_0) \geq -C_3/a$ and (2.87) follows.

Now suppose that $\lambda < 0$ (expanders). In this case, (2.89) and (2.91) give

$$\begin{aligned}
0 &\leq \frac{2|F|}{n\phi}\left(F - \frac{n\lambda\phi}{2} + \frac{n(1+\theta)^2}{a^2 r_0^2} + \frac{nC_2}{2a} + \frac{n\lambda\varphi' r}{4ar_0}\right) \\
&\leq \frac{2|F|}{n\phi}\left(F + \frac{C_3}{a} - \frac{n\lambda}{2}\left(\varphi - \frac{\varphi' r}{2ar_0}\right)\right) \\
&\leq \frac{2|F|}{n\phi}\left(F + \frac{C_3}{a} - \frac{n\lambda}{2} + \frac{n\lambda}{2}\left(1 - \varphi + \frac{\varphi' r}{2ar_0}\right)\right).
\end{aligned}$$

However, by our construction of φ, specifically by (2.85), we have

$$1 - \varphi\left(\frac{r}{ar_0}\right) + \frac{r}{2ar_0}\varphi'\left(\frac{r}{ar_0}\right) \geq -\varepsilon(\delta)$$

at x_0, so (2.87) follows in this case as well.

From the lower bound on F, we immediately obtain that

$$R(p) = F_{\delta,a}(p) \geq \begin{cases} -C_2/a & \text{if } \lambda \geq 0, \\ (1+\varepsilon)\frac{\lambda n}{2} - \frac{C_1}{a}\lambda & \text{if } \lambda < 0 \end{cases}$$

on $B_{\delta ar_0}(x)$ for all $0 < \delta < 1/10$ and $a > 1/\delta$. Sending $a \to \infty$ for any arbitrary $0 < \delta < 1/10$ and then sending $\delta \to 0$ completes the proof of the scalar curvature lower bounds in Theorem 2.14.

Next, we prove the characterization of the equality case. If R achieves one of these minimum values at some point, that is, if $R(p) = 0$ when $\lambda \geq 0$ or $R(p) = n\lambda/2$ when $\lambda < 0$, then R must coincide everywhere with this minimum value by the strong maximum principle. But then the equation for $\Delta_X R$ implies $|\text{Ric} - (R/n)g|^2 \equiv 0$, and the claim follows.

Finally, suppose in addition that $\lambda > 0$ and the shrinker is gradient. Then we have that $\nabla^2 f = \frac{1}{2}g > 0$ and $f = |\nabla f|^2 \geq 0$. Hence $\inf_{\mathcal{M}} f = f(o) = 0$, where o is the unique critical point of f (which exists by Theorem 4.3 below). Defining $\rho := 2\sqrt{f}$, we have on $\mathcal{M}^n \setminus \{o\}$ that

$$(2.92) \qquad \nabla^2(\rho^2) = 2g \quad \text{and} \quad |\nabla\rho|^2 = 1.$$

It now follows from the proof of Proposition 2.9 that $(\mathcal{M}^n, g)$ is isometric to Euclidean space. This completes the proof of the theorem. $\qquad\square$

Regarding the lower bound for the scalar curvature, more generally one may consider a solution to the Ricci flow $(\mathcal{M}^n, g(t))$. Then

$$(2.93) \qquad \frac{\partial R}{\partial t} = \Delta R + 2\left|\mathrm{Ric}\right|^2 \geq \Delta R + \frac{2}{n}R^2 \geq \Delta R.$$

Recall from Definition 1.11 that an ancient solution is a solution to the Ricci flow which exists on an interval of the form $(-\infty, \omega)$. The following result for complete ancient solutions is due to B.-L. Chen; see [86] for the proof.

Theorem 2.23. *Any complete ancient solution to the Ricci flow must have nonnegative scalar curvature. If the solution has zero scalar curvature at some point and time, then the solution is Ricci-flat at all earlier times.*

Chen's theorem in particular applies to both shrinking and steady Ricci solitons.

2.8. Completeness of the soliton vector field

The equivalence of Ricci solitons and self-similar solutions to the Ricci flow is a fundamental heuristic principle and one that is at least *morally* true. However, the correspondence established in Proposition 2.2 falls short of realizing a true equivalence between the two concepts since the self-similar solution it produces from a Ricci soliton need only be defined locally. In order to properly leverage this correspondence, we will need to know when the two concepts are really the same. The crucial issue is the *completeness* of the Ricci soliton vector field.

Definition 2.24. A vector field X on a manifold $\mathcal{M}^n$ is said to be **complete** if for all $p \in \mathcal{M}^n$ the maximal integral curve $\sigma(t)$ of X with $\sigma(0) = p$ is defined for all $t \in \mathbb{R}$.

In this section, we will present two criteria which guarantee the completeness of the Ricci soliton vector field which together show that in the situations of greatest interest for singularity analysis, the concepts of Ricci solitons and self-similar solutions are indeed equivalent.

The first criterion is completely elementary.

Theorem 2.25 (Completeness of the soliton field, I). *Suppose $(\mathcal{M}^n, g, X, \lambda)$ is a Ricci soliton for which $(\mathcal{M}^n, g)$ is complete and of bounded Ricci curvature. Then X is complete.*

Proof. Fix any point $p \in \mathcal{M}^n$ and let $\sigma : (A, \Omega) \to \mathcal{M}^n$ be the maximal integral curve of X with $\sigma(0) = p$. The completeness of $(\mathcal{M}^n, g)$ and the local theory of ODEs imply that $-\infty \leq A < 0 < \Omega \leq \infty$ and — given the maximality of σ — that if either $A > -\infty$ or $\Omega < \infty$, then $d(p, \sigma(t)) \to \infty$ as $t \searrow A$ or $t \nearrow \Omega$, respectively.

Using the Ricci soliton equation, we compute that the function $t \mapsto |X|^2(\sigma(t))$ satisfies

$$\frac{d}{dt}|X|^2 = 2\langle \nabla_X X, X \rangle = \lambda |X|^2 - 2\mathrm{Ric}(X, X)$$

for all $t \in (A, \Omega)$. Hence, since the Ricci curvature is bounded, there is a constant C such that

$$-2C|X|^2 \le \frac{d}{dt}|X|^2 \le 2C|X|^2$$

along σ, and thus

$$\mathrm{e}^{-Ct}|X|(0) \le |X|(\sigma(t)) \le \mathrm{e}^{Ct}|X|(\sigma(0))$$

for all $t \in (A, \Omega)$.

From this we see that, if $\Omega < \infty$, then $|X|(\sigma(t)) \le C'$ for all $t \in [0, \Omega)$. But then, along any sequence $0 \le t_i \nearrow \Omega$, we would have

$$d(p, \sigma(t_i)) \le \mathrm{L}(\sigma|_{[0,t_i]}) = \int_0^{t_i} |X|(\sigma(t)) \, dt < C'\Omega,$$

contradicting the maximality of σ; here, L denotes the Riemannian length. Thus we must have $\Omega = \infty$. A similar argument shows that $A - -\infty$ and hence that $\sigma(t)$ is defined for all $t \in \mathbb{R}$. It follows that X is complete. $\square$

Remark 2.26. Since Theorem 2.14 implies that the scalar curvature of a complete Ricci soliton is bounded below, the two-sided bound on the Ricci curvature in the theorem above may be replaced with merely an upper bound.

The assumption that $(\mathcal{M}^n, g)$ be complete in Theorem 2.25 is certainly necessary: If $(\mathcal{M}^n, g, X, \Lambda)$ is a complete Ricci soliton with a nontrivial (i.e., not identically zero) vector field and $p \in \mathcal{M}^n$ is such that $X(p) \ne 0$, then the restriction of X to $\mathcal{M}^n \setminus \{p\}$ will not be complete. However, the necessity of the assumption of bounded Ricci curvature is less clear. The following result of Z.-H. Zhang [303] shows that, at least for *gradient* Ricci solitons, the completeness of the manifold alone is enough to ensure the completeness of the vector field.

Theorem 2.27 (Completeness of the soliton field, II). *Suppose $(\mathcal{M}^n, g, f, \lambda)$ is a gradient Ricci soliton for which $(\mathcal{M}^n, g)$ is complete. Then ∇f is a complete vector field.*

The key to the proof is Hamilton's identity (2.44) and the universal lower bound for scalar curvature proven in Theorem 2.14.

Proof of Theorem 2.27. By combining Theorem 2.14 and (2.44), we have

$$(2.94) \qquad\qquad |\nabla f|^2 \leq \lambda f + C$$

for some $C = C(\lambda, n) \geq 0$. Fix $p \in \mathcal{M}^n$ and let $r(x) = d(x, p)$.

When $\lambda \neq 0$, (2.94) implies that $h = \lambda f + C$ satisfies $h \geq 0$ and $|\nabla h|^2 \leq |\lambda|^2 h$; that is,

$$|\nabla \sqrt{h}| \leq |\lambda|/2.$$

Choosing $q \in \mathcal{M}^n$ and integrating along any minimizing unit speed geodesic $\gamma : [0, r(q)] \to \mathcal{M}^n$, we find

$$\sqrt{h}(q) - \sqrt{h}(p) = \int_0^{r(q)} \left\langle \nabla \sqrt{h}(\gamma(s)), \gamma'(s) \right\rangle ds \leq \int_0^{r(q)} \left| \nabla \sqrt{h} \right| ds \leq \frac{|\lambda|}{2} r(q).$$

Hence there is a constant $C' > 0$ such that

$$(2.95) \qquad\qquad |\nabla f|(q) \leq |\lambda| r(q) + C'$$

on all of $\mathcal{M}^n$. On the other hand, when $\lambda = 0$, (2.94) says that $|\nabla f| \leq \sqrt{C}$, so, after possibly enlarging C', estimate (2.95) is valid for all λ. The theorem is now a consequence of the following lemma, which says that the vector field X is complete. $\qquad\square$

Lemma 2.28. *Let X be a smooth vector field on $\mathcal{M}^n$. If there is a complete metric g on $\mathcal{M}^n$ relative to which $|X|_g(q) \leq C(d(p, q) + 1)$ for some constant C and $p \in \mathcal{M}^n$, then X is complete.*

Proof. Suppose g is a complete metric on $\mathcal{M}^n$ relative to which the growth of $|X| = |X|_g$ is no more than linear relative to the distance $r(q) = d(p, q)$ from some fixed $p \in \mathcal{M}^n$. Fix an arbitrary $q_0 \in \mathcal{M}^n$ and let $\sigma : (A, \Omega) \to \mathcal{M}^n$, $-\infty \leq A < 0 < \Omega \leq \infty$, be any maximal integral curve of X with $\sigma(0) = q_0$.

Now, by assumption, there is a constant $C \geq 0$ such that, for any $t \in [0, \Omega)$, we have

$$r(\sigma(t)) \leq r(q_0) + d(q_0, \sigma(t))$$
$$\leq r(q_0) + \int_0^t |X|(\sigma(s)) \, ds$$
$$\leq r(q_0) + Ct + C \int_0^t r(\sigma(s)) \, ds,$$

and hence by Grönwall's inequality,

$$r(\sigma(t)) \leq e^{Ct}(r(q_0) + Ct)$$

for all $t < \Omega$. This shows that $\lim_{t \to \Omega} r(\sigma(t)) = \infty$ only if $\Omega = \infty$. The same argument, applied to the integral curve $t \to \sigma(-t)$ of $-X$, shows that $A = -\infty$, and it follows that X is complete. $\qquad\square$

2.9. Compact steadies and expanders are Einstein

On closed manifolds, nonshrinking Ricci solitons are trivial. We have the following result of Ivey [**192**].

Theorem 2.29. *Any steady or expanding Ricci soliton on a closed manifold is Einstein; i.e.,* $\mathrm{Ric} = \frac{r}{n} g$, *where* $r = R_{\mathrm{avg}}$.

Proof. Let $(\mathcal{M}^n, g, X, \lambda)$ be a compact Ricci soliton with $\lambda \leq 0$. Integrating the equation $R + \mathrm{div}\, X = n\lambda/2$, we see that $r = n\lambda/2 \leq 0$. By taking the divergence of the Ricci soliton equation (2.1), we obtain

$$(2.96) \qquad \Delta X + \mathrm{Ric}(X) = 0.$$

From the equation

$$(2.97) \qquad \Delta_X R - \lambda R + 2 \left| \mathrm{Ric} \right|^2 = 0$$

we see that

$$(2.98) \qquad \Delta_X (R - r) + 2 \left| \mathrm{Ric} - \frac{r}{n} \right|^2 + \frac{2r}{n}(R - r) = 0.$$

Since $\mathcal{M}^n$ is compact, R achieves its minimum value R_{min} at some $x_0 \in \mathcal{M}^n$, and at any such point

$$2 \left| \mathrm{Ric} - \frac{r}{n} \right|^2 + \frac{2r}{n}(R - r) \leq 0.$$

Both terms are nonnegative and thus vanish. In particular, $R_{\mathrm{min}} = R(x_0) = r$, so $R(x) = r$ for all $x \in \mathcal{M}^n$. But then every term in (2.98) must vanish identically on $\mathcal{M}^n$, including $|\mathrm{Ric} - (r/n)g|^2$. $\qquad \square$

The theorem is also true in the nongradient case; see Exercise 2.30 for a proof.

2.10. Notes and commentary

The mathematical theory of Ricci solitons was first rigorously developed by Hamilton [**175–177, 179**], laying the foundations of the theory and exhibiting its deep connection to Ricci flow singularity analysis. Bryant, Cao, Ivey, and Koiso made important contributions to the early development of this theory. In the physics literature, the Ricci soliton equation first appeared in Friedan [**151**]. A widely cited survey is by Cao [**61**]. Expository accounts include [**111**, Chapter 4], [**101**, Chapter 1], and [**104**, Chapter 27]. See the extensive references therein on Ricci solitons. Additionally, a selection of papers on Riemannian Ricci solitons and Kähler Ricci solitons, not cited elsewhere in this book, are referenced in the notes and commentary sections of Chapters 4 and 3, respectively.

2.11. Exercises

2.11.1. Scalings and pullbacks of solitons.

Exercise 2.1 (Curvature under scaling). Prove the elementary curvature scaling properties: If α is a positive real number, then

$$(2.99) \qquad \mathrm{Rm}(\alpha g) = \alpha \mathrm{Rm}(g), \quad \mathrm{Ric}(\alpha g) = \mathrm{Ric}(g), \quad R(\alpha g) = \alpha^{-1} R(g).$$

Exercise 2.2 (Pullback of curvatures). Let ϕ be a local diffeomorphism. Prove that:

(1) $\mathrm{Rm}_{\phi^* g} = \phi^* \mathrm{Rm}_g$.

(2) $\mathrm{Ric}_{\phi^* g} = \phi^* \mathrm{Ric}_g$.

(3) $R_{\phi^* g} = R_g \circ \phi$.

Exercise 2.3 (Pullback of Lie derivative). Prove that if $\phi : \mathcal{N}^n \to \mathcal{M}^n$ is a diffeomorphism, X is a vector field on $\mathcal{M}^n$, and α is a (covariant) tensor on $\mathcal{M}^n$, then

$$(2.100) \qquad \qquad \phi^* (\mathcal{L}_X \alpha) = \mathcal{L}_{\phi^* X} (\phi^* \alpha).$$

Exercise 2.4 (Lie derivative of the metric). Prove the Lie derivative of the metric identity (2.28). Generalize this to

$$(2.101) \qquad \qquad (\mathcal{L}_X g)_{ij} = \nabla_i X_j + \nabla_j X_i.$$

Exercise 2.5 (Lie derivative of the volume form). Prove that the Lie derivative of the volume form is given by

$$(2.102) \qquad \qquad \mathcal{L}_X d\mu = \mathrm{div}(X) \, d\mu.$$

Exercise 2.6 (Diffeomorphism-invariance of solitons). Prove the diffeomorphism-invariance property (2) for Ricci solitons: If $(\mathcal{M}^n, g, X, \lambda)$ satisfies (2.1) and if $\varphi : \mathcal{M}^n \to \mathcal{M}^n$ is a diffeomorphism, then

$$(2.103) \qquad \qquad \mathrm{Ric}_{\varphi^* g} + \frac{1}{2} \mathcal{L}_{\varphi^* X} \varphi^* g = \frac{\lambda}{2} \varphi^* g.$$

2.11.2. Product solitons.

Exercise 2.7. Let $(\mathcal{M}_i^{n_i}, g_i)$, $i = 1, 2$, be Riemannian manifolds with Levi-Civita connections ∇_i. Show that the Riemannian product $(\mathcal{M}_1^{n_1}, g_1) \times (\mathcal{M}_2^{n_2}, g_2)$ has Levi-Civita connection ∇ given by

$$(2.104) \qquad \nabla_{X_1 + X_2} (Y_1 + Y_2) = (\nabla_1)_{X_1} Y_1 + (\nabla_2)_{X_2} Y_2$$

for $X_i, Y_i \in T\mathcal{M}_i$, $i = 1, 2$.

Exercise 2.8. Denote the Riemann, Ricci, and scalar curvatures of $(\mathcal{M}_i^{n_i}, g_i)$ by Rm_i, Ric_i, and R_i, respectively.

(1) Prove that the Riemann curvature tensor Rm of the Riemannian product $(\mathcal{M}_1^{n_1}, g_1) \times (\mathcal{M}_2^{n_2}, g_2)$ is given by

$$(2.105) \quad \mathrm{Rm}(X_1 + X_2, Y_1 + Y_2, Z_1 + Z_2, W_1 + W_2)$$
$$= \mathrm{Rm}_1(X_1, Y_1, Z_1, W_1) + \mathrm{Rm}_2(X_2, Y_2, Z_2, W_2).$$

(2) Prove (2.16), that the Ricci tensor Ric of the Riemannian product satisfies $\mathrm{Ric} = \mathrm{Ric}_1 + \mathrm{Ric}_2$; that is,

$$(2.106) \qquad \mathrm{Ric}(X_1 + X_2, Y_1 + Y_2) = \mathrm{Ric}_1(X_1, Y_1) + \mathrm{Ric}_2(X_2, Y_2).$$

(3) Prove that the scalar curvature R of the Riemannian product satisfies

$$(2.107) \qquad R(x_1, x_2) = R_1(x_1) + R_2(x_2)$$

for $x_1 \in \mathcal{M}_1^{n_1}, x_2 \in \mathcal{M}_2^{n_2}$.

2.11.3. Nongradient Ricci solitons.

Exercise 2.9 (The Topping–Yin expanding soliton [**278**]). Prove that the quadruple $(\mathbb{R}^2, g, X, -1)$ in Example 2.4 satisfies the expanding Ricci soliton equation (2.1) with $\lambda = -1$.

Exercise 2.10. Let $(\mathcal{M}^n, g, X, \lambda)$ be a Ricci soliton. Prove (2.96):

$$\Delta X + \mathrm{Ric}(X) = 0.$$

By taking the divergence of the equation above, prove (2.98):

$$\Delta_X (R - r) + 2 \left| \mathrm{Ric} - \frac{r}{n} \right|^2 + \frac{2r}{n}(R - r) = 0.$$

Exercise 2.11 (Compact case of R lower bound). Prove Theorem 2.14 in the case where $\mathcal{M}^n$ is compact. Observe how the proof is simpler than in the noncompact case. The parabolic version of this fact is that on a closed manifold, under the Ricci flow the minimum of the scalar curvature is nondecreasing.

2.11.4. Level sets of the potential function.

Exercise 2.12 (Level sets as evolving hypersurfaces). Let $F : \mathcal{M}^n \to \mathbb{R}$ be a smooth function with $\nabla F(x) \neq 0$ for all $x \in \mathcal{M}^n$. Show that each level set $\Sigma_c := \{F = c\}$ is a smooth hypersurface. Define a 1-parameter group of diffeomorphisms $\phi_t : \mathcal{M}^n \to \mathcal{M}^n$ by $\partial_t \phi_t = \frac{\nabla F}{|\nabla F|^2} \circ \phi_t$, where we assume that $(\mathcal{M}^n, g)$ is complete and the vector field on the right-hand side is complete. Prove that $\phi_t(\Sigma_c) = \Sigma_{c+t}$.

Exercise 2.13. Prove that the second fundamental form, defined by (2.35), is symmetric:

$$(2.108) \qquad \mathrm{II}(Y, X) = \mathrm{II}(X, Y) \quad \text{for } X, Y \in T_x \Sigma_c, \ x \in \Sigma_c.$$

HINT: We may extend the vectors X, Y to vector fields defined in a neighborhood $\mathcal{U}$ of x in $\mathcal{M}^n$ so that X, Y are tangent to $\Sigma_c \cap \mathcal{U}$. Note that then $[X, Y]$ is tangent to $\Sigma_c \cap \mathcal{U}$.

Exercise 2.14. Prove the **Gauss equation** for a hypersurface $\Sigma \subset \mathcal{M}^n$ with unit normal vector field ν (if you like, you may assume that Σ is a level set, but this doesn't simplify things): For $X, Y, Z, W \in T_x \Sigma$,

$$(2.109) \quad \mathrm{Rm}_{\mathcal{M}}(X, Y, Z, W) = \mathrm{Rm}_{\Sigma}(X, Y, Z, W)$$
$$- \mathrm{II}(X, W)\, \mathrm{II}(Y, Z) + \mathrm{II}(X, Z)\, \mathrm{II}(Y, W).$$

HINT: Extend X, Y, Z, W to vector fields defined in a neighborhood of x and tangent to Σ. Use the formula

$$(2.110) \qquad \nabla_X^{\mathcal{M}} Y = \nabla_X^{\Sigma} Y - \mathrm{II}(X, Y)\nu.$$

Take the tangential component of the defining equation for $\mathrm{Rm}_{\mathcal{M}}$.

Remark 2.30. The interested reader may take the normal component and derive the **Codazzi equation**:

$$(2.111) \qquad (\nabla_X^{\Sigma} \mathrm{II})(Y, Z) - (\nabla_Y^{\Sigma} \mathrm{II})(X, Z) = -\langle \mathrm{Rm}_{\mathcal{M}}(X, Y)Z, \nu \rangle.$$

2.11.5. Special solitons.

Exercise 2.15 (Manifolds with trace-free Ricci tensor). Use the contracted second Bianchi identity (1.60) to prove that if $(\mathcal{M}^n, g)$ satisfies $\mathrm{Ric} = \frac{1}{n}Rg$ and $n \geq 3$, then R is a constant. In particular, $(\mathcal{M}^n, g)$ is an Einstein manifold.

Exercise 2.16. Suppose that a quadruple $(\mathcal{M}^n, g, f, \lambda)$ satisfies $\nabla^2 f = \frac{\lambda}{2}g$. Prove that, by adding a constant to f if necessary, we have

$$(2.112) \qquad |\nabla f|^2 = \lambda f.$$

Exercise 2.17. Hypothesize as in the previous exercise, now assuming that $\lambda = 1$ and $f > 0$. Define $\rho := 2\sqrt{f}$. Show that $|\nabla \rho| = 1$ and $\nabla_{\nabla \rho}\nabla \rho = 0$. Prove that

$$\mathcal{L}_{\nabla \ln \rho}\left(\frac{g}{\rho^2}\right) = -\frac{4}{\rho^2} d\ln \rho \otimes d\ln \rho.$$

2.11.6. Properties of solitons.

Exercise 2.18 (Critical points of f and R). Prove that for any GRS with positive Ricci curvature, if x is a critical point of R, then x is a critical point of f. Does this result hold for negative Ricci curvature?

Exercise 2.19 (Steady GRS have bounded R). Prove that the scalar curvature of any steady GRS is uniformly bounded. Prove that for any steady GRS, if $R \geq 0$ (which is proved later), then $|\nabla f|$ is uniformly bounded.

2.11.7. The f-divergence.

Exercise 2.20. Prove the f-contracted second Bianchi identity:

$$(2.113) \qquad \operatorname{div}_f \left(\operatorname{Ric} + \nabla^2 f \right) = \frac{1}{2} \nabla R_f,$$

where div_f is defined by (2.62). Derive from this that $R_f + \lambda f$ is constant on a gradient Ricci soliton (for a normalized gradient Ricci soliton we have (2.49)).

Exercise 2.21 (f-divergence theorem). Prove that on a compact Riemannian manifold $(\mathcal{M}^n, g)$ with boundary, for any vector field V we have

$$(2.114) \qquad \int_{\mathcal{M}} \operatorname{div}_f(V) e^{-f} d\mu = \int_{\partial \mathcal{M}} \langle V, \nu \rangle \, e^{-f} d\sigma,$$

where ν denotes the outward unit normal and where $d\sigma$ is the induced volume element of $\partial \mathcal{M}$. A useful special case is when V is a gradient vector field. For example, we obtain

$$(2.115) \qquad \int_{\mathcal{M}} |\nabla f|^2 e^{-f} d\mu = \int_{\mathcal{M}} \Delta f \, e^{-f} d\mu$$

on a closed manifold.

2.11.8. Variation of arc length and Laplacian comparison.

Exercise 2.22. Prove the first variation of arc length formula (2.70). HINT: Define the map $\Gamma(r, v) := \gamma_v(r)$. Use the formula

$$(2.116) \qquad \partial_v |\gamma'(r)|^2 = 2 \langle \nabla_V^\Gamma \gamma'(r), \gamma'(r) \rangle,$$

where ∇^Γ denotes the covariant derivative along the map Γ.

Exercise 2.23. Prove the second variation of arc length formula (2.71). HINT: Calculate

$$\partial_v|_{v=0} \left\langle \frac{\gamma_v'(r)}{|\gamma_v'(r)|}, \nabla_{\partial_r}^\Gamma V \right\rangle,$$

while using the formula

$$\operatorname{Rm}(V, \gamma_v'(r))V = \nabla_{\partial_v}^\Gamma (\nabla_{\partial_r}^\Gamma V) - \nabla_{\partial_r}^\Gamma (\nabla_{\partial_v}^\Gamma V).$$

Exercise 2.24. Denote $r(x) := d(x, p)$. Prove that, in the strong barrier sense,

$$(2.117) \qquad \Delta r(x) \le \frac{1}{r(x)} - \frac{1}{r(x)^2} \int_0^{r(x)} r^2 \operatorname{Ric}\left(\gamma'(r), \gamma'(r)\right) dr.$$

Exercise 2.25. Let $k \in \mathbb{R}$. Choose $\zeta(r) = \frac{\operatorname{sn}_k(r)}{\operatorname{sn}_k(r_x)}$ in the inequality (2.72) for the Laplacian of the distance function, where

$$(2.118) \qquad \operatorname{sn}_k(r) := \begin{cases} \frac{1}{\sqrt{-k}} \sinh\left(r\sqrt{-k}\right) & \text{if } k < 0, \\ r & \text{if } k = 0, \\ \frac{1}{\sqrt{k}} \sin\left(r\sqrt{k}\right) & \text{if } k > 0. \end{cases}$$

What upper bound do you obtain for $\Delta r(x)$?

Exercise 2.26. Let $r_0 \le r(x)/2$. What second variation inequality do you obtain if you replace $\zeta(r)$ in (2.76) by the slightly more general

$$(2.119) \qquad \zeta(r) = \begin{cases} \frac{r}{r_0} & \text{if } 0 \le r \le r_0, \\ 1 & \text{if } r_0 < r \le r(x) - r_0, \\ \frac{r(x)-r}{r_0} & \text{if } r(x) - r_0 < r \le r(x) \, ? \end{cases}$$

2.11.9. Maximum principles.

Exercise 2.27 (Elliptic maximum principle). Suppose that a function h with compact support on a complete Riemannian manifold $(\mathcal{M}^n, g)$ satisfies

$$(2.120) \qquad \Delta h + V \cdot \nabla h \ge ah^2 + bh,$$

where $a \in \mathbb{R}^+$, $b \in \mathbb{R}$, and V is a vector field. What is the best upper bound for h that you can obtain?

Exercise 2.28 (Weak maximum principle). Prove Lemma B.1 on the elliptic weak and strong maximum principles in Appendix B.
HINT: See Theorem 4 on p. 333 of Evan's book [**145**], which implies that part (2) holds locally on a manifold. Use part (2) to prove parts (1) and (3) by contradiction.

Exercise 2.29. Prove that for a shrinking gradient Ricci soliton $(\mathcal{M}^n, g, f)$, at any minimum point o of f we have $f(o) \le \frac{n}{2}$.
HINT: Apply the elliptic maximum principle (Lemma B.1) to the equation (2.54) for $\Delta_f f$.

Exercise 2.30 (Formulas for Ricci solitons). Prove that for a Ricci soliton $(\mathcal{M}^n, g, X, \lambda)$:

(1) The function $S := R - \frac{n\lambda}{2}$ satisfies

$$(2.121) \qquad \Delta S - \langle X, \nabla S \rangle + 2 \left| \operatorname{Ric} - \frac{\lambda}{2} g \right|^2 + \lambda S = 0.$$

(2) Prove Theorem 2.29 for Ricci solitons that are not necessarily gradient.

HINT: When $\lambda \leq 0$, deduce that S is constant by applying the strong maximum principle to (2.121).

The 2-Dimensional Classification

The Ricci soliton equation simplifies dramatically in two dimensions, to the extent that it is possible to classify all complete (and even some incomplete) gradient solitons. We will present the details of the classification in the complete case here and return to a general study of the GRS equation in the next chapter.

3.1. Rotationally symmetric 2-dimensional solitons

One of the main types of Ricci solitons in dimension 2 consists of the rotationally symmetric solitons.

3.1.1. Killing vector fields for 2-dimensional solitons.

The key to the classification of 2-dimensional GRS in general is the observation of Hamilton that a 2-dimensional GRS with nonconstant Gauss curvature has a nontrivial (i.e., not identically zero) Killing vector field. More generally, all gradient Kähler Ricci solitons with nonconstant potential admit a nontrivial local Killing vector field. In dimension 2, the nontrivial local symmetry reduces the Ricci soliton equation to an ODE for a single scalar equation. See §3.4.3 for a short review of Kähler manifolds, and see Exercise 3.2 for a description of the proof of the existence of these Killing vector fields that avoids Kähler geometry.

Lemma 3.1. *Suppose that* $(\mathcal{M}^{2m}, g, f, \lambda)$ *is a Kähler GRS of complex dimension* m *with complex structure* J. *Then* $J(\nabla f)$ *is a Killing vector field; that is,* $\mathcal{L}_{J(\nabla f)} g = 0$.

Proof. Let X, Y be smooth vector fields on $\mathcal{M}^{2m}$. Since $g(JX, Y) = -g(X, JY)$ (a Hermitian metric) and $\nabla J = 0$ (a Kähler metric), we have by (2.28) that

$$\langle \nabla_X(J\nabla\phi), Y \rangle = \langle J(\nabla_X \nabla\phi), Y \rangle = -\nabla^2\phi(X, JY)$$

for any function ϕ, and so, on the one hand we have that

$$(\mathcal{L}_{J\nabla\phi}g)(X, Y) = -\left(\nabla^2\phi(X, JY) + \nabla^2\phi(JX, Y)\right).$$

On the other hand, from the GRS equation (2.3), we see that for the potential function f of the gradient Ricci soliton,

$$\nabla^2 f(X, JY) = -\mathrm{Ric}(X, JY) + \frac{\lambda}{2}g(X, JY)$$

$$= \mathrm{Ric}(JX, Y) - \frac{\lambda}{2}g(JX, Y)$$

$$= -\nabla^2 f(JX, Y),$$

where we used that g is Kähler. Thus $\mathcal{L}_{J\nabla f}g(X, Y) = 0$. $\qquad\square$

When $m = 1$, that is, $\mathcal{M}^2$ is an oriented surface, then J produces a counterclockwise rotation by the angle $90°$.

When the integral curves of the Killing vector fields are closed, we have rotational symmetry.

Example 3.2. The standard almost complex structure on $\mathbb{R}^{2m}$ is

$$(3.1) \qquad\qquad J = \begin{pmatrix} 0_m & -I_m \\ I_m & 0_m \end{pmatrix},$$

where 0_m and I_m are the $m \times m$ zero and identity matrices, respectively. Let $\{x^1, \ldots, x^m, y^1, \ldots, y^m\}$ denote the corresponding standard Euclidean coordinates. The gradient of the potential function for the Gaussian soliton is

$$(3.2) \qquad\qquad \nabla f = \sum_{i=1}^{m} \frac{\lambda x^i}{2} \frac{\partial}{\partial x^i} + \sum_{j=1}^{m} \frac{\lambda y^j}{2} \frac{\partial}{\partial y^j}.$$

The covariant derivative of J acting on the exterior derivative of f is

$$(3.3) \qquad\qquad \nabla(J(df)) = -\sum_{j=1}^{m} \frac{\lambda}{2} dy^j \otimes dx^j + \sum_{i=1}^{m} \frac{\lambda}{2} dx^i \otimes dy^i.$$

Since the antisymmetrization of this 2-tensor vanishes, we conclude that $\mathcal{L}_{J(\nabla f)}g_{\mathrm{Euc}} = 0$; that is, $J(\nabla f)$ is a Killing vector field of the Euclidean metric.

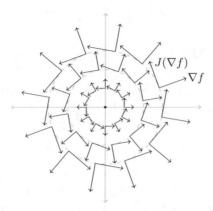

Figure 3.1. The gradient of the potential function of the Gaussian shrinker ∇f and its counterclockwise 90 degree rotation $J(\nabla f)$.

3.1.2. The cigar soliton.

Under the rotationally symmetric ansatz for a steady GRS, we obtain the **cigar soliton [175]** (a.k.a. **Witten's black hole [287]**) $\left(\mathbb{R}^2, g_\Sigma, f_\Sigma\right)$, which is defined by

$$(3.4) \qquad g_\Sigma(x, y) := \frac{4 g_{\mathrm{Euc}}}{1 + x^2 + y^2} \quad \text{and} \quad f_\Sigma(x, y) := -\ln\left(1 + x^2 + y^2\right).$$

The gradient of the potential function is given by

$$(3.5) \qquad \nabla_{g_\Sigma} f_\Sigma(x, y) = \left(-\frac{x}{2}, -\frac{y}{2}\right).$$

In polar coordinates (r, θ), this is

$$(3.6) \qquad g_\Sigma(r, \theta) = \frac{4\left(dr^2 + r^2 d\theta^2\right)}{1 + r^2} \quad \text{and} \quad f_\Sigma(r, \theta) = -\ln(1 + r^2).$$

If we replace the radial coordinate r by the arc length parameter (distance to the origin) $s := 2\sinh^{-1} r = 2\ln(r + \sqrt{1 + r^2})$, then we obtain

$$(3.7)$$
$$g_\Sigma(s, \theta) = ds^2 + 4\tanh^2\left(\frac{s}{2}\right) d\theta^2 \quad \text{and} \quad f_\Sigma(s, \theta) = -2\ln\left(\cosh\left(\frac{s}{2}\right)\right).$$

Thus we see that the cigar soliton is asymptotic to a cylinder of circumference 4π. In particular, the metric g_Σ is complete.

For a metric of the form $g = ds^2 + \phi(s)^2 d\theta^2$, where ϕ is a positive function, we have[1]

$$(3.8) \qquad R(s, \theta) = -2\frac{\phi''(s)}{\phi(s)}.$$

---

[1] A nice way to compute the Gauss curvature $K = \frac{1}{2}R$ is to use the moving coframe $\omega^1 = ds$, $\omega^2 = \phi(s)\, d\theta$. See Exercise 3.16.

Hence

$$(3.9) \qquad R_{g_\Sigma} = \operatorname{sech}^2\left(\frac{s}{2}\right) = \frac{1}{1+r^2}.$$

From this we can verify that $\operatorname{Ric}_{g_\Sigma} + \nabla^2_{g_\Sigma} f_\Sigma = 0$, so that the cigar is a steady GRS. Note also that $R_{g_\Sigma} + |\nabla f_\Sigma|^2 = 1$, that

$$(3.10) \qquad\qquad R_{g_\Sigma} = e^{f_\Sigma},$$

and that the curvature, which is a radial decreasing function, decays exponentially as a function of the distance to the origin. We also have that for any $\kappa > 0$ the cigar soliton is not κ-noncollapsed on all scales in the sense of Definition 1.12; see Exercise 3.5. Intuitively, this is because by (3.7) we have that outside a sufficiently large ball centered at the origin 0^2, the cigar soliton is arbitrarily close to the cylinder $\mathbb{S}^1 \times (0,\infty)$ with the metric $ds^2 + 4d\theta^2$; cf. §1.2.4.4.

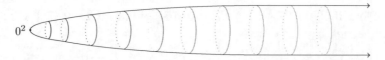

Figure 3.2. Under the Ricci flow, the cigar soliton metric $g_\Sigma(t)$ evolves purely by conformal diffeomorphisms (scalings of the conformally equivalent Euclidean metric). Geometrically, a point on a lighter circle evolves to a point on a darker circle. That is, if (x, t_1) is on a circle, then (x, t_2), for $t_2 > t_1$, is on a darker circle. Note that for any $x \in \mathbb{R}^2 \setminus \{0^2\}$, the function $t \mapsto R_\Sigma(x, t)$ is strictly increasing.

Note that for any integer $p \geq 2$, the finite cyclic group $\mathbb{Z}_p$ acts on $(\mathbb{R}^2, g_\Sigma)$ by isometries, generated by (counterclockwise) rotation by the angle $2\pi/p$:

$$(x, y) \mapsto \big(\cos(2\pi/p)x + \sin(2\pi/p)y, -\sin(2\pi/p)x + \cos(2\pi/p)y\big).$$

Consequently, the quotient $(\mathbb{R}^2, g_\Sigma)/\mathbb{Z}_p$ is a steady GRS on the orbifold $\mathbb{R}^2/\mathbb{Z}_p$.

3.1.3. Rotationally symmetric shrinkers on the teardrop and football orbifolds.

Hamilton [**175**, §10] proved the following; cf. Remark 3.9 below.

Lemma 3.3. *For each bad orbifold (on a topological 2-sphere with one or two cone points), there exists a unique shrinking GRS. This shrinker is rotationally symmetric and has positive curvature.*

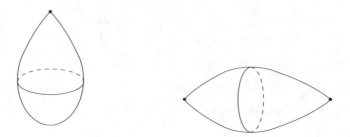

Figure 3.3. The teardrop and football solitons.

3.1.4. The rotationally symmetric noncompact expander.

In this section we describe the 2-dimensional rotationally symmetric expanding gradient Ricci solitons of Gutperle, Headrick, Minwalla, and Schomerus [**169**, Appendix A]. We will call these the **GHMS expanders**.

Consider a rotationally symmetric Riemannian metric g on the punctured plane $\mathbb{R}^2 \setminus \{0\} \cong (0, \infty) \times S^1(F_0)$, where F_0 is the radius of the circle, of the form

$$(3.11) \qquad g = F(r)^2\, dr^2 + r^2\, d\theta^2,$$

where $F : (0, \infty) \to (0, \infty)$, $r \in (0, \infty)$, and $\theta \in S^1(F_0) = \mathbb{R}/(2\pi F_0 \mathbb{Z})$ with $F_0 \subset (0, 2)$.[2] The scalar curvature of g is given by (see Exercise 3.16)

$$(3.12) \qquad R = \frac{2F'(r)}{rF(r)^3}.$$

Note that $g_{\mathrm{Euc}} := F_0^2\, dr^2 + r^2\, d\theta^2$ is the Euclidean metric since $\theta \in \mathbb{R}/(2\pi F_0 \mathbb{Z})$.

Define on $\mathbb{R}^2 \setminus \{0\}$ the radial vector field

$$(3.13) \qquad X := -\frac{r}{F(r)}\frac{\partial}{\partial r}.$$

If ϕ is any radial function, then $\mathrm{grad}_g\phi = \frac{1}{F(r)^2}\phi'(r)\frac{\partial}{\partial r}$. Hence X is a gradient vector field:

$$(3.14) \qquad X = \nabla f := \mathrm{grad}_g f,$$

where f is defined (up to an additive constant) by

$$(3.15) \qquad f'(r) := -rF(r).$$

---

[2] $\mathbb{R}/(2\pi F_0 \mathbb{Z})$ is the quotient space $\mathbb{R}/\sim$, where $\sim$ is the equivalence relation on $\mathbb{R}$ defined by $x \sim y$ if $x - y$ is an integer multiple of $2\pi F_0$. We identify this space with a circle of circumference $2\pi F_0$ and hence radius F_0.

We now search for F (which determines f) such that $(\mathbb{R}^2 \setminus \{0\}, g, f)$ is an expanding gradient Ricci soliton, i.e.,

$$(3.16) \qquad Rg + \mathcal{L}_{\nabla f} g + g = Rg + \mathcal{L}_X g + g = 0,$$

that extends smoothly over the origin.

Define an orthonormal coframe by

$$\omega^1 = F(r)\, dr, \quad \omega^2 = r\, d\theta,$$

which satisfies $g = \omega^1 \otimes \omega^1 + \omega^2 \otimes \omega^2$. The 1-form dual to X is

$$X^\flat := -rF(r)\, dr = -r\omega^1.$$

From the first Cartan structure equation (see e.g. Hicks's book [**187**, §5.2]), the connection 1-forms $\{\omega_i^j\}$ satisfy

$$\left(\nabla_V \omega^j\right)(e_i) = -\omega^j(\nabla_V e_i) = -\omega_i^j(V)$$

for any vector V, so that $\nabla \omega^j = -\omega_i^j \otimes \omega^i$. Using this and the formula $\omega_2^1 = -\frac{1}{rF(r)}\omega^2$, we compute that

$$\mathcal{L}_X g = -2\operatorname{Sym}\left(\nabla X^\flat\right) = -2\operatorname{Sym}\left(dr \otimes \omega^1 + r\nabla\omega^1\right) = -\frac{2}{F(r)}g,$$

where Sym denotes the symmetrization. Thus, the expanding GRS equation (3.16) is equivalent to the first-order ODE:

$$(3.17) \qquad F'(r) = rF(r)^2 \left(1 - \frac{F(r)}{2}\right).$$

By (3.15) and (3.17), we have

$$f'(r) = -rF(r) = -\frac{F'(r)}{F(r)\left(1 - \frac{F(r)}{2}\right)}.$$

Integrating this, we obtain the relation (setting a constant of integration to be zero)

$$(3.18) \qquad f(r) = \ln\left(\frac{2}{F(r)} - 1\right).$$

Assume that $0 < F(r) < 2$, so that the right-hand side of (3.18) is well-defined.

By (3.12) and (3.17),

$$(3.19) \qquad R(r) = \frac{2}{F(r)} - 1 = e^{f(r)} > 0$$

is the scalar curvature, which is positive. Solving the separable ODE (3.17), we have the relation

$$(3.20) \qquad R(r) + \ln R(r) = -r^2 + C$$

for some constant C. Let $R_0 = \lim_{r \to 0} R(r)$, provided the limit exists and is positive. Then $C = R_0 + \ln R_0$. By taking the exponential of (3.20), we have

$$R(r) e^{R(r)} = R_0 e^{R_0} e^{-r^2}.$$

Thus,

(3.21) $$R(r) = W\left(R_0 e^{R_0 - r^2}\right),$$

where $W : \mathbb{R}^+ \to \mathbb{R}^+$ is the **Lambert W function**, defined as the inverse of the function $x \mapsto x e^x$.

Let $F_0 = \ln R_0$. In terms of $F(r)$, the equation above says that

(3.22) $$F(r) = \frac{2}{W\left(\left(\frac{2}{F_0} - 1\right) \exp\left(\frac{2}{F_0} - 1 - r^2\right)\right) + 1}.$$

Conversely, given any $F_0 \in (0, 2)$, the metric g in (3.11) with $F(r)$ defined by (3.22) satisfies

$$\lim_{r \to 0} F'(r) = F_0$$

and hence extends to a smooth metric g on $\mathbb{R}^2$. Note that this implies that $R(r)$ satisfies (3.21), where $R_0 := \lim_{r \to 0} R(r) = e^{F_0}$. We have proved that $(\mathbb{R}^2, g, f)$ is an expanding gradient Ricci soliton.

It is easy to see that the metric g defined by (3.11) is complete. By the form of g, the **cone angle** $\mathrm{CA}(g)$ of g at infinity is

(3.23) $$\mathrm{CA}(g) = \frac{2\pi F_0}{\lim_{r \to \infty} F(r)} = \pi F_0.$$

Since $F_0 \in (0, 2)$, the set of possible cone angles at infinity is the whole interval $(0, 2\pi)$. Now (3.19) and (3.22) imply that

$$R(r, \theta) = W\left(\left(\frac{2}{F_0} - 1\right) \exp\left(\frac{2}{F_0} - 1 - r^2\right)\right).$$

Finally, we note from (3.18) that we have

(3.24) $$f(r, \theta) = \ln R(r, \theta) = \ln W\left(R_0 e^{R_0} e^{-r^2}\right).$$

Summarizing, we have proved the following (see Exercise 3.7).

Theorem 3.4 (Existence of GHMS expanders). *For each $\alpha \in (0, 2\pi)$ there exists a complete 2-dimensional rotationally symmetric expanding gradient Ricci soliton $(\mathcal{M}^2, g, f)$, where $\mathcal{M}^2$ is diffeomorphic to $\mathbb{R}^2$, g has positive, exponentially decaying curvature and is asymptotically conical with cone angle α.*

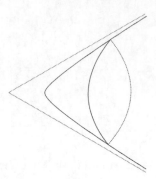

Figure 3.4. A 2-dimensional rotationally symmetric expanding soliton and its asymptotic cone.

Observe that taking the Laplacian of (3.24) and using the trace of (3.16) yields

$$(3.25) \qquad \Delta \ln R = \Delta f = -R - 1,$$

so that Hamilton's Harnack quantity in (1.41) is constant on expanding solitons. This consideration helped Hamilton discover in all dimensions the matrix Harnack quadratic form for solutions to the Ricci flow (see Subsection 4.5.2), which he proved is nonnegative provided that the solution is complete and has bounded nonnegative curvature operator (see [**177**]).

In all complex dimensions there are U(m)-invariant expanding **Kähler-Ricci solitons** on $\mathbb{C}^m$ due to H.-D. Cao [**60**]; see also Feldman, Ilmanen, and Knopf [**147**] for further U(m)-invariant examples.

3.2. Classification of compact 2-dimensional Ricci solitons

In this section we show that all Ricci solitons on closed surfaces necessarily have constant curvature.

3.2.1. The Ricci soliton equation on surfaces and its consequences.

Let $\left(\mathcal{M}^2, g, X\right)$ be a Ricci soliton on a surface. By (2.1) and $\mathrm{Ric} = \frac{1}{2}Rg$, we have

$$(3.26) \qquad (\lambda - R)g = \mathcal{L}_X g,$$

which implies that X is a *conformal vector field* (see (1.71)). Tracing, we have

$$(3.27) \qquad \lambda - R = \mathrm{div}\, X.$$

Taking the divergence of (3.26) and commuting covariant derivatives, we have

$$-\nabla_j R = \nabla^i(\nabla_i X_j + \nabla_j X_i) = \Delta X_j + \nabla_j \operatorname{div} X + R_{jk} X^k.$$

Thus (this is a special case of (2.96)),

$$(3.28) \qquad \Delta X + \frac{1}{2} R X = 0.$$

Taking a second divergence, we obtain (see (2.97) or Exercise 3.8)

$$(3.29) \qquad \Delta R = \langle \nabla R, X \rangle + \lambda R - R^2.$$

Now, consider the case where $X = \nabla f$ is a gradient vector field; i.e., $(\mathcal{M}^2, g, f)$ is a gradient Ricci soliton. By (3.26) and (3.27), we have

$$(3.30) \qquad (\lambda - R)g = 2\nabla^2 f \quad \text{and} \quad \lambda - R = \Delta f.$$

Taking the divergence of the first equation and commuting covariant derivatives yields (see Exercise 3.9)

$$(3.31) \qquad \nabla R = R \nabla f.$$

By taking a second divergence, we obtain (this also follows immediately from (3.29))

$$(3.32) \qquad \Delta R = \langle \nabla R, \nabla f \rangle + \lambda R - R^2.$$

We further rederive the following identity which is especially important and remains unchanged in higher dimensions:

$$\nabla R = 2\operatorname{Ric}(\nabla f) = (\lambda \nabla f - 2\nabla^2 f)\nabla f = \lambda \nabla f - \nabla |\nabla f|^2,$$

and hence (cf. (2.44))

$$(3.33) \qquad R + |\nabla f|^2 - \lambda f = C$$

for some constant C.

Next assume that $(\mathcal{M}^2, g, X)$ is a compact Ricci soliton. By integrating (3.27) and applying the divergence theorem, we obtain that $\lambda = r$, where r is the average scalar curvature. Thus

$$(3.34) \qquad (r - R)g = \mathcal{L}_X g \quad \text{and} \quad \operatorname{div} X = r - R.$$

In particular, if $(\mathcal{M}^2, g, f)$ is a gradient Ricci soliton on a compact surface, then we have

$$(3.35a) \qquad Rg + 2\nabla^2 f = rg,$$

$$(3.35b) \qquad R + \Delta f = r,$$

$$(3.35c) \qquad \Delta R - \langle \nabla R, \nabla f \rangle + R^2 = rR.$$

3.2.2. Ricci solitons on closed surfaces must have constant curvature.

From the work of Hamilton [**175**], Bryant [**54**], Ivey [**195**], Ramos [**259**], and Bernstein and Mettler [**33**], there is a complete classification of 2-dimensional complete GRS. We begin with a result on closed surfaces originally due to Hamilton. Ricci solitons on closed surfaces are the same as constant curvature metrics:

Theorem 3.5. *If $\left(\mathcal{M}^2, g, X\right)$ is a Ricci soliton on a closed surface, then g has constant curvature. Hence:*

(1) *A shrinking Ricci soliton must be a constant positive curvature metric on the 2-sphere or the real projective plane.*

(2) *A steady Ricci soliton must be a flat metric on the torus or the Klein bottle.*

(3) *An expanding Ricci soliton must be a constant negative curvature metric on a surface of genus at least 2.*

Figure 3.5. A round 2-sphere and a flat torus.

Proof. Observe that the Gauss–Bonnet formula implies that the Euler characteristic χ of $\mathcal{M}^2$ has the same sign as the average scalar curvature r.

Case 1 (nonshrinking): $r \le 0$. We may rewrite (3.29) as

$$(3.36) \qquad \Delta R - \langle \nabla R, X \rangle + (R - r)^2 + r(R - r) = 0.$$

Since $\mathcal{M}^2$ is compact, there exists a point $x_0 \in \mathcal{M}^2$ at which R attains its minimum. By the elliptic maximum principle (i.e., by the first and second derivative tests from calculus), we have

$$\nabla R(x_0) = 0 \quad \text{and} \quad \Delta R(x_0) \ge 0.$$

Applying this to (3.36) yields

$$(R - r)^2(x_0) \le r(r - R)(x_0) \le 0,$$

where the second inequality follows from $r \le 0$ and $R(x_0) = R_{\min} \le r$. Therefore $R_{\min} = r$, which implies that $R \equiv r$ on $\mathcal{M}^2$.

Case 2 (shrinking): $r > 0$. Since then $\chi(\mathcal{M}^2) > 0$, by passing to a 2-fold covering if $\mathcal{M}^2$ is nonorientable, we may assume that $\mathcal{M}^2$ is diffeomorphic to $\mathbb{S}^2$. Now, since X is a conformal vector field, by the Kazdan–Warner identity (A.1) in Appendix A and by integrating by parts, we obtain

$$(3.37) \qquad 0 = \int_{\mathbb{S}^2} \langle \nabla R, X \rangle \, d\mu = - \int_{\mathbb{S}^2} R \operatorname{div} X \, d\mu = \int_{\mathbb{S}^2} (r - R)^2 d\mu.$$

We conclude that $R \equiv r$. $\qquad\square$

Remark 3.6. Hamilton's original proof that shrinking GRS on closed surfaces must have constant curvature did not use the Kazdan–Warner identity, but rather the fact that any Riemann 2-sphere after removing a point is conformal to the Euclidean plane. However this fact also relies on the uniformization theorem.

The uniformization theorem in complex analysis implies the following.

Theorem 3.7. *If* (M^2, g) *is a simply connected Riemannian surface, then it is conformally equivalent to either the unit sphere* S^2, *the Euclidean plane* $\mathbb{R}^2$, *or the unit disk* D^2.

The differential geometric version of the uniformization theorem says that for any closed Riemannian surface $(\mathcal{M}^2, g)$, there exists a function u on $\mathcal{M}^2$ such that the metric $e^u g$ has constant scalar curvature. This is a consequence of the uniformization theorem in complex analysis.

In view of the fact that the proof of the Kazdan–Warner identity and Hamilton's proof both use the uniformization theorem, the following result of Chen, Lu, and Tian [93] in the gradient case is of importance in that it enables the Ricci flow to give another proof of the differential geometric version of the uniformization theorem.

Theorem 3.8. *One can prove that any GRS on* $\mathbb{S}^2$ *has constant curvature without using the uniformization theorem.*

Proof. Let $(\mathcal{M}^2, g, f, \lambda)$ be a GRS on an oriented closed surface with compatible complex structure J. By $\lambda = r$ and the Gauss–Bonnet formula, we have $\lambda > 0$. Without loss of generality, assume that $\lambda = 1$. For a contradiction, suppose that g does not have constant curvature. By Lemma 3.1, $J(\nabla f)$ is a nontrivial (i.e., not identically zero) Killing vector field. Since $\mathcal{M}^2$ is closed, f attains its minimum at some point o, so that $J(\nabla f)(o) = 0$. In fact, by (3.35a) we have that $J(\nabla^2 f(V)) = \nabla^2 f(J(V))$ for any vector V, so that $J(\nabla f)$ is a nonconstant real holomorphic vector field on a surface. This implies that $J(\nabla f)$ has isolated zeroes. Since $J(\nabla f)$ is also a Killing vector field, the index of each such zero must be equal to 1 (since the 1-parameter group of isometries that $J(\nabla f)$ generates is asymptotically

comprised of rotations fixing the zero). We may now appeal to the Poincaré–
Hopf theorem, which says that the sum of the indices of a vector field with
isolated zeroes is equal to the Euler characteristic of $\mathbb{S}^2$, which is equal to
2. We conclude that $J(\nabla f)$ has exactly 2 zeroes, one which must be the
maximum of f and one which must be a minimum of f.

Let $\phi_t : \mathcal{M}^2 \to \mathcal{M}^2$, $t \in \mathbb{R}$, be the 1-parameter group of orientation
preserving isometries generated by $J(\nabla f)$. Then o is a fixed point of each
ϕ_t and hence
$$d\phi_t : (T_o\mathcal{M}, g_o) \to (T_o\mathcal{M}, g_o)$$
is a rotation.

Since f is not constant, the map $\mathbb{R} \to \mathrm{SO}(T_o\mathcal{M}, g_o)$ defined by $t \mapsto d\phi_t$ is
a nontrivial homomorphism. Hence there exists a smallest $t_0 > 0$ such that
$d\phi_{t_0} = \mathrm{id}_{T_o\mathcal{M}}$, which implies that the isometry ϕ_{t_0} is the identity map $\mathrm{id}_\mathcal{M}$.
Hence we have the effective $\mathbb{S}^1$-action by isometries defined by $e^{i\frac{2\pi t}{t_0}} \mapsto \phi_t$.
This proves that $(\mathcal{M}^2, g)$ is rotationally symmetric and that f is rotationally
invariant, which reduces the GRS equation to an ODE.

Thus we may write $g = ds^2 + w^2(s)d\theta^2$, where $0 \le s \le L$ for some
$L < \infty$ and $0 \le \theta \le 2\pi$, and we have that f is a function of s only. The
shrinking GRS equation $\frac{R}{2}g + \nabla^2 f = \frac{1}{2}g$ then says that
$$-\frac{w''}{w}\left(ds^2 + w^2 d\theta^2\right) + f'' ds^2 + ww' f' d\theta^2 = \frac{1}{2}ds^2 + \frac{1}{2}w^2 d\theta^2.$$
This is equivalent to

(3.38) $$\frac{w''}{w} + \frac{1}{2} = f'' = \frac{w' f'}{w}.$$

Integrating $\frac{f''}{f'} = \frac{w'}{w}$, we obtain that $f'(s) = aw(s)$, where $a \in \mathbb{R}$. Thus

(3.39) $$w'' + \frac{1}{2}w = aww'.$$

Multiplying this equation by w' and integrating yields

(3.40) $$\frac{1}{2}(w')^2(s) + \frac{1}{4}w^2(s)\Big|_{s=0}^{L} = a\int_0^L w(s)(w')^2(s)ds.$$

Since the metric g is C^∞ on $\mathbb{S}^2$ by assumption, we have $w(0) = w(L) = 0$
and $w'(0) = -w'(L) = 1$. This implies that the left-hand side of (3.40) is
zero. Since $w(s) > 0$ for $s \in (0, L)$, we conclude that $a = 0$. Hence $f'(s) \equiv 0$
and consequently $R \equiv 1$. $\square$

Remark 3.9. The method of Chen, Lu, and Tian [**93**] to analyze Ricci
solitons applies to 2-dimensional closed orbifolds with conical singularities
to yield a new proof of Hamilton's result that in this case any shrinking GRS
is rotationally symmetric and has at most two cone points; see Lemma 3.3.

When the orbifold is bad (i.e., there is one cone point, a teardrop, or there are two cone points of different orders, a football), its curvature is no longer constant.

We can further analyze the ODE (3.39) for a rotationally symmetric 2-dimensional shrinking GRS. We rewrite it as a first-order system:

(3.41a) $$w' =: x,$$

(3.41b) $$x' = \left(ax - \tfrac{1}{2}\right)w.$$

By replacing x by ax and w by aw, we obtain the normalized system (3.41) where $a = 1$, which we henceforth assume.

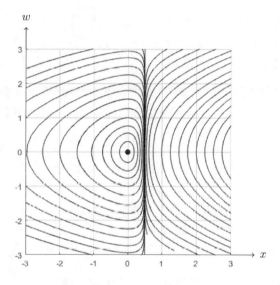

Figure 3.6. Phase-plane portrait of the ODE system (3.41) with $a = 1$ in the xw-plane. In the plot, the horizontal axis is the x-axis and the vertical axis is the w-axis.

In Figure 3.6, the line $x = \frac{1}{2}$ corresponds to a Gaussian shrinker on a cone. Here, $x(s) \equiv \frac{1}{2}$ and $w(s) = \frac{1}{2}s + C$. If we choose the angle θ to lie in the circle of length L, then at $w = 0$ we have a cone angle equal to $L/2$. So the usual Gaussian shrinker on $\mathbb{R}^2$ is given by $x(s) \equiv \frac{1}{2}$ with $L = 4\pi$.

Each point (x, w) in the phase-plane portrait represents the pair of the radius w of a circle in the rotationally symmetric shrinking GRS and the derivative x of the radius. Observe the symmetry of the solution curves with respect to the reflection map $(x, w) \mapsto (x, -w)$. This is because the inverse slope is equal to $\frac{x'}{w'} = \left(ax - \frac{1}{2}\right)\frac{w}{x}$ at points where $x \neq 0$. The line $w = 0$ represents cone points, where the cone angle is given by $|x|L$. Since the warping function is w^2, we only consider the upper half-plane $w \geq 0$.

Now, we see that any curve to the right of the line $x = \frac{1}{2}$ describes a shrinking GRS on $\mathbb{R}^2$ with a single cone point of cone angle xL, where x is the x-value at which the curve intersects the x-axis. On the other hand, any curve to the left of the line $x = \frac{1}{2}$ describes a shrinking GRS on $\mathbb{S}^2$, with two cone points, corresponding to each point of intersection of the curve with the x-axis. Let $-\infty < x_- < 0 < x_+ < \frac{1}{2}$ denote the x-values of the points of intersection. Then the cone angles are $-x_- L$ and $x_+ L$. One can show that $-x_- > x_+$, where the ratio $\frac{x_+}{-x_-}$ may take any value in the interval $(0, 1)$. In particular, for any positive integers $p < q$, there exists a solution $(x(t), w(t))$ to the ODE system (3.41) with $a = 1$ satisfying $\frac{x_+}{-x_-} = \frac{p}{q}$. We may then choose $L = \frac{2\pi}{x_+ q} = \frac{2\pi}{-x_- p}$, so that the cone angles are $2\pi/p$ and $2\pi/q$, and we are hence on a teardrop or football orbifold.

3.3. Classification of noncompact 2-dimensional GRS

In this section we determine the classification of 2-dimensional complete steady and shrinking GRS. We also state the classification result for 2-dimensional expanding GRS.

3.3.1. Complete 2-dimensional steady GRS.

In this subsection we extend the classification of 2-dimensional steady GRS from the compact case to the complete case. Let $\left(\mathcal{M}^2, g, f, \lambda\right)$ be a complete GRS on a surface.

The gradient formula (3.31) implies that $\nabla \left(e^{-f} R\right) = 0$. Therefore

$$(3.42) \qquad\qquad R = ce^f, \quad \text{where } c \in \mathbb{R}.$$

In particular, the scalar curvature of *any* GRS on a surface always has a fixed sign (we always assume that our manifolds are connected). If we are on a shrinking or steady GRS, then R cannot be negative everywhere (see Theorem 2.14). Hence either $R > 0$ everywhere or $R \equiv 0$. Furthermore, by (3.33), on a steady GSR we have that

$$R \leq R + |\nabla f|^2 = C,$$

so that the curvature is bounded on a steady GRS.

Given a complete steady GRS $(\mathcal{M}^2, g, f)$, we may define a solution to the Ricci flow as follows. Let $\varphi_t : \mathcal{M}^2 \to \mathcal{M}^2$ be the 1-parameter group of diffeomorphisms generated by ∇f; that is,

$$\partial_t \varphi_t(x) = \nabla f(\varphi_t(x)),$$
$$\varphi_0 = \mathrm{id}_{\mathcal{M}}.$$

Since $|\nabla f|^2 = C - R \leq C$ by (3.33), φ_t is defined for all $t \in (-\infty, \infty)$. Define $g(t) = \varphi_t^* g$ and $f(t) = f \circ \varphi_t$. Then

$$(3.43) \qquad \partial_t g(t) = -R(t)g(t) = 2\nabla^2_{g(t)} f(t),$$

where $0 \leq R(t) \leq C$ for all t.

Thus, by Shi's local derivative estimates (Theorem 1.6), we have:

Lemma 3.10. *If $(\mathcal{M}^2, g, f)$ is a complete steady GRS, then there exist constants C_k such that $|\nabla^k R| \leq C_k$ on $\mathcal{M}^2$ for $k \geq 0$.*

We have the following result of Hamilton [175], [179, Theorem 26.3].

Theorem 3.11. *If $(\mathcal{M}^2, g, f)$ is a complete steady GRS, then it is either the cigar soliton or flat.*

Proof. If $\mathcal{M}^2$ is compact, then Theorem 3.5 implies that $R \equiv 0$. Thus, if g is not flat, then $R > 0$ by the strong maximum principle and $\mathcal{M}^2$ is noncompact.

STEP 1. *The scalar curvature attains its maximum.* This follows from the claim that

$$R(x) \to 0 \quad \text{as } x \to \infty.$$

Suppose for a contradiction that there exists a sequence $p_i \to \infty$ such that $R(p_i) \geq c > 0$ for all i. By Lemma 3.10, there exist constants C_k such that $|\nabla^k R| \leq C_k$ on $\mathcal{M}^2$. Moreover, by the Gromoll–Meyer theorem (see Theorem 1.18), we have that $\mathcal{M}^2$ is diffeomorphic to $\mathbb{R}^2$ and that $\mathrm{inj}_g(p_i) \geq \delta$ for some positive constant δ. Thus we may apply the Cheeger–Gromov compactness theorem (see Theorem 1.2) to obtain that there exists a subsequence such that $(\mathcal{M}^2, g, p_i)$ converges to a complete Riemannian surface $(\mathcal{M}^2_\infty, g_\infty, p_\infty)$ with bounded nonnegative curvature. Observe that g_∞ is nonflat since $R_{g_\infty}(x_\infty) \geq c > 0$. On the other hand, since the sectional curvatures are nonnegative and $p_i \to \infty$ we may apply the splitting Theorem 1.19 to obtain that $(\mathcal{M}^2_\infty, g_\infty)$ is the metric product of a line with a 1-manifold, which must be flat. Hence we have a contradiction.

STEP 2. *Rotational symmetry.* Let $o \in \mathcal{M}^2$ be a point at which R attains its maximum. It follows from $\nabla R = R\nabla f$ that $\nabla f(o) = 0$. Thus, as observed by Chen, Lu, and Tian [93], we may apply the same proof as in Theorem 3.8 to conclude that any 2-dimensional complete nonflat steady GRS must be rotationally symmetric. So we may let $g = ds^2 + w^2(s)d\theta^2$ and assume that the potential function is radial: $f = f(s)$. Then $R = -2\frac{w''}{w}$ and $\nabla^2 f = f'' ds^2 + ww' f' d\theta^2$. Thus $\frac{R}{2}g + \nabla^2 f = 0$ yields

$$(3.44) \qquad w'' = w' f' = w f''.$$

The second equality implies $f'(s) = aw(s)$ for some constant a. This and the first equality in (3.44) imply that $w'(s) - \frac{1}{2}aw^2(s) = b$ for some constant b.

Since $\mathcal{M}^2$ is diffeomorphic to $\mathbb{R}^2$, without loss of generality, we may assume the closure conditions $w(0) = 0$ and $w'(0) = 1$. Thus $b = 1$. If $a = 0$, then $w(s) = s$, which is the Euclidean metric. If $a > 0$, then $w(s) = \frac{1}{a}\tan(as)$, which is incomplete. Since g is nonflat and complete, we conclude that $a < 0$. Normalizing so that $a = -\frac{1}{2}$, we obtain $w(s) = 2\tanh\left(\frac{s}{2}\right)$, which yields the cigar soliton as defined by (3.6). $\qquad\square$

3.3.2. Complete 2-dimensional shrinking GRS.

Another result of Hamilton is the following [**175**], [**179**, §26].

Theorem 3.12. *If $\left(\mathcal{M}^2, g, f\right)$ is a complete nonflat shrinking GRS, then $\mathcal{M}^2$ is closed; hence g has constant positive curvature. Moreover, the only complete noncompact shrinking GRS is the Euclidean plane.*

Proof. Recall that, since g is nonflat, we have $R > 0$. Normalize the potential function so that $f = R + |\nabla f|^2 > 0$. Now, by (3.42) we have that $R = ce^f > c$, where $c > 0$. Since $\mathrm{Ric} = \frac{1}{2}Rg > \frac{c}{2}g$, by Myers's theorem $\mathcal{M}^2$ is compact. That g has constant positive curvature then follows from either Theorem 3.5 or Theorem 3.8. $\qquad\square$

3.3.3. Complete 2-dimensional expanding GRS.

It remains to classify complete 2-dimensional expanding GRS. To this end, the following is useful; cf. [**101**, Lemma 1.18].

Lemma 3.13. *Any GRS $(\mathcal{M}^2, g, f, \lambda)$ on a surface with $\nabla f \neq 0$ is locally a warped product.*

Proof. Let $x \in \mathcal{M}^2$, and define $\gamma : (-\varepsilon, \varepsilon) \to \mathcal{M}^2$ by $\gamma'(s) = \frac{\nabla f}{|\nabla f|}(\gamma(s))$, $\gamma(0) = x$. Define $\alpha_s : (-\varepsilon, \varepsilon) \to \mathcal{M}^2$ by

$$\alpha_s'(\theta) = J(\nabla f)(\alpha_s(\theta)), \quad \alpha_s(0) = \gamma(s),$$

where J is the almost complex structure on $(\mathcal{M}^2, g)$ (see (2.23)), and define $\gamma_\theta : (-\varepsilon, \varepsilon) \to \mathcal{M}^2$ by

$$\gamma_\theta'(s) = \frac{\nabla f}{|\nabla f|}(\gamma_\theta(s)), \quad \gamma_\theta(0) = \alpha_0(\theta).$$

Note that $\gamma_0 = \gamma$. We compute that

$$\left[J\nabla f, \frac{\nabla f}{|\nabla f|}\right] = \frac{1}{|\nabla f|}[J\nabla f, \nabla f] - \frac{1}{2}\frac{1}{|\nabla f|^3}J\nabla f(|\nabla f|^2)\nabla f = 0,$$

where we used $[J\nabla f, \nabla f] = 0$ and $J\nabla f\,|\nabla f|^2 = 0$. From this we obtain that

$$\alpha_s(\theta) = \gamma_\theta(s).$$

In particular, (s, θ) are local coordinates on a neighborhood of x. Since $|J\nabla f| = |\nabla f|$ is constant on each α_s, we may define $\phi(s) := |\nabla f|(\alpha_s)$. We conclude that the metric g is given in a neighborhood of x by

$$g = \left\langle \frac{\nabla f}{|\nabla f|}, \frac{\nabla f}{|\nabla f|} \right\rangle ds^2 + 2 \left\langle J\nabla f, \frac{\nabla f}{|\nabla f|} \right\rangle ds\, d\theta + \langle J\nabla f, J\nabla f \rangle d\theta^2$$
$$= ds^2 + \phi(s)^2 \, d\theta^2.$$

This completes the proof of the lemma. $\qquad\square$

The following theorem was obtained by Bernstein and Mettler [**33**] and independently by Ramos [**259**]; see §3.4.2 below for the definition of a (topological) **end** of a noncompact manifold. Intuitively, an end of a manifold is a connected component of its "infinity". Given $\alpha \in (0, \infty)$, the flat Riemannian cone with cone angle α is $(0, \infty) \times (\mathbb{R}/\alpha\mathbb{Z})$ with the metric $dr^2 + r^2 d\theta^2$.

Theorem 3.14. *If $\left(\mathcal{M}^2, g, f \right)$ is a complete expanding GRS not of constant curvature, then g and f are rotationally symmetric and either:*

(i) *g has positive curvature and has one end, which is asymptotic to a flat Riemannian cone with cone angle in $(0, 2\pi)$. These are the GHMS expanding GRS on $\mathbb{R}^2$ given by Theorem 3.4.*

(ii) *g has negative curvature and has one end, which is asymptotic to a flat Riemannian cone with cone angle in $(2\pi, \infty)$.*

(iii) *g has negative curvature and has two ends, where one end is asymptotic to a hyperbolic cusp and the other end is asymptotic to a flat Riemannian cone with cone angle in $(0, \infty)$. There is an "infinite cone angle" version of this described in the proof below.*

The proof of this theorem will occupy the rest of this subsection. We assume that g is not flat. Recall from formula (3.42) that then either $R > 0$ or $R < 0$ on all of $\mathcal{M}^2$. Recall also from Lemma 3.1 that $J(\nabla f)$ is a nontrivial (i.e., not identically zero) Killing vector field. Note that as in the argument by Chen, Lu, and Tian in the proof of Theorem 3.8, if the potential function f has a critical point somewhere on $\mathcal{M}^2$, then one can deduce that both f and the metric g are rotationally symmetric; see Exercise 3.18. Since $\mathcal{M}^2$ is noncompact, we have that if f has a critical point, then the critical point is unique.

So we assume that f has no critical points and prove rotational symmetry of f and g in this case too. We first take a local point of view, where the arguments also apply to the case where f has a critical point.

Since f has no critical points and by Lemma 3.13, in a neighborhood of any point $x \in \mathcal{M}^2$, there exist local coordinates (s, θ) and a positive function $\phi(s)$ such that we can write the metric g as

$$(3.45) \qquad\qquad g = ds^2 + \phi(s)^2 d\theta^2,$$

where $\phi(s) > 0$. If f has a critical point, then this result is true in the complement of the critical point. Recall that by (3.8) the scalar curvature is then given by

$$R(s, \theta) = -2\frac{\phi''(s)}{\phi(s)}.$$

Since $R(x) \neq 0$, we have $\phi''(s_0) \neq 0$ at the coordinate s_0 of x. If $\phi'(s_0) = 0$, then we have $\phi'(s) \neq 0$ for s near but not equal to s_0. With this in mind, we assume henceforth that we are in a region in $\mathcal{M}^2$ where $\phi'(s) \neq 0$. Later, we will see that $\phi' \neq 0$ actually holds at all points of $\mathcal{M}^2$.

We change variables by setting $t := \frac{1}{4}\phi(s)^2$, so that $dt = \frac{1}{2}\phi(s)\phi'(s)ds$. Then we may put the locally expressed warped product metric g in the form

$$(3.46) \qquad\qquad g = \frac{\psi(t)^2}{t}dt^2 + 4t\,d\theta^2,$$

where $\psi(t) = \frac{1}{\phi'(s)}$ and $s = \phi^{-1}(2\sqrt{t})$. That is,

$$(3.47) \qquad\qquad \psi\left(\frac{\phi(s)^2}{4}\right)\phi'(s) = 1.$$

We proceed to derive an ODE for ψ. By (3.8) again, we have that

$$(3.48) \qquad\qquad R(t, \theta) = -\frac{2}{\sqrt{t}}\frac{d^2\sqrt{t}}{ds^2} = \frac{\psi'(t)}{\psi(t)^3},$$

where we used that $\frac{d}{ds} = \frac{\sqrt{t}}{\psi(t)}\frac{d}{dt}$. In particular, R has the same sign as ψ'.

On the other hand, by (3.31) and Lemma 3.1,

$$J(\nabla f) = J\nabla \ln|R| = J\left(\frac{\partial \ln|R|}{\partial s}\frac{\partial}{\partial s}\right) = \frac{\sqrt{t}}{\psi(t)}(\ln|R|)'(t)J\left(\frac{\partial}{\partial s}\right)$$

is a nonzero Killing vector field. Moreover, since $J(\nabla f)$ is orthogonal to $\frac{\partial}{\partial s}$ and the metric g is a warped product by (3.46) and since $J(\nabla f)$ is a Killing vector field, we must have that $J(\nabla f) = \frac{a}{2}\frac{\partial}{\partial \theta}$ for some nonzero constant a. Therefore, since we may assume that $J\left(\frac{\partial}{\partial s}\right) = \frac{1}{2\sqrt{t}}\frac{\partial}{\partial \theta}$ (which is a unit vector), we obtain

$$(3.49) \qquad\qquad a = \frac{1}{\psi(t)}(\ln|R|)'(t) = \frac{1}{\psi(t)}\left(\ln\left|\frac{\psi'(t)}{\psi(t)^3}\right|\right)',$$

where we used (3.48) for the last equality.

Recall from (3.32), with $\lambda = -1$, that for a 2-dimensional expanding GRS, we have

$$\Delta R = \langle \nabla R, \nabla f \rangle - R - R^2.$$

Substituting into this equation the formula $\nabla R = R \nabla f$, we obtain (cf. (3.25))

$$(3.50) \qquad \qquad \Delta \ln |R| + R + 1 = 0.$$

Now, the Laplacian with respect to g of any radial function $h(s)$ is given by the formula

$$(3.51) \quad \Delta h = \frac{d^2 h}{ds^2} + \frac{\phi'(s)}{\phi(s)} \frac{dh}{ds} = \frac{1}{\phi(s)} \frac{d}{ds} \left(\phi(s) \frac{dh}{ds} \right) = \frac{1}{\psi(t)} \frac{d}{dt} \left(\frac{t}{\psi(t)} \frac{dh}{dt} \right).$$

By applying this to $h = \ln |R|$, we have

$$-1 - \frac{\psi'(t)}{\psi(t)^3} = \frac{1}{\psi(t)} \left(\frac{t}{\psi(t)} \left(\ln \left| \frac{\psi'(t)}{\psi(t)^3} \right| \right)' \right)'.$$

By substituting (3.49) into this equation, we obtain

$$(3.52) \qquad \qquad -1 - \frac{\psi'(t)}{\psi(t)^3} = \frac{1}{\psi(t)} (at)' = \frac{a}{\psi(t)}.$$

We conclude that any nonflat 2-dimensional expanding GRS locally has the form

$$(3.53) \qquad \qquad g = \frac{\psi(t)^2}{t} dt^2 + 4t \, d\theta^2,$$

where $\psi(t)$ satisfies

$$(3.54) \qquad \qquad \psi'(t) = -\psi(t)^2 (a + \psi(t))$$

and where $a \in \mathbb{R} \setminus \{0\}$. This is an autonomous first-order ODE for ψ.

Note that if $\psi(t)$ is a solution to (3.54), then its negative satisfies

$$(-\psi)'(t) = -\big(- \psi(t) \big)^2 \big(- a + (-\psi(t)) \big),$$

which has the effect of changing the sign of a. So, without loss of generality, we may assume that $a < 0$.

Note that by (3.48) and (3.52), we have that the scalar curvature satisfies

$$(3.55) \qquad \qquad R(t, \theta) = -1 - \frac{a}{\psi(t)}.$$

From this we see that R is bounded if and only if ψ is bounded from below by a positive constant.

Note that since g is assumed not to have constant curvature, $\frac{\partial}{\partial \theta}$ is the unique, up to scaling, Killing vector field of the metric g. Now we invoke the completeness assumption on g to see the global form of the metric. By this assumption and the local warped product form of g and by the uniqueness of the Killing vector field, we have a global warped product form for g. Therefore, one of the following possibilities must hold:

(1) (Real line level sets) The level sets of the function t are diffeomorphic to lines and $\theta \in \mathbb{R}$.

(2) (Circle level sets) The level sets of the function t are diffeomorphic to circles and $\theta \in \mathbb{R}/2\pi L\mathbb{Z}$, where $L > 0$. We call L the θ-**radius**.

Now we consider the symmetries of the ODE (3.54). Since the ODE is autonomous, it is translation invariant. This does not change the isometry classes of the corresponding metrics. We also have the following scaling symmetry: Let $\tilde{\psi}(t) = c\psi(c^2 t)$, where c is a positive constant. Then $\tilde{\psi}(t)$ satisfies

$$(3.56) \qquad\qquad \tilde{\psi}'(t) = -\tilde{\psi}(t)^2(ac + \tilde{\psi}(t)),$$

which is (3.54) with a replaced by ac. Note that

$$\tilde{g} := \frac{\tilde{\psi}(t)^2}{t}dt^2 + 4t\,d\theta^2 = \frac{\psi(\tilde{t})^2}{\tilde{t}}d\tilde{t}^2 + 4\tilde{t}\,d\tilde{\theta}^2,$$

where $\tilde{t} := c^2 t$ and $\tilde{\theta} := \frac{\theta}{c}$. So the scaling symmetry of the ODE discussed above corresponds geometrically to changing, in the circle level set case, the θ-radius from L to L/c.

In view of the symmetry discussed above, since $a < 0$, without loss of generality we may assume that $a = -2$, and the ODE is

$$(3.57) \qquad\qquad \psi'(t) = \psi(t)^2\big(2 - \psi(t)\big).$$

If $\psi \neq 0$ and $\psi \neq 2$, then this is equivalent to

$$\left(\frac{1}{\psi} + \frac{2}{\psi^2} + \frac{1}{2 - \psi}\right)\psi' = 4.$$

Thus, the solution $\psi(t)$ satisfies

$$\ln|\psi(t)| - \frac{2}{\psi(t)} - \ln|2 - \psi(t)| = 4t + C,$$

where C is a real constant. In any case, it is not difficult to see that solutions to the ODE have the following properties. (Compare with Figure 3.7, which pictures the general solutions to the ODE in (3.57).) Let $t_0 \in \mathbb{R}_+$ and assume that ψ is defined at t_0. This implies that $\psi(t)$ is defined for all $t \geq t_0$. Then:

   (1) If $\psi(t_0) > 0$, then $\psi(t) \to 2$ as $t \to \infty$.
       (a) If $\psi(t_0) = 2$, then $\psi(t) \equiv 2$ for all $t \in \mathbb{R}$.
       (b) If $\psi(t_0) \in (0, 2)$, then $\psi(t)$ is defined for all $t \in \mathbb{R}$, $t \mapsto \psi(t)$ is increasing, and $\psi(t) \to 0$ as $t \to -\infty$.

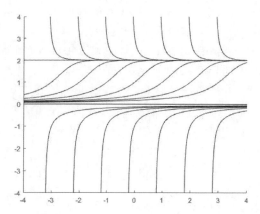

Figure 3.7. Graphs of solutions $\psi(t)$ to the ODE in (3.57). The horizontal axis pictured is the t-axis.

      (c) If $\psi(t_0) \in (2, \infty)$, then there exists $t_* \in (-\infty, t_0)$ such that $\psi(t)$ is defined for $t \in (t_*, \infty)$, $t \mapsto \psi(t)$ is decreasing, and $\psi(t) \to \infty$ as $t \to (t_*)_+$.

(2) If $\psi(t_0) = 0$, then $\psi(t) \equiv 0$ for all $t \in \mathbb{R}$.

(3) If $\psi(t_0) < 0$, then there exists $t_* \in (-\infty, t_0)$ such that $\psi(t)$ is defined for $t \in (t_*, \infty)$, $t \mapsto \psi(t)$ is increasing, and $\psi(t) \to -\infty$ as $t \to (t_*)_+$.

In general, the solution $\psi(t)$, $t \in \big(\max\{t_*, 0\}, \infty \big)$, to the ODE (3.57) defines an expanding GRS metric $g = \frac{\psi(t)^2}{t} dt^2 + 4t \, d\theta^2$. We determine below in which cases g is complete or extends to a complete metric.

The constant solution $\psi(t) \equiv 2$ defines the Euclidean metric. On the other hand, the trivial solution $\psi(t) \equiv 0$ does not define a metric. Firstly, suppose that $t_* \leq 0$. Then $\psi(0) \in [-\infty, \infty]$ is defined. We consider solutions satisfying:

(1)(b) $\psi(0) \in (0, 2)$,

(1)(c) $\psi(0) \in (2, \infty]$, or

(3) $\psi(0) \in [-\infty, 0)$.

In case (1)(b), we have that $\psi(0) =: F_0 \in (0, 2)$. The solution $\psi(t)$ is defined for $t \in (0, \infty)$. The metric g extends to a smooth metric at $t = 0$ if and only if $\theta \in \mathbb{R}/2\pi F_0 \mathbb{Z} = S^1(F_0)$.[3] This is the GHMS expanding GRS on $\mathbb{R}^2$ given by Theorem 3.4 as well as by (3.60) in Remark 3.16 below. Recall

---

[3]This includes the fact that if $\theta \in \mathbb{R}$, then g does not extend to a smooth metric at $t = 0$.

that g has positive curvature and g is asymptotic to a cone with cone angle $\mathrm{CA}(g) = F_0\pi \in (0, 2\pi)$ at infinity. This proves case (i) of Theorem 3.14.

In case (1)(c), the solution $\psi(t)$, $t \geq 0$, defines a smooth complete metric on $\mathbb{R}^2$ if and only if $\theta \in \mathbb{R}/2\pi F_0\mathbb{Z}$, where $\psi(0) = F_0 \in (2, \infty)$. Since $\lim_{t\to\infty} \psi(t) = 2$, we have that g is asymptotic to a cone with cone angle $\mathrm{CA}(g) = F_0\pi \in (2\pi, \infty)$ at infinity. Recall from (3.48) that $R(t, \theta) = \frac{\psi'(t)}{\psi(t)^3}$, so that the curvature is negative everywhere since $\psi'(t) < 0$ and $\psi(t) > 0$. This proves case (ii) of Theorem 3.14.

The remaining possibility in case (1)(c) is that

$$(3.58) \qquad\qquad\qquad \lim_{t\to 0_+} \psi(t) = +\infty.$$

Here, we may define (i) $\theta \in \mathbb{R}/2\alpha\mathbb{Z}$ for any $\alpha \in (0, \infty)$ or (ii) $\theta \in \mathbb{R}$. In case (i), then g is asymptotic on one end to a cone with cone angle α as $t \to \infty$. The other end is where $t \to 0_+$. By (3.57) and (3.58), for $t \approx 0$ we have $\psi'(t) \approx -\psi(t)^3$, so that $(\psi(t)^{-2})' \approx 2$, which implies that $\psi(t) \approx (2t)^{-1/2}$. Thus, for $t \approx 0$ we have that

$$g \approx \frac{2}{t^2} dt^2 + 4t\, d\theta^2 = 2ds^2 + 4\mathrm{e}^{-s} d\theta^2,$$

where $s := -\ln t$; note that $s \to \infty$ as $t \to 0_+$. We also have that $R(s, \theta) \to -1$ as $s \to \infty$, and g is asymptotic to a hyperbolic cusp on this end. The metric g is also complete in case (ii), which we can think of as corresponding to $\alpha = \infty$. This proves case (iii) of Theorem 3.14.

In case (3), we have $\psi(t) < 0$ and $\lim_{t\to\infty} \psi(t) = 0$. So for $t \gg 1$, we have $\psi(t) \approx 0$, and hence by (3.57) we have that $\psi'(t) \approx 2\psi(t)^2$, so that $\psi(t) \approx -\frac{1}{2t}$. From this we can deduce that the metric g is incomplete by considering $t \to \infty$.

Finally, we suppose that $t_* > 0$. In this case, the metric g is incomplete unless $\psi(0) := \lim_{t\to 0_+} \psi(t) = \pm\infty$. As in case (3) above, if $\psi(0) = -\infty$, then again the metric g is incomplete by considering $t \to \infty$. Finally, if $\psi(0) = \infty$, then $\psi(t) \approx (2(t - t_*))^{-1/2}$ and we see that the metric g is incomplete by considering $t \to (t_*)_+$.

This completes the proof of Theorem 3.14.

Remark 3.15. Note that if $\psi(t) < 0$, then the metric g is incomplete because of its behavior as the coordinate $t \to \infty$. On the other hand, if $\psi(t) > 0$ and g is incomplete, then $\psi(t) > 0$. Furthermore, any cases where the metrics g are incomplete are not because of $\psi(t) \to \infty$ as $t \to t_*$. Hence, there are no incomplete cases that correspond to $\phi'(s) \to 0$. This justifies our assumption that $\phi'(s) \neq 0$ on all of $\mathcal{M}^2$.

Remark 3.16. If we make the change of variables $r = 2\sqrt{t}$, then (3.46) becomes

$$(3.59) \qquad\qquad g = \psi(r^2/4)^2 dr^2 + r^2 d\theta^2.$$

In the ansatz (3.11) this corresponds to $F(r) = \psi(r^2/4)$. The GHMS expanders are given by (3.22) as

$$(3.60) \qquad\qquad \psi(t) = \frac{2}{W\left(A_0 e^{A_0} e^{-4t}\right) + 1},$$

where W is the inverse of $x \mapsto xe^x$ and $A_0 := \frac{2}{F_0} - 1 > 0$. The radii of the level set circles are identically equal to $\frac{2}{A_0 + 1} = F_0 \in (0, 2)$. Recall that $W'(u) = \frac{W(u)}{u(W(u)+1)}$, so that $uW'(u) = 1 - \frac{1}{W(u)+1}$. Thus,

$$\begin{aligned} \psi'(t) &= 2\psi(t)^2 A_0 e^{A_0} e^{-4t} W'\left(A_0 e^{A_0} e^{-4t}\right) \\ &= -\psi(t)^2 \left(-2 + \psi(t)\right), \end{aligned}$$

which is (3.54) with $a = -2$. Note that $\psi(0) = F_0 \in (0, 2)$.

By [**101**, Lemma A.2] (the cylinder-to-ball rule), the Riemannian metric

$$g = ds^2 + \phi(s)^2 d\theta^2 = \frac{\psi(t)^2}{t} dt^2 + 4t\, d\theta^2, \quad s, t > 0, \ \theta \in \mathbb{R}/2\pi F_0 \mathbb{Z},$$

extends smoothly over the origin if and only if $\phi(s)$ satisfies $\lim_{s\to 0_+} \phi(s) = 0$, $\lim_{s\to 0_+} \phi'(s) = \frac{1}{F_0}$, and $\lim_{s\to 0_+} \phi^{(2k)}(s) = 0$ for $k \geq 1$. Recall that $t = \frac{1}{4}\phi(s)^2$ and from (3.47) that $\psi\left(\frac{\phi(s)^2}{4}\right) \phi'(s) - 1$. By translation, we may assume that $\phi(0) = 0$. We have that $\phi'(0) = \frac{1}{\psi(0)} = \frac{1}{F_0}$. One also checks that $\lim_{s\to 0_+} \phi^{(2k)}(s) = 0$ for $k \geq 1$, so that g does extend smoothly over the origin. For example

$$\phi''(0) = -\psi^{-2}\left(\frac{\phi(0)^2}{4}\right) \frac{1}{2} \phi(0)\phi'(0) = 0.$$

3.3.4. The bounded curvature question.

As a consequence of their classification, Bernstein and Mettler obtained:

Theorem 3.17. *Any 2-dimensional complete GRS must have bounded curvature.*

One may ask: Must any Ricci soliton on a surface be gradient? By Theorem 3.17, an *a priori* weaker question is: Must any (nongradient) complete Ricci soliton on a surface have bounded curvature?

The answer to the first question is no. Topping and Yin proved that there exists a nongradient expanding Ricci soliton on $\mathbb{R}^2$; see Example 2.4. Their example has bounded curvature.

Cheng and Zhang [**96**] proved that any complete shrinking *breather solution* to the Ricci flow with Ricci curvature bounded from below must be a shrinking GRS. **Breather solutions** are periodic orbits of the Ricci flow. That is, a Ricci flow $g(t)$ is a breather solution if there exist $t_1 \neq t_2$ such that $g(t_2)$ is isometric to $cg(t_1)$ for some positive number c. In particular, Ricci solitons are breather solutions.

3.4. Notes and commentary

3.4.1. Noncollapsed 2-dimensional ancient solutions.

Definition 3.18. An ancient solution $(\mathcal{M}^n, g(t))$, $t \in (-\infty, 0)$, to the Ricci flow is **Type I** if

$$(3.61) \qquad \sup_{\mathcal{M}^n \times (-\infty, 0)} |\mathrm{Rm}||t| < \infty.$$

The following result of Hamilton (see [**179**, Theorem 26.1]) extends Theorem 3.5 in the case of shrinking Ricci solitons.

Theorem 3.19. *If* $(\mathcal{M}^2, g(t))$, $t \in (-\infty, 0)$, *is an ancient solution to the Ricci flow on a surface that is a Type I ancient solution, i.e.,*

$$\sup_{\mathcal{M}^2 \times (-\infty, 0)} R|t| < \infty,$$

or is κ-noncollapsed on all scales in the sense of Definition 1.12, then its Riemannian universal cover is isometric to either a round shrinking $\mathbb{S}^2$ or the flat $\mathbb{R}^2$.

3.4.2. Ends of noncompact manifolds.

Given a compact subset $\mathcal{K}$ of a noncompact differentiable manifold $\mathcal{M}^n$, define $\mathcal{E}(\mathcal{K})$ to be the set of connected components E of $\mathcal{M}^n \setminus \mathcal{K}$ where $\bar{E}$ is noncompact, i.e., where E is an unbounded component of $\mathcal{M}^n \setminus \mathcal{K}$. We call E an **end with respect to $\mathcal{K}$**.

The number $N_{\mathcal{K}}(\mathcal{M}^n)$ of unbounded components of $\mathcal{M}^n \setminus \mathcal{K}$, i.e.,

$$N_{\mathcal{K}}(\mathcal{M}^n) := |\mathcal{E}(\mathcal{K})| := \text{cardinality of } \mathcal{E}(\mathcal{K}),$$

is called the **number of ends with respect to $\mathcal{K}$**. For compact subsets $\mathcal{K}_1 \subset \mathcal{K}_2$, we have a natural map $\phi : \mathcal{E}(\mathcal{K}_2) \to \mathcal{E}(\mathcal{K}_1)$ defined by $\phi(E_2) := E_1$, where E_1 is the connected component of $\mathcal{M}^n \backslash \mathcal{K}_1$ containing $E_2 \in \mathcal{E}(\mathcal{K}_2)$. Since ϕ is a surjection, we have $N_{\mathcal{K}_1}(\mathcal{M}^n) \leq N_{\mathcal{K}_2}(\mathcal{M}^n)$.

Let $\{\mathcal{K}_i\}$ be a collection of smooth compact n-manifolds with boundary in $\mathcal{M}^n$ such that $\bigcup_i \mathcal{K}_i = \mathcal{M}^n$ and $\mathcal{K}_i \subset \text{int}(\mathcal{K}_{i+1})$ for all i; that is, $\{\mathcal{K}_i\}$ is a **compact exhaustion** of $\mathcal{M}^n$.

Definition 3.20. The **number of ends of** $\mathcal{M}^n$, defined by

$$N(\mathcal{M}^n) := \lim_{i \to \infty} N_{\mathcal{K}_i}(\mathcal{M}^n),$$

is independent of the choice of $\{\mathcal{K}_i\}$. Since $\mathcal{M}^n$ is noncompact, $N(\mathcal{M}^n) \in [1, \infty]$.

We say that a compact set $\mathcal{K} \subset \mathcal{M}^n$ is **end-complementary** if for every compact set $\mathcal{K}' \supset \mathcal{K}$ the map $\phi : \mathcal{E}(\mathcal{K}') \to \mathcal{E}(\mathcal{K})$ is a bijection. If $\mathcal{K}$ is an end-complementary compact set, then we call each element of $\mathcal{E}(\mathcal{K})$ a **topological end**. In this case, $N_{\mathcal{K}}(\mathcal{M}^n) = N(\mathcal{M}^n)$. If $N(\mathcal{M}^n) < \infty$, then there exists $\mathcal{K}$ such that $N_{\mathcal{K}}(\mathcal{M}^n) = N(\mathcal{M}^n)$ and any such $\mathcal{K}$ is end-complementary.

Definition 3.21. If $N(\mathcal{M}^n) = 1$, we say that $\mathcal{M}^n$ is **connected at infinity** or that $\mathcal{M}^n$ has only one (topological) end.

3.4.3. Kähler–Ricci solitons.

Recall that an **almost complex structure** on $\mathcal{M}^{2m}$ is an automorphism $J : T\mathcal{M} \to T\mathcal{M}$ satisfying $J^2 = -\operatorname{id}$. A Riemannian manifold $(\mathcal{M}^{2m}, g)$ with an almost complex structure J is **Kähler** if g is J-invariant and J is parallel; i.e., $g(JX, JY) = g(X, Y)$ and $\nabla J = 0$. Every orientable Riemannian surface is Kähler, and therefore every 2-dimensional gradient Ricci soliton is locally Kähler.

A short partial list of works in the field of Kähler–Ricci solitons, roughly in chronological order, includes the following: Koiso [**203**], Cao [**60**], Ivey [**194**], Tian and Zhu [**272–274**], Cao and Hamilton [**65**], Zhu [**307**], Chen and Zhu [**89**], Feldman, Ilmanen, and Knopf [**147**], Wang and Zhu [**283**], Cao, Tian, and Zhu [**67**], Chau and Tam [**81, 82**], Cao and Šešum [**68**], Bryant [**55**], Chan [**76**], Chen, Sun, and Tian [**94**], Dancer and Wang [**124**], Tian and Zhang [**271**], Yang [**293**], Munteanu and Wang [**238**], Drugan, Lu, and Yuan [**141**], Munteanu and Wang [**240**], Phong, Song, and Sturm [**257**], Takahashi [**268**], Chodosh and Fong [**98**], Kotschwar [**207**], Conlon, Deruelle, and Sun [**122**], Conlon and Deruelle [**121**], Deng and Zhu [**131**], Guo, Phong, Song, and Sturm [**166**], Bamler, Cifarelli, Conlon, and Deruelle [**25**], and Li and Wang [**221**]. For surveys, see Cao [**61, 62**].

3.5. Exercises

Exercise 3.1. Show that the Killing vector field $J(\nabla f)$ on $\mathbb{R}^{2n}$ in Example 3.2 generates the circle action given by $(\mathbf{x}, \mathbf{y}) \mapsto \left(\cos\left(\frac{\lambda}{2}t\right)\mathbf{x}, \sin\left(\frac{\lambda}{2}t\right)\mathbf{y} \right)$.

Exercise 3.2. This exercise allows us to avoid Kähler geometry for the proof of Lemma 3.1 in the case where $n = 1$. Let $(\mathcal{M}^2, g, f, \lambda)$ be a 2-dimensional GRS, and let J be the almost complex structure defined by (2.23), so that $J^2 = -\operatorname{id}$ and $\nabla J = 0$. Given a 2-tensor α, define $J(\alpha)(X, Y) := \alpha(X, J(Y))$. Using the formula $\mathcal{L}_X g = 2\operatorname{Sym}(\nabla X)$ and the GRS equation, show that

$$\mathcal{L}_{J(\nabla f)}g = (\lambda - R)\operatorname{Sym}(J(g)) = 0.$$

Exercise 3.3. By making the change of variable $\rho := M\cosh^2 s$, show that the cigar soliton metric defined by (3.4) may be expressed as

$$(3.62) \qquad g = \left(1 - \frac{M}{\rho}\right)d\theta^2 + \left(1 - \frac{M}{\rho}\right)^{-1}\frac{d\rho^2}{4\rho^2},$$

where $\rho > M$.

Exercise 3.4. Verify that the cigar soliton is a steady GRS. You may use your favorite coordinate system to do this.

Exercise 3.5. Prove that for any $\kappa > 0$ the cigar soliton is not κ-noncollapsed on all scales.

Exercise 3.6. Prove that a bad 2-dimensional closed orbifold (i.e, a 2-sphere with one cone point or two cone points with different cone angles) cannot admit a metric with constant curvature. In particular, the shrinking GRS metric in Lemma 3.3 does not have constant curvature.

Exercise 3.7. Show that the expanding soliton metric g defined in §3.1.4 is asymptotic at infinity to a flat cone and its curvature decays exponentially as a function of the distance to the origin.

Exercise 3.8. Prove (3.29).

Exercise 3.9. Use (3.30) and the commutator formula

$$\operatorname{div}(\nabla^2 f)_j = \nabla_i\nabla_j\nabla_i f = \nabla_j\nabla_i\nabla_i f - R_{ijik}\nabla_k f$$

to prove (3.31) for a GRS on a surface.

3.5.1. A primer on moving frames.

A reference for moving frames is Hick's book [**187**]. Two mathematicians central to the development of moving frames are Élie Cartan and Shiing-Shen Chern. Let $(\mathcal{M}^n, g)$ be a Riemannian manifold. A way to calculate locally is to assume that we have an orthonormal frame field $\{e_i\}_{i=1}^n$ on an open set $U \subset \mathcal{M}^n$; that is, $g(e_i, e_j) = \delta_{ij}$. The dual **coframe field** is the collection of 1-forms $\{\omega^j\}_{j=1}^n$ on U defined by $\omega^j(e_i) = \delta_{ij}$ for all $i, j = 1, \dots, n$. The **connection 1-forms** $\{\omega_i^j\}_{i,j=1}^n$ are the components of the Levi-Civita connection:

$$(3.63) \qquad \nabla_X e_i := \sum_{j=1}^n \omega_i^j(X)\,e_j,$$

for all vector fields X on U. The **curvature 2-forms** $\{\mathrm{Rm}_i^j\}_{i,j=1}^n$ are defined by

$$(3.64) \qquad \mathrm{Rm}_i^j\,(X,Y)\,e_j := \frac{1}{2}\,\mathrm{Rm}\,(X,Y)\,e_i.$$

Exercise 3.10. Prove that

$$(3.65) \qquad g = \sum_{j=1}^n \omega^j \otimes \omega^j.$$

Exercise 3.11. Prove that the connection 1-forms are antisymmetric:

$$(3.66) \qquad \omega_i^j = -\omega_j^i$$

for all i,j. In particular, $\omega_i^i = 0$.
HINT: Calculate $X\langle e_i, e_j\rangle$ for an arbitrary tangent vector X.

Exercise 3.12. Prove

$$(3.67) \qquad \nabla_X \omega^j = -\sum_{i=1}^n \omega_i^j\,(X)\,\omega^i.$$

Exercise 3.13. Prove the **first and second Cartan structure equations**:

$$(3.68) \qquad d\omega^j = \sum_{i=1}^n \omega^i \wedge \omega_i^j,$$

$$(3.69) \qquad \mathrm{Rm}_i^j = d\omega_i^j - \sum_{k=1}^n \omega_i^k \wedge \omega_k^j.$$

HINT: Use

$$d\omega^j\,(X,Y) = \frac{1}{2}\left(\nabla_X\omega^j\right)(Y) - \frac{1}{2}\left(\nabla_Y\omega^j\right)(X),$$

$$\mathrm{Rm}_i^j\,(X,Y) = \frac{1}{2}\left\langle \nabla_{X,Y}^2 e_i - \nabla_{Y,X}^2 e_i, e_j \right\rangle.$$

Exercise 3.14. Prove that if $n = 2$, i.e., we are on a surface, then we have

$$(3.70a) \qquad d\omega^1 = \omega^2 \wedge \omega_2^1,$$

$$(3.70b) \qquad d\omega^2 = \omega^1 \wedge \omega_1^2,$$

$$(3.70c) \qquad \mathrm{Rm}_2^1 = d\omega_2^1.$$

In particular, the Gauss curvature K is given by

$$K := \langle \mathrm{Rm}\,(e_1,e_2)\,e_2, e_1 \rangle = 2d\omega_2^1\,(e_1, e_2).$$

Exercise 3.15. Prove (1.20):

$$R_{e^u g} = e^{-u}(R_g - \Delta_g u).$$

HINT: If e_1, e_2 is a moving frame for g, then $e^{u/2}e_1, e^{u/2}e_2$ is a moving frame for $e^u g$. Also, $R = 2K$.

Exercise 3.16. Consider the rotationally symmetric metric

$$g = h(r)^2 dr^2 + f(r)^2 g_{S^{n-1}}$$

and an orthonormal frame field $\{e_i\}_{i=1}^n$ with $e_n = \frac{1}{h(r)}\frac{\partial}{\partial r}$. Show that the curvature 2-forms satisfy

$$\operatorname{Rm}(g)_i^j = \frac{1-(e_n f)^2}{f^2}\omega^j \wedge \omega^i$$

and

$$\operatorname{Rm}(g)_n^j = -\frac{e_n e_n f}{f}\omega^j \wedge \omega^n.$$

For example, if $n = 2$ and $f(r) = r$, then

$$\operatorname{Rm}(g)_2^1 = \frac{h'(r)}{rh^3(r)}\omega^1 \wedge \omega^2$$

and the Gauss curvature is $K = 2\operatorname{Rm}(g)_2^1(e_1, e_2) = \frac{h'(r)}{rh^3(r)}$.

Exercise 3.17. Prove Lemma 3.3 using the hint in Remark 3.9.

Exercise 3.18. Let $(\mathcal{M}^2, g, f, \lambda)$ be a 2-dimensional complete noncompact GRS with the property that f has a critical point. Prove that g and f are rotationally symmetric and that the critical point of f is unique.

3.5.2. Ancient solutions. Recall from (1.17) that the Ricci flow on surfaces is given by $\partial_t g = -Rg$ and from (1.18) that the normalized Ricci flow on surfaces is given by $\partial_t g = (r - R)g$.

Exercise 3.19. Let $(\mathcal{M}^2, g(t))$ be a solution to the Ricci flow on compact surfaces with positive curvature. Prove that the normalized surface entropy defined by $\tilde{N}(g) := \int_\mathcal{M} R\ln(RA)d\mu$, where $A := \operatorname{Area}(g)$, satisfies
(3.71)

$$d_t\tilde{N}(g(t)) = -2\int_\mathcal{M}\left|\operatorname{Ric}+\nabla^2 f - \frac{r}{2}g\right|^2 d\mu - \int_\mathcal{M}\frac{|\nabla R - 2\operatorname{Ric}(\nabla f)|^2}{R}d\mu,$$

which is manifestly the same equation as (3.71).

(a) Prove that if the solution satisfies $\tilde{N}(g(t_1)) = \tilde{N}(g(t_2))$ for some $t_1 \neq t_2$, then $g(t)$ is a shrinking GRS.

(b) Prove that if the solution satisfies $d_t\tilde{N}(g(t_0)) = 0$ for some t_0, then $g(t)$ is a shrinking GRS.

Exercise 3.20. Let $(\mathcal{M}^2, g(t))$, $t \in \mathbb{R}$, be a nonflat complete eternal solution to the Ricci flow on surfaces,[4] where R attains its space-time maximum at some point (x_0, t_0). Prove that $g(t)$ is a cigar soliton solution.

HINT: Prove under the Ricci flow on surfaces that

$$Q := \Delta \ln R + R = \partial_t \ln R - |\nabla \ln R|^2$$

satisfies

$$(3.72) \qquad (\partial_t - \Delta - 2\nabla \ln R \cdot \nabla) Q = 2 \left| \mathrm{Ric} + \nabla^2 \ln R \right|^2.$$

Cf. (1.70) for the normalized Ricci flow on surfaces. Show that $Q \geq 0$. Apply the strong maximum principle.

Exercise 3.21. Let $(\mathcal{M}^2, g(t))$, $t \in (-\infty, 0)$, be an ancient solution to the Ricci flow on a compact surface satisfying $-tR(x, t) \leq C$ for all $x \in \mathcal{M}^2$ and $t \in (-\infty, -1]$. Prove that $\lim_{t \to -\infty} \tilde{N}(g(t))$ exists and is finite.

Exercise 3.22. Continuing the previous exercise, let $t_i \to -\infty$, and choose x_i so that $K_i := R(x_i, t_i) = \max_{x \in \mathcal{M}} R(x, t_i)$. Prove that the pointed sequence of solutions $(\mathcal{M}^2, g_i(t), x_i)$, where $g_i(t) := K_i g(t_i + K_i^{-1} t)$, subconverges to a compact ancient solution to the Ricci flow with constant normalized surface entropy. Conclude that $g(t)$ has constant curvature.

HINT: Show that the $g_i(0)$ satisfy uniform injectivity radius and diameter estimates.

---

[4]Recall from Definition 1.11 that an eternal solution is a solution which exists for all time $t \in \mathbb{R}$.

Chapter 4

Estimates for Shrinking Ricci Solitons

The methods discussed in this chapter for studying GRS may be roughly categorized as addressing the following issues:

(1) what the potential function f says about the metric g,

(2) conversely, what g says about f,

(3) mutually, what f and g say about each other.

By exploiting the interaction between f and g, we can prove some crucial estimates for GRS, including the lower bound for the potential function due to Cao and Zhou [**69**]. These estimates lay the foundation for the further study of GRS, which is contained in the ensuing chapters. We also discuss some relations between GRS and the Ricci flow.

4.1. Sharp lower bound for the potential function of shrinkers

Recall that the Gaussian shrinking GRS on $\mathbb{R}^n$ has $f(x) = \frac{|x|^2}{4}$. Hence the upper bound for the potential function afforded by (2.63) is qualitatively sharp. Complementing this, there is a qualitatively sharp lower bound. This is the first indication that the geometries of noncompact shrinking GRS may be rather rigid. Intuitively, with scaling considered, the rigidity of a GRS correlates positively with the increase of the expansion constant λ. In any case, control of the potential function is clearly important.

4.1.1. f-Laplacian comparison on gradient Ricci solitons.

Now let $(\mathcal{M}^n, g, f, \lambda)$ be a GRS. Since the f-Laplacian $\Delta_f = \Delta - \nabla f \cdot \nabla$ is its naturally associated elliptic operator, we consider the f-Laplacian of the distance function: $\Delta_f r$.

Adopt the notation of Proposition 2.18. Suppose that x is not in the cut locus of p. Since γ is a geodesic, by applying the fundamental theorem of calculus to $(f \circ \gamma)'(r)$ while using the GRS equation, we calculate that

(4.1)

$$
\begin{aligned}
\langle \nabla f, \nabla r \rangle (x) - \langle \nabla f(p), \gamma'(0) \rangle &= \int_0^{r_x} (f \circ \gamma)''(r) dr \\
&= \int_0^{r_x} \nabla^2 f \left(\gamma'(r), \gamma'(r) \right) dr \\
&= - \int_0^{r_x} \operatorname{Ric} \left(\gamma'(r), \gamma'(r) \right) dr + \frac{\lambda}{2} r(x).
\end{aligned}
$$

By combining this with the Laplacian upper bound (2.72), we obtain

(4.2)
$$
\begin{aligned}
\Delta_f r(x) \le (n-1) \int_0^{r_x} \left(\zeta' \right)^2 (r) dr - \langle \nabla f(p), \gamma'(0) \rangle - \frac{\lambda}{2} r(x) \\
+ \int_0^{r_x} \left(1 - \zeta^2(r) \right) \operatorname{Ric} \left(\gamma'(r), \gamma'(r) \right) dr.
\end{aligned}
$$

Let $\zeta(r) = r$ for $0 \le r \le 1$ and $\zeta(r) = 1$ for $1 < r \le r_x$. We conclude:

Lemma 4.1. *If* $(\mathcal{M}^n, g, f, \lambda)$ *is a GRS and* $p \in \mathcal{M}^n$, *then*

(4.3)
$$
\Delta_f r(x) + \frac{\lambda}{2} r(x) \le A
$$

for x not contained in $B_1(p)$ union the cut locus of p, where

(4.4)
$$
A := n - 1 + |\nabla f|(p) + \frac{2}{3} S(p),
$$

and where $S(p)$ is defined by (2.78). In fact, (4.3) holds in the strong barrier sense on $\mathcal{M}^n \setminus B_1(p)$.

Observe that (4.3) implies that

(4.5)
$$
\Delta_f(r^2) + \lambda r^2 \le 2Ar + 2.
$$

4.1.2. A qualitatively sharp lower bound.

Recall:

Definition 4.2. A function is **proper** if the preimages of compact sets are compact.

A continuous function is proper if and only if it has the property that $x_i \to \infty$ implies $f(x_i) \to \infty$.

There are many applications of the following original result of Cao and Zhou [69] in the study of shrinking GRS. (The constant inside the brackets of the formula was sharpened by Haslhofer and Müller [185].) The rough idea of the proof is that the stability inequality for minimal geodesics says something about the Ricci curvature, which may be converted by the GRS equation to say something about the potential function.

Theorem 4.3. *For any shrinking GRS $(\mathcal{M}^n, g, f)$ and $p \in \mathcal{M}^n$,*

$$(4.6) \qquad f(x) \geq \frac{1}{4} \left(\left(d(x,p) - 2\sqrt{f(p)} - 5n \right)_+ \right)^2,$$

where for any real number u, we denote $u_+ := \max\{u, 0\}$. Hence f is a proper function, the minimum of f is attained, at any minimum point $o \in \mathcal{M}^n$ we have $f(o) \leq \frac{n}{2}$, and the potential function has the lower bound

$$(4.7) \qquad f(x) \geq \frac{1}{4} \left((d(x,o) - 5n)_+ \right)^2.$$

Example 4.4. The Gaussian shrinker satisfies $f(x) = \frac{1}{4} d_{\mathrm{Euc}}(x,0)^2$, so (4.7) is qualitatively sharp.

Proof. We may assume that $x \in \mathcal{M}^n \setminus B_2(p)$. Let $\gamma : [0, r(x)] \to \mathcal{M}^n$ be a unit speed minimal geodesic joining p to x, let $f(r) := f(\gamma(r))$, and let $\zeta(r)$ be as in (2.76). By applying the shrinking GRS equation (2.5) to the geodesic stability inequality (2.77),

$$2(n-1) \geq \int_0^{r(x)} \zeta^2(r) \, \mathrm{Ric}\left(\gamma'(r), \gamma'(r) \right) dr,$$

we obtain that

$$(4.8) \qquad 2(n-1) \geq \int_0^{r(x)} \zeta^2(r) \left(\frac{1}{2} - f''(r) \right) dr$$

$$= \frac{1}{2} r(x) - \frac{2}{3} + 2 \int_0^1 r f'(r) dr - 2 \int_{r(x)-1}^{r(x)} \zeta(r) f'(r) dr$$

$$\geq \frac{1}{2} r(x) - \frac{2}{3} - 2 \int_0^1 r \left(\sqrt{f(p)} + \frac{r}{2} \right) dr$$

$$\quad - 2 \int_{r(x)-1}^{r(x)} \zeta(r) \left(\sqrt{f(x)} + \frac{\zeta(r)}{2} \right) dr,$$

where for the last inequality we used by (2.63) that

$$|f'(r)| \leq |\nabla f|(\gamma(r)) \leq \sqrt{f(p)} + \frac{r}{2}.$$

for $0 \le r \le 1$ and $|f'(r)| \le \sqrt{f}(x) + \frac{\zeta(r)}{2}$ for $r(x) - 1 \le r \le r(x)$. Hence

$$2(n-1) \ge \frac{1}{2}r(x) - \frac{4}{3} - \sqrt{f}(p) - \sqrt{f}(x).$$

That is,

(4.9)
$$\sqrt{f}(x) + \sqrt{f}(p) \ge \frac{1}{2}d(x,p) - 2n + \frac{2}{3}.$$

This implies that the minimum of f is attained. If $o \in \mathcal{M}^n$ is a minimum point of f, then we have $f(o) \le \frac{n}{2}$ by applying the elliptic maximum principle to (2.54). Hence $\sqrt{f}(x) \ge \frac{1}{2}d(x,o) - \frac{35}{16}n$. The theorem follows. $\qquad\square$

Define
$$\rho(x) = 2\sqrt{f}(x).$$

The function ρ is a proxy for the distance to a fixed point. Namely, by (2.63) and (4.6), we have $|\nabla\rho| \le 1$ and

(4.10)
$$d(x,p) - \rho(p) - 4n + \frac{4}{3} \le \rho(x) \le d(x,p) + \rho(p).$$

Example 4.5. For the product soliton of two shrinking GRS $(\mathcal{M}_i^{n_i}, g_i, f_i)$, $i = 1, 2$, the potential function f on $\mathcal{M}_1^{n_1} \times \mathcal{M}_2^{n_2}$ is defined by (2.18) as $f(x_1, x_2) = f_1(x_1) + f_2(x_2)$. The distance function d of the product metric $g_1 + g_2$ satisfies

$$d_g^2\big((x_1, x_2), (y_1, y_2)\big) = d_{g_1}^2(x_1, y_1) + d_{g_2}^2(x_2, y_2).$$

Note that this is consistent with the previously stated results on upper and lower bounds for potential functions of shrinkers.

4.2. Shrinkers have at most Euclidean volume growth

In this and the next section we shall see two approaches towards proving that shrinkers have at most Euclidean volume growth. Each approach has its own advantage. In essence, we consider GRS versions of Riemannian geometry constructs used in the study of Ricci curvature. Since the GRS equation is the Einstein metric (constant Ricci curvature) equation modified by adding the Hessian of f, we will use the prefix f- to denote these GRS constructs.

4.2.1. Volume estimate via the Riccati equation.

The Bishop–Gromov volume comparison theorem is dominant in the study of Ricci curvature. One approach toward volume comparison is the Riccati equation for the mean curvature of geodesic spheres.

4.2.1.1. *Jacobi fields and geodesic spherical coordinates.* Let $p \in \mathcal{M}^n$ and let V be a unit tangent vector at p. In the following discussion we will assume that we are within the cut locus of p. Let

$$\gamma := \gamma_V : [0, \infty) \to \mathcal{M}^n$$

be the unit-speed geodesic with $\gamma'(0) = V$. Choose $e_i \in T_p\mathcal{M}$, $1 \leq i \leq n - 1$, so that $\{e_1, \ldots, e_{n-1}, V\}$ forms a positively oriented orthonormal frame. Define

$$J_i(r) := J_i(r, V)$$

to be the **Jacobi field**, i.e., the variation field of a 1-parameter family of geodesics, along $\gamma(r)$ with

(4.11) $$J_i(0) = \vec{0} \quad \text{and} \quad (\nabla_V J_i)(0) = e_i,$$

where $\vec{0}$ denotes the origin in $T_p\mathcal{M}$. We have that the J_i are tangent to the distance spheres $S_r(p) := \partial B_r(p)$ at smooth points since we are within the cut locus.

The vector fields J_i along γ, since they are Jacobi fields, satisfy the **Jacobi equation**

(4.12) $$\nabla_{\gamma'(r)}(\nabla_{\gamma'(r)} J_i) + \operatorname{Rm}(J_i, \gamma'(r))\gamma'(r) = 0.$$

Recall that this follows from the calculation

$$\begin{aligned}
0 &= \nabla_{J_i}(\nabla_{\gamma'(r)}\gamma'(r)) \\
&= \nabla_{\gamma'(r)}(\nabla_{J_i}\gamma'(r)) + \operatorname{Rm}(J_i, \gamma'(r))\gamma'(r) \\
&= \nabla_{\gamma'(r)}(\nabla_{\gamma'(r)} J_i) + \operatorname{Rm}(J_i, \gamma'(r))\gamma'(r),
\end{aligned}$$

where the covariant derivatives are along the path $\gamma(r)$. The Jacobi equation shows that the spread of geodesics is governed by the Riemann curvature tensor.

The Jacobi fields J_i are coordinate tangent vectors for geodesic spherical coordinates, which we now discuss. Let $\mathcal{S}_p^{n-1}$ denote the unit sphere in $T_p\mathcal{M}$. On $\mathcal{S}_p^{n-1}$ define positively oriented local coordinates $\{\theta^i\}_{i=1}^{n-1}$ in a neighborhood $\mathcal{N}$ of V with $\frac{\partial}{\partial\theta^i} = e_i$ at V. Local spherical coordinates $\{s^i\}_{i=1}^n$ on the cone over $\mathcal{N}$, which is an open subset of $T_p\mathcal{M} \setminus \{\vec{0}\}$, are given by

(4.13) $$s^n(W) := |W| \quad \text{and} \quad s^i(W) := \theta^i(W/|W|)$$

for $W \in T_p\mathcal{M} \setminus \{\vec{0}\}$ and $1 \leq i \leq n - 1$. The Euclidean volume form $d\mu_{g_p}$ of $(T_p\mathcal{M}, g_p)$ is given by

(4.14) $$(d\mu_{g_p})_W = |W|^{n-1} ds^1 \wedge \cdots \wedge ds^n.$$

Let $D_p := \{V \in T_p\mathcal{M} : d(\exp_p(V), p) = |V|\}$. Recall that

$$\exp_p : D_p^\circ \to \mathcal{M}^n \setminus \mathrm{Cut}(p)$$

is a diffeomorphism, where $\mathrm{Cut}(p)$ is the **cut locus** of p and where the **injectivity domain** D_p° of the exponential map denotes the interior of D_p. So we may define local **geodesic spherical coordinates** $\{x^i\}_{i=1}^n$ on an open subset of $\mathcal{M}^n \setminus \mathrm{Cut}(p)$ by

(4.15) $$x^i := s^i \circ \exp_p^{-1} \quad \text{for } 1 \le i \le n.$$

Note that $x^n(x) = r(x) := d(x, p)$.

4.2.1.2. *The first and second fundamental forms of distance spheres.* Let $T = \gamma'$, which is the unit outward normal field to the distance spheres $S_r(p)$. We have $\frac{\partial}{\partial x^n}(\gamma(r)) = T$ and $\frac{\partial}{\partial x^i}(\gamma(r)) = J_i(r)$ for $1 \le i \le n-1$. The components of the metric g with respect to $\{x^i\}_{i=1}^n$ are

(4.16) $$g_{ij} := g\left(\frac{\partial}{\partial x^i}, \frac{\partial}{\partial x^j}\right) \quad \text{for } 1 \le i \le n.$$

We have $g_{nn} = 1$, $g_{in} = g_{ni} = 0$, and the components of the first fundamental form I (a.k.a. the induced metric) of $S_r(p)$ are

(4.17) $$\mathrm{I}_{ij} = g_{ij} = g(J_i, J_j) \quad \text{for } 1 \le i, j \le n-1.$$

Recall that the J_i are tangent to $S_r(p)$.

Since T is the unit outward normal to $S_r(p)$, the components of the second fundamental form II of $S_r(p)$, defined by (2.35), are given by

(4.18) $$\mathrm{II}_{ij} := \mathrm{II}(J_i, J_j) = \langle \nabla_{J_i} T, J_j \rangle = \langle \nabla_T J_i, J_j \rangle \quad \text{for } 1 \le i \le n-1.$$

For the third equality we used that since T and the J_i are coordinate tangent vectors of $\{x^i\}$, we have $\vec{0} = [T, J_i] = \nabla_T J_i - \nabla_{J_i} T$.

From (4.18) we obtain that the evolution of the first fundamental form along γ is given by

(4.19) $$T(g_{ij})(r) = \langle \nabla_T J_i, J_j \rangle + \langle J_i, \nabla_T J_j \rangle = 2\,\mathrm{II}(J_i(r), J_j(r)).$$

4.2.1.3. *The volume density in geodesic spherical coordinates.* The **Jacobian of the exponential map in spherical coordinates** for g is

(4.20) $$\mathrm{J}(r) := \mathrm{J}_V(r) := \sqrt{\det(g_{ij}(r))}.$$

Note that $\lim_{r \to 0} \frac{g_{ij}(r)}{r^2} = \lim_{r \to 0} \frac{g(J_i(r), J_j(r))}{r^2} = \delta_{ij}$ by (4.11), and hence

(4.21) $$\lim_{r \to 0} \frac{\mathrm{J}(r)}{r^{n-1}} = 1.$$

Since $(d\exp_p)_{rV}(re_i) = J_i(r)$, we have

(4.22) $$\exp_p^*(d\mu_g)(rV) = \frac{\mathrm{J}(r)}{r^{n-1}} d\mu_{g_p} \quad \text{in } D_p^\circ.$$

Equivalently, with respect to the geodesic spherical coordinates $\{x^i\}$,

$$d\mu_g\left(\gamma(r)\right) = \mathrm{J}\left(r\right)dx^1 \wedge \cdots \wedge dx^n.$$

That is, J is the volume density in geodesic spherical coordinates.

4.2.1.4. *Radial evolution of the volume density.* We compute using (4.19) that
(4.23)
$$T(\ln \mathrm{J})(r) = \frac{1}{2}T\left(\ln \det(g_{ij})\right)(r) = \frac{1}{2}g^{ij}T(g_{ij})(r) = g^{ij}\,\mathrm{II}(J_i, J_j) = H\left(\gamma(r)\right),$$

where H is the **mean curvature** of $S_r\left(p\right)$. That is, the mean curvature is the radial derivative of the logarithm of the volume density.

4.2.1.5. *Radial evolution of the mean curvature: The Riccati equation.* In terms of the distance function $r(x) = d(x, p)$, the second fundamental form and the mean curvature of the distance spheres are given by (2.35) as

(4.24)
$$\mathrm{II} = \nabla^2 r \quad \text{and} \quad H = \Delta r.$$

Therefore, the fundamental Bochner formula (see (1.55))

(4.25)
$$0 = \frac{1}{2}\Delta|\nabla r|^2 = \langle \nabla r, \nabla \Delta r\rangle + |\nabla^2 r|^2 + \mathrm{Ric}(\nabla r, \nabla r),$$

where $\nabla r = T$, may be expressed as the **Riccati equation**:

Corollary 4.6. *The radial derivative of the mean curvature of the geodesic spheres satisfies*

(4.26)
$$T(H) = -\,\mathrm{Ric}(T, T) - |\mathrm{II}|^2.$$

Since $|\mathrm{II}|^2 \geq \frac{H^2}{n-1}$, *we obtain the* **Riccati inequality**:

(4.27)
$$T(H) \leq -\,\mathrm{Ric}(T, T) - \frac{H^2}{n-1}.$$

4.2.1.6. *The f-Riccati equation.* Now we consider the GRS versions of the Riemannian geometry constructs above. Let $f(r) := f(\gamma(r))$, $H(r) = H(\gamma(r))$, etc. Note that $f(0) = f(p)$. Then (4.27) implies that

(4.28)
$$T\left(r^2 H(r)\right) \leq 2rH(r) - \frac{r^2 H^2(r)}{n-1} - r^2\,\mathrm{Ric}(T, T)$$
$$\leq n - 1 + r^2(f''(r) - \lambda/2).$$

The f-**mean curvature** is

(4.29)
$$H_f := H - \langle \nabla f, \nabla r\rangle = \Delta_f r$$

since $\Delta r = H$. By (4.28), we have

(4.30)
$$T\left(r^2 H_f(r)\right) \leq n - 1 - 2rf'(r) - \frac{\lambda}{2}r^2.$$

Integrating this along $\gamma|_{[0,r]}$ while using $\lim_{r\searrow 0} r^2 H_f(r) = 0$, we obtain

$$(4.31) \qquad H(r) - f'(r) = H_f(r) \leq \frac{n-1}{r} - \frac{\lambda}{6}r - \frac{2}{r^2}\int_0^r sf'(s)ds.$$

The f-mean curvature is a useful quantity for the analysis of GRS.

Alternatively, the f-**Bochner formula** says that

$$(4.32) \qquad 0 = \frac{1}{2}\Delta_f|\nabla r|^2 = |\nabla^2 r|^2 + \langle \nabla r, \nabla\Delta_f r\rangle + \mathrm{Ric}_f(\nabla r, \nabla r).$$

This implies the f-**Riccati equation**:

$$(4.33) \qquad T(H_f) = -\mathrm{Ric}_f(T,T) - |\mathrm{II}|^2 \leq -\frac{\lambda}{2} - \frac{H^2}{n-1}.$$

Thus

$$T(r^2 H_f) \leq -\frac{\lambda}{2}r^2 - \frac{r^2 H^2}{n-1} + 2r(H - f') \leq n - 1 - 2rf'(r) - \frac{\lambda}{2}r^2,$$

which is (4.30) again.

4.2.1.7. *The f-volume.* The GRS version of the Jacobian of the exponential map is defined as follows. Let

$$(4.34) \qquad \mathrm{J}_f(r) := e^{-f(\gamma(r))}\,\mathrm{J}(r)$$

denote the f-**Jacobian**. Using (4.23), we compute that

$$(4.35) \qquad T(\ln \mathrm{J}_f)(r) = T(\ln \mathrm{J})(r) - f'(r) = H_f(\gamma(r)).$$

Hence

$$(4.36) \qquad \frac{\mathrm{J}_f(r_2)}{\mathrm{J}_f(r_1)} = e^{\int_{r_1}^{r_2} H_f(\gamma(r))dr}.$$

Now consider the case where $\lambda = 0$, i.e., $\mathrm{Ric}_f \geq 0$. At points for which we have the bound $|\nabla f| \leq B$, where $B \geq 0$, we have

$$(4.37) \qquad H_f(r) \leq \frac{n-1}{r} + B.$$

By substituting this into (4.36), we obtain

$$\frac{\mathrm{J}_f(r_2)}{\mathrm{J}_f(r_1)} \leq e^{\int_{r_1}^{r_2}(\frac{n-1}{r}+B)dr} = e^{B(r_2-r_1)}\left(\frac{r_2}{r_1}\right)^{n-1}.$$

Taking $r_1 \to 0$ and calling $r_2 = r$, we conclude that

$$(4.38) \qquad \mathrm{J}_f(r) \leq e^{-f(p)+Br}r^{n-1}$$

since $\lim_{r\to 0} r^{n-1}\mathrm{J}_f(r) = e^{-f(p)}$ by (4.21).

Let

$$(4.39) \quad C(r) := \{V \in T_p\mathcal{M} : |V| = 1 \text{ and } \gamma_V|_{[0,r]} \text{ is minimizing}\} \subset \mathcal{S}_p^{n-1},$$

where $\mathcal{S}_p^{n-1}$ is the unit sphere in $T_p\mathcal{M}$. Observe that $\bigcup_{r>0} r \cdot C(r) = D_p \setminus \{\vec{0}\}$. The function $r \mapsto C(r)$ is nonincreasing in the sense that if $r_1 \leq r_2$, then $C(r_2) \subset C(r_1)$. Since C_p and $\mathrm{Cut}(p)$ have measure zero, for any integrable function ϕ on a geodesic ball $B_r(p)$ we have

$$(4.40) \qquad \int_{B_r(p)} \phi(x) d\mu_g(x) = \int_{B_r(\vec{0}) \cap D_p^\circ} \phi(\exp_p(sV)) \frac{\mathrm{J}_V(s)}{s^{n-1}} d\mu_{g_p}(sV)$$

$$= \int_0^r \int_{C(s)} \phi(\exp_p(sV)) \, \mathrm{J}_V(s) \, d\Theta(V) ds,$$

where $d\Theta$ is the volume $(n-1)$-form of $\mathcal{S}_p^{n-1}$. In particular,

$$(4.41) \qquad \mathrm{Vol}_g B_r(p) = \int_{B_r(\vec{0}) \cap D_p^\circ} \frac{\mathrm{J}_V(s)}{s^{n-1}} d\mu_{g_p}(sV).$$

By (4.40), the *f-volume* of $B_r(p)$ defined by

$$\mathrm{Vol}_f(B_r(p)) := \int_{B_r(p)} \mathrm{e}^{-f(x)} d\mu_g(x)$$

may be expressed as

$$(4.42) \qquad \mathrm{Vol}_f(B_r(p)) = \int_{B_r(0) \cap D_p^\circ} \mathrm{J}_f(s, V) d\mu_{g_p}(sV).$$

4.2.1.8. *The Bakry–Emery volume comparison theorem.* Integrating (4.38) and using (4.42) yields the Bakry–Emery volume comparison theorem; see Wei and Wylie [**285**].

Theorem 4.7. *Let $f : \mathcal{M}^n \to \mathbb{R}$. If $p \in \mathcal{M}^n$ and $r > 0$ are such that $\mathrm{Ric}_f \geq 0$ and $|\nabla f| \leq B$ in $B_r(p)$ for some $B > 0$, then*

$$(4.43) \qquad \mathrm{Vol}_f(B_r(p)) \leq \omega_n \mathrm{e}^{-f(p)+Br} r^n,$$

where ω_n denotes the volume of the n-dimensional Euclidean unit ball. In particular, the f-volume estimate above holds for steady and shrinking GRS.

4.2.1.9. *Munteanu–Wang proof of the Cao–Zhou volume estimate.* We start with this proof since it is shorter and it gives a better multiplicative constant, although it is qualitatively equivalent to the original Cao and Zhou estimate. Starting from the estimate (4.31) for the f-mean curvature and by a clever combination, we can estimate the Jacobian of the exponential map. This leads to the following result, whose proof is by Munteanu and Wang [**237**] and which sharpens an earlier result of Cao and Zhou [**69**] by a different method, which we discuss in the next section.

Theorem 4.8. *For any shrinking GRS and $p \in \mathcal{M}^n$,*

(4.44) $\operatorname{Vol} B_r(p) \le \omega_n e^{f(p)} r^n$ *for $r > 0$.*

In particular, if $o \in \mathcal{M}^n$ is a minimum point of f, then

(4.45) $\operatorname{Vol} B_r(o) \le \omega_n e^{n/2} r^n$.

Proof. By (4.23) and (4.31) with $\lambda = 1$, we have

(4.46) $$\frac{\partial}{\partial r} \ln \left(\frac{\mathrm{J}(r)}{r^{n-1}} \right) = H(r) - \frac{n-1}{r}$$

$$\le -\frac{r}{6} + f'(r) - \frac{2}{r^2} \int_0^r s f'(s)\, ds.$$

Since $\lim_{r \to 0} \frac{\mathrm{J}(r)}{r^{n-1}} = 1$, we obtain

(4.47) $$\ln \left(\frac{\mathrm{J}(t)}{t^{n-1}} \right) \le \int_0^t \left(-\frac{r}{6} + f'(r) - \frac{2}{r^2} \int_0^r s f'(s) ds \right) dr$$

$$= -\frac{t^2}{12} + f(t) - f(0) + 2 \int_0^t \frac{d}{dr} \left(r^{-1} \right) \int_0^r s f'(s)\, ds\, dr$$

$$= -\frac{t^2}{12} - f(t) + f(0) + \frac{2}{t} \int_0^t r f'(r)\, dr,$$

where we integrated by parts and used $\lim_{r \to 0} \frac{1}{r} \int_0^r s f'(s)\, ds = 0$ to obtain the last line. Note that $f(p) = f(0)$. Another version of this formula is

(4.48) $$\ln \left(\frac{\mathrm{J}(t)}{t^{n-1}} \right) \le -\frac{t^2}{12} + f(0) + f(t) - \frac{2}{t} \int_0^t f(r)\, dr.$$

By combining (4.46) and (4.47), we obtain

(4.49) $$\frac{\partial}{\partial t} \left(t \ln \frac{\mathrm{J}(t)}{t^{n-1}} \right) = \ln \frac{\mathrm{J}(t)}{t^{n-1}} + t \frac{\partial}{\partial t} \left(\ln \frac{\mathrm{J}(t)}{t^{n-1}} \right)$$

$$\le -\frac{t^2}{4} - f(t) + f(0) + t f'(t)$$

$$= f(0) - \left(f'(t) - \frac{t}{2} \right)^2 - \left(f(t) - (f'(t))^2 \right).$$

Let $(\nabla f)^T$ be the tangential component to $S_r(p)$ of ∇f. Then $f = |\nabla f|^2 + R = (f')^2 + |(\nabla f)^T|^2 + R$. By integrating (4.49) from $t = 0$ to $t = r$, we obtain

(4.50)

$$\ln \frac{\mathrm{J}_V(r)}{r^{n-1}} \le f(0) - \frac{1}{r} \int_0^r \left(\left(f'(t) - \frac{t}{2} \right)^2 + |(\nabla f)^T|^2 (\gamma_V(t)) + R(\gamma_V(t)) \right) dt$$

$$\le f(0).$$

Therefore, since $f(0) = f(p)$, we conclude that

$$(4.51) \qquad \qquad \mathrm{J}(r) \leq e^{f(p)} r^{n-1}.$$

Integrating this over $B_r(\vec{0}) \cap D_p^\circ$ in $T_p \mathcal{M}$ yields (4.44). □

From (4.6) and (4.51), we obtain:

Corollary 4.9 (Finiteness of weighted volume). *For any complete noncompact shrinking GRS $(\mathcal{M}^n, g, f)$, we have:*

(1) *If $p > \frac{n}{2}$, then*

$$(4.52) \qquad \qquad \int_\mathcal{M} f^{-p} d\mu < \infty.$$

(2) *The **f-volume** of $(\mathcal{M}^n, g)$ is finite:*

$$(4.53) \qquad \qquad \mathrm{Vol}_f(g) := \int_\mathcal{M} e^{-f} d\mu < \infty.$$

4.3. The asymptotic volume ratio

Assume that $\mathcal{M}^n$ is noncompact and let $p \in \mathcal{M}^n$. The **asymptotic volume ratio** (AVR) is defined by

$$(4.54) \qquad \qquad \mathrm{AVR}(g) := \lim_{r \to \infty} \frac{\mathrm{Vol}_g(B_r(p))}{\omega_n r^n},$$

provided that the limit exists. If this limit exists for some p, then it exists for all p and is independent of p (we leave this to the reader to check). If $\mathrm{Ric} \geq 0$, then the $\mathrm{AVR}(g)$ exists and $0 \leq \mathrm{AVR}(g) \leq 1$ by the Bishop volume comparison theorem (see Exercise 4.6).

In this section we show that the AVR of a noncompact shrinking GRS exists and reprove that it is bounded. With the good control on the potential function, we have a proxy for the distance function that is natural from the point of view of the shrinking GRS structure. The methods of this section are due to Cao and Zhou.

4.3.1. Volume ratios of sublevel sets.

We shall need the co-area formula (see e.g. Lin and Yang's book [**222**]):

Proposition 4.10. *Let $(\mathcal{M}^n, g)$ be a compact manifold with or without boundary. If $\phi : \mathcal{M}^n \to \mathbb{R}$ is Lipschitz and if $h : \mathcal{M}^n \to \mathbb{R}$ is L^1 or nonnegative and measurable, then*

$$(4.55) \qquad \qquad \int_\mathcal{M} h \, |\nabla \phi| \, d\mu = \int_{-\infty}^\infty \int_{\{\phi = c\}} h \, d\sigma \, dc,$$

where $d\sigma$ is the induced measure on $\{\phi = c\}$.

Define the functions $V : \mathbb{R} \to (0, \infty)$ and $R : \mathbb{R} \to (0, \infty)$ by

(4.56a) $$V(t) := \int_{\{f<t\}} d\mu = \mathrm{Vol}\{f < t\},$$

(4.56b) $$R(t) := \int_{\{f<t\}} R \, d\mu$$

for $t \in \mathbb{R}$. Since $R \geq 0$, we have $R(t) \geq 0$. By (4.10) we have

$$\frac{1}{4}(d(x,p) - C_1)^2 \leq f(x) \leq \frac{1}{4}(d(x,p) + C_1)^2$$

for $d(x,p) \geq C_1$, where $C_1 := f(p) + 4n - \frac{4}{3}$. Hence:

Lemma 4.11. *The limit of the volume ratios* $\lim_{t \to \infty} \frac{V(t)}{t^{n/2}}$ *exists if and only if the* $\mathrm{AVR}(g)$ *exists, in which case*

(4.57) $$\lim_{t \to \infty} \frac{V(t)}{2^n \omega_n t^{n/2}} = \mathrm{AVR}(g).$$

To analyze $V(t)$ and $R(t)$, we compute their derivatives:

Lemma 4.12.

(4.58a) $$V'(t) = \int_{\{f=t\}} \frac{1}{|\nabla f|} d\sigma,$$

(4.58b) $$R'(t) = \int_{\{f=t\}} \frac{R}{|\nabla f|} d\sigma,$$

where $d\sigma$ is the volume element of $\{f = t\}$.

Proof. Since $|\nabla f|^{-1}$ is measurable (see the discussion after Theorem 4.28), we may apply the co-area formula (4.55) with both $h = |\nabla f|^{-1}$ and $h = R|\nabla f|^{-1}$ to the compact manifold with boundary $\{f \leq \bar{t}\}$ to conclude that

$$V(\bar{t}) = \int_0^{\bar{t}} dt \int_{\{f=t\}} \frac{1}{|\nabla f|} d\sigma,$$

$$R(\bar{t}) = \int_0^{\bar{t}} dt \int_{\{f=t\}} \frac{R}{|\nabla f|} d\sigma,$$

respectively. The lemma follows. $\qquad\square$

Integrating the formula (2.39) over $\{f < t\}$ yields

(4.59) $$\frac{n}{2} V(t) - R(t) = \int_{\{f<t\}} \Delta f \, d\mu = \int_{\{f=t\}} \frac{\partial f}{\partial \nu} d\sigma = \int_{\{f=t\}} |\nabla f| \, d\sigma,$$

where $\nu = \frac{\nabla f}{|\nabla f|}$ is the unit outward normal to $\partial\{f < t\} = \{f = t\}$. In particular,

(4.60) $$R(t) \leq \frac{n}{2} V(t).$$

By (2.53), we have wherever $|\nabla f| \neq 0$ that $|\nabla f| = \frac{f-R}{|\nabla f|}$. Applying this to (4.59) yields:

Lemma 4.13.

$$(4.61) \qquad \frac{n}{2} \, \mathrm{V}(t) - t \, \mathrm{V}'(t) = \mathrm{R}(t) - \mathrm{R}'(t).$$

4.3.2. The AVR exists and is bounded.

We modify the volume ratio by defining the quantity

$$(4.62) \qquad \mathrm{P}(t) := \frac{\mathrm{V}(t)}{t^{n/2}} - \frac{\mathrm{R}(t)}{t^{(n+2)/2}}.$$

For convenience, let $\mathrm{N}(t) := \frac{\mathrm{R}(t)}{t \, \mathrm{V}(t)}$, so that $\mathrm{N}(t) \leq \frac{n}{2t}$ and

$$(4.63) \qquad \left(1 - \frac{n}{2t}\right) \frac{\mathrm{V}(t)}{t^{n/2}} \leq \mathrm{P}(t) = (1 - \mathrm{N}(t)) \frac{\mathrm{V}(t)}{t^{n/2}} \leq \frac{\mathrm{V}(t)}{t^{n/2}}.$$

The existence of the AVR in the following result is due to Lu, Yang, and the author [**113**].

Theorem 4.14. *For any shrinking GRS, $t \mapsto \mathrm{P}(t)$ is monotonically nonincreasing and its asymptotic limit is*

$$(4.64) \qquad \lim_{t \to \infty} \mathrm{P}(t) = 2^n \omega_n \, \mathrm{AVR}(g).$$

In particular, the asymptotic volume ratio $\mathrm{AVR}(g)$ exists. By Theorem 4.8, the $\mathrm{AVR}(g)$ is bounded above by a constant depending only on n.

Proof. Using the ODE (4.61), which motivated the definition of $\mathrm{P}(t)$, we compute that

$$(4.65) \qquad \mathrm{P}'(t) = \frac{\mathrm{V}'(t)}{t^{n/2}} - \frac{n}{2} \frac{\mathrm{V}(t)}{t^{(n+2)/2}} - \frac{\mathrm{R}'(t)}{t^{(n+2)/2}} + \frac{n+2}{2} \frac{\mathrm{R}(t)}{t^{(n+4)/2}}$$

$$= -\left(1 - \frac{n+2}{2t}\right) \frac{\mathrm{R}(t)}{t^{(n+2)/2}}$$

$$= -\frac{\left(1 - \frac{n+2}{2t}\right) \mathrm{N}(t)}{1 - \mathrm{N}(t)} \mathrm{P}(t).$$

Hence we have the monotonicity formula

$$(4.66) \qquad \mathrm{P}'(t) \leq 0$$

for $t \geq \frac{n+2}{2}$. Therefore, by (4.57) and (4.63), we have

$$(4.67) \qquad 2^n \omega_n \, \mathrm{AVR}(g) = \lim_{t \to \infty} \frac{\mathrm{V}(t)}{t^{n/2}} = \lim_{t \to \infty} \mathrm{P}(t),$$

where the limits exist by (4.66).

Finally, we show that the $\mathrm{AVR}(g)$ is bounded above by a constant depending only on n. We have for all $t \geq \frac{n+2}{2}$,

$$\mathrm{P}(t) \leq \mathrm{P}\left(\frac{n+2}{2}\right) \leq \frac{\mathrm{V}(\frac{n+2}{2})}{(\frac{n+2}{2})^{n/2}} \leq \frac{\mathrm{Vol}\, B_{C_n}(p)}{(\frac{n+2}{2})^{n/2}},$$

where $C_n := \sqrt{2(n+2)} + 5n$ and the last inequality follows from (4.7). Since the $\mathrm{AVR}(g)$ is independent of the basepoint, we may choose p to be a minimum point o of f. The theorem follows from:

Claim. *There exists $C(n) < \infty$ such that $\mathrm{Vol}\, B_{C_n}(o) \leq C(n)$. Hence, by (4.67) we have the $\mathrm{AVR}(g) \leq \frac{C(n)}{\omega_n(2(n+2))^{n/2}}$.*

Proof of the claim. By Corollary 2.15(1), for $x \in B_{C_n}(o)$ we have that the potential function f satisfies

$$|\nabla f|(x) \leq \sqrt{f}(x) \leq \frac{1}{2}(C_n + \sqrt{2n}) \leq 3.4n.$$

Since $\mathrm{Ric}_f = \frac{1}{2}g \geq 0$, by Theorem 4.7,

$$(4.68) \qquad \mathrm{Vol}_f(B_{C_n}(o)) = \int_{B_{C_n}(o)} \mathrm{e}^{-f} d\mu \leq \omega_n \mathrm{e}^{3.4nC_n} C_n^n.$$

The claim and hence the theorem follow from this and

$$\mathrm{Vol}(B_{C_n}(o)) \leq \mathrm{e}^{(3.4n)^2} \mathrm{Vol}_f(B_{C_n}(o)). \qquad \square$$

Recall from (4.45) that $\mathrm{Vol}\, B_r(o) \leq \omega_n \mathrm{e}^{n/2} r^n$. In particular, we have $\mathrm{AVR}(g) \leq \mathrm{e}^{n/2}$.

Conjecture 4.15. *For any shrinking GRS we have $\mathrm{AVR}(g) \leq 1$.*

4.3.3. A characterization of when the AVR is positive.

Let $R_{\mathrm{avg}}(r) = \frac{1}{\mathrm{Vol}\, B_r(p)} \int_{B_r(p)} R d\mu$ denote the average scalar curvature on $B_r(p)$. Lu, Yang, and the author proved the following.

Proposition 4.16. *On a shrinking GRS we have that*

$$(4.69) \qquad \mathrm{AVR}(g) > 0 \quad \textit{if and only if} \quad \int_1^\infty R_{\mathrm{avg}}(r) \frac{dr}{r} < \infty.$$

Proof. Integrating (4.65) yields

$$(4.70) \qquad \mathrm{P}(t) = \mathrm{P}(n+2)\, \mathrm{e}^{-\int_{n+2}^t \left(1 - \frac{n+2}{2s}\right) \frac{\mathrm{N}(s)}{1-\mathrm{N}(s)} ds}$$

for $t \geq n+2$. On the other hand, from $\mathrm{N}(s) \leq \frac{n}{2s}$ we have that for any $t \geq n+2$,

$$(4.71) \quad \frac{1}{2} \int_{n+2}^t \mathrm{N}(s)\, ds \leq \int_{n+2}^t \left(1 - \frac{n+2}{2s}\right) \frac{\mathrm{N}(s)}{1-\mathrm{N}(s)} ds \leq 2 \int_{n+2}^t \mathrm{N}(s)\, ds.$$

Hence

$$\mathrm{P}(n+2)\mathrm{e}^{-2\int_{n+2}^{t}\frac{\mathrm{R}(s)}{s\,\mathrm{V}(s)}ds} \leq \mathrm{P}(t) \leq \mathrm{P}(n+2)\mathrm{e}^{-\frac{1}{2}\int_{n+2}^{t}\frac{\mathrm{R}(s)}{s\,\mathrm{V}(s)}ds}.$$

From (4.67) we conclude that the $\mathrm{AVR}(g) > 0$ if and only if $\int_{n+2}^{\infty}\frac{\mathrm{R}(s)}{s\,\mathrm{V}(s)}ds < \infty$. Finally, we observe from (2.63) and (4.6) that $\int_{n+2}^{\infty}\frac{\mathrm{R}(s)}{s\,\mathrm{V}(s)}ds < \infty$ if and only if $\int_{1}^{\infty}R_{\mathrm{avg}}(r)\frac{dr}{r} < \infty$. $\qquad\square$

4.3.4. The scalar curvature and volume growth.

Roughly, the larger the scalar curvature is near infinity, the slower the volume growth. The following is due to S. J. Zhang [**301**].

Corollary 4.17. *For a shrinking GRS with $R \geq \delta \geq 0$ and for $p \in \mathcal{M}^n$, we have*

$$(4.72) \qquad \mathrm{Vol}(B_r(p)) \leq C\,r^{n-2\delta} \quad \text{for } r \geq 1,$$

where C depends on the shrinker and p.

Proof. Since $R \geq \delta$, we have $\mathrm{N}(s) \geq \frac{\delta}{s}$. Hence, combining this inequality with (4.70), for $t \geq n+2$ we have

$$\mathrm{P}(t) \leq \mathrm{P}(n+2)\,\mathrm{e}^{\int_{n+2}^{t}\left(1-\frac{n+2}{2s}\right)\frac{\delta}{s}ds} \leq \mathrm{P}(n+2)\,(n+2)^{\delta}\,\mathrm{e}^{\frac{\delta}{2}}t^{-\delta}.$$

Because $\mathrm{P}(t) \geq \frac{\mathrm{V}(t)}{2\,t^{n/2}}$ for $t \geq n+2$ by (4.63), we obtain

$$(4.73) \qquad \mathrm{Vol}(B_{2\sqrt{t}-C}(p)) \leq \mathrm{V}(t) \leq 2\,\mathrm{P}(n+2)\,(n+2)^{\delta}\,\mathrm{e}^{\frac{\delta}{2}}t^{\frac{n}{2}-\delta}.$$

The corollary follows. $\qquad\square$

Proposition 4.18. *For a shrinking GRS with $\frac{\mathrm{R}(t)}{\mathrm{V}(t)} \leq \delta$ for t sufficiently large, where $\delta \in (0,\frac{n}{2})$, and for $p \in \mathcal{M}^n$, we have*

$$\mathrm{Vol}(B_r(p)) \geq c\,r^{n-2\delta} \quad \text{for } r \geq 1,$$

where $c > 0$ depends on the shrinker and p.

Proof. There exists $t_0 \geq \frac{n}{2}+1$ such that $\mathrm{R}(t) \leq \delta\,\mathrm{V}(t)$ for $t \geq t_0$. By this, (4.61), and $R \geq 0$, we have

$$t\,\mathrm{V}'(t) = \frac{n}{2}\,\mathrm{V}(t) - \mathrm{R}(t) + \mathrm{R}'(t) \geq \left(\frac{n}{2}-\delta\right)\mathrm{V}(t) \quad \text{for } t \geq t_0.$$

Integrating this, we obtain $\mathrm{V}(t) \geq \mathrm{V}(t_0)\left(\frac{t}{t_0}\right)^{n/2-\delta}$ for $t \geq t_0$. $\qquad\square$

4.4. GRS as fixed points of the Ricci flow—the canonical form revisited

GRS can be interpreted as fixed points of the Ricci flow in the space of metrics modulo the actions of diffeomorphisms and scalings.

4.4.1. Existence of a canonical dynamical form for GRS.

Equation (2.29) says that infinitesimally a GRS changes by scaling and the pullback by diffeomorphisms when evolved by Ricci flow. By Theorem 2.27, this may be integrated to obtain the following result, which gives a **canonical form** for the associated time-dependent version of a GRS.

Proposition 4.19 (Canonical form, II). *If $(\mathcal{M}^n, g, f, \lambda)$ is a complete GRS, then there exist a solution $g(t)$ of the Ricci flow with $g(0) = g$, diffeomorphisms $\varphi_t : \mathcal{M}^n \to \mathcal{M}^n$ with $\varphi_0 = \mathrm{id}$, and $f(t) : \mathcal{M}^n \to \mathbb{R}$ with $f(0) = f$, all of the above defined for all t with $\tau(t) := 1 - \lambda t > 0$, such that the following hold:*

(1) $\frac{\partial}{\partial t} \varphi_t(x) = \frac{1}{\tau(t)} (\nabla_g f)(\varphi_t(x))$,

(2) $g(t) = \tau(t) \varphi_t^* g$,

(3) $f(t) = f \circ \varphi_t$.

Moreover,

$$(4.74) \qquad \mathrm{Ric}_{g(t)} + \nabla_{g(t)}^2 f(t) = \frac{\lambda}{2\tau(t)} g(t)$$

and

$$(4.75) \qquad \frac{\partial f}{\partial t}(t) = |\nabla f(t)|_{g(t)}^2.$$

*We call $(\mathcal{M}^n, g(t), f(t), \frac{\lambda}{\tau(t)})$ the **canonical form** or **evolving GRS associated to** $(\mathcal{M}^n, g, f, \lambda)$.*

Proof. Since the vector field $\nabla_g f$ is complete by Theorem 2.27, there exists a 1-parameter family of diffeomorphisms φ_t generated by the vector fields $\tau(t)^{-1} \nabla_g f$, defined for t such that $\tau(t) > 0$. We can now define $f(t) = f \circ \varphi_t$ and $g(t) = \tau(t) \varphi_t^* g$. We then have

$$\frac{\partial}{\partial t} g(t) = -\frac{\lambda}{\tau(t)} g(t) + \tau(t) \frac{\partial}{\partial t}(\varphi_t^* g).$$

We compute that

$$\tau(t) \frac{\partial}{\partial t}(\varphi_t^* g) = \tau(t) \mathcal{L}_{(\varphi_t^{-1})_* \frac{\partial}{\partial t} \varphi_t}(\varphi_t^* g) = \mathcal{L}_{\nabla_{g(t)} f(t)} g(t),$$

where we used $\frac{\partial}{\partial t}\varphi_t = \frac{1}{\tau(t)}\nabla_g f = (\varphi_t)_*(\nabla_{g(t)}f(t))$. Hence

$$\frac{\partial}{\partial t}g(t) = -\frac{\lambda}{\tau(t)}g(t) + \mathcal{L}_{\nabla_{g(t)}f(t)}g(t).$$

We have

$$-2\operatorname{Ric}_{g(t)} = -2\varphi_t^*(\operatorname{Ric}_g) = \varphi_t^*(-\lambda g + \mathcal{L}_{\nabla_g f}g)$$
$$= -\frac{\lambda}{\tau(t)}g(t) + \mathcal{L}_{\nabla_{g(t)}f(t)}g(t).$$

Hence

$$\frac{\partial}{\partial t}g(t) = -\frac{\lambda}{\tau(t)}g(t) + \mathcal{L}_{\nabla_{g(t)}f(t)}g(t) = -2\operatorname{Ric}_{g(t)}.$$

Finally we calculate that

$$\frac{\partial f}{\partial t}(x, t) = df\left(\frac{\partial \varphi_t}{\partial t}\right)(x) = \frac{1}{\tau(t)}|\nabla f|_g^2(\varphi_t(x)) = |\nabla f(t)|_{g(t)}^2(x). \qquad \square$$

Note that Proposition 4.19(2) says that the solution $g(t)$ to the Ricci flow is isometric to $\tau(t)$ times g. Observe that on the Gaussian shrinking GRS we have $f(x) = \frac{1}{4}|x|^2$, $\varphi_t(x) = \frac{1}{\sqrt{1-t}}x$, and $\varphi_t^* g_{\mathrm{Euc}} = \frac{1}{1-t}g_{\mathrm{Euc}}$. For $t \in (0, 1)$ the pullback homothetically expands the metric, so from that perspective the static metric looks like it is shrinking. Exercise 4.7 considers the canonical form of the Gaussian soliton and the cigar soliton.

Remark 4.20. The application of the above dynamical version of GRS to the study of GRS, one of which is discussed for 3-dimensional steady GRS in §8.7 of Chapter 8, seems to be essential.

By applying Theorem 1.6 to the canonical form, we have the following global derivative estimates for GRS.

Corollary 4.21. *Given a constant C, there exist constants C_m depending only on m, n, and C such that if $(\mathcal{M}^n, g, f, \lambda)$ is a complete GRS with $|\mathrm{Rm}_g| \leq C$ on $\mathcal{M}^n$, then for $m \geq 1$ we have $|\nabla^m \mathrm{Rm}| \leq C_m$ on $\mathcal{M}^n$.*

Proof. By hypothesis, we have $|\mathrm{Rm}_g| \leq C$. Recall $g(t) = (1 - \lambda t)\varphi_t^* g$, so that $|\mathrm{Rm}_{g(t)}|(x) = \frac{1}{1-\lambda t}|\mathrm{Rm}_g|(\varphi_t(x))$. If $\lambda \geq 0$, then $|\mathrm{Rm}_{g(t)}| \leq C$ for $t \in [-1, 0]$. If $\lambda < 0$, then we have $|\mathrm{Rm}_{g(t)}| \leq 2C$ for $t \in [\frac{1}{2\lambda}, 0]$. In either case we may apply Theorem 1.6 to obtain $|\nabla_g^m \mathrm{Rm}_g| \leq C_m$. $\qquad \square$

In the case of steady GRS one can obtain local derivative estimates at distance scales less than 1.

Lemma 4.22. *Given a constant C, there exist constants C_m depending only on m, n, and C such that if $(\mathcal{M}^n, g, f)$ is a non-Ricci-flat complete steady GRS with $|\mathrm{Rm}|(x) \leq Cr^{-2}$ for $x \in B_{2r}(p)$, where $r \in (0, 1]$, then $|\nabla^m \mathrm{Rm}| \leq C_m r^{-2-m}$ in $B_r(p)$.*

Proof. Let $(\mathcal{M}^n, g(t), f(t))$ be the associated canonical form. Suppose that $|\mathrm{Rm}|(x, 0) \leq Cr^{-2}$ for $x \in B_{2r}^{g(0)}(p)$. Since $\frac{\partial}{\partial t}\varphi_t(x) = (\nabla_{g(0)}f(0))(\varphi_t(x))$, by combining with (2.65), we have the inequality $\left|\frac{\partial}{\partial t}\varphi_t(x)\right|_{g(0)} \leq 1$. Hence, if $x \in B_{3r/2}^{g(0)}(p)$ and $t \in [-\frac{r^2}{2}, 0]$, then

$$d_{g(0)}(\varphi_t(x), x) \leq \int_t^0 \left|\frac{\partial}{\partial t}\varphi_{\bar{t}}(x)\right|_{g(0)} d\bar{t} \leq \frac{r^2}{2} \leq \frac{r}{2}$$

since $r \leq 1$. Thus $\varphi_t(x) \in B_{2r}^{g(0)}(p)$ so that $|\mathrm{Rm}|(x, t) = |\mathrm{Rm}|(\varphi_t(x), 0) \leq Cr^{-2}$. By Theorem 1.6, $|\nabla^m \mathrm{Rm}|(x, 0) \leq C_m r^{-2-m}$ for $x \in B_r^{g(0)}(p)$. $\qquad\square$

Remark 4.23. Observe that for scales $r > 1$, if we impose the additional condition that $|\nabla f|(x) \leq r^{-1}$ for $x \in B_{2r}(p)$, then we obtain $|\nabla^m \mathrm{Rm}| \leq C_m r^{-2-m}$ in $B_r(p)$.

For shrinking GRS we have the following.

Lemma 4.24. *For any C_0, there exist constants C_m depending only on n, m, and $o \in \mathcal{M}^n$ such that if $(\mathcal{M}^n, g, f)$ is a complete noncompact shrinking GRS satisfying $|\mathrm{Rm}|(x) \leq C_0(d(x, o) + 1)^\alpha$ on $\mathcal{M}^n$, where $\alpha \geq 0$, then*

$$(4.76) \qquad |\nabla^m \mathrm{Rm}|(x) \leq C_m(d(x, o) + 1)^{\alpha(1 + \frac{m}{2})} \quad \text{on } \mathcal{M}^n \text{ for } m \geq 1.$$

Proof. From (2.64) and (4.6), we have that the potential function $f(x)$ is comparable to $(d(x, o) + 1)^2$. Let $(\mathcal{M}^n, g(t), f(t))$, $t \in (-\infty, 1)$, be the canonical form of $(\mathcal{M}^n, g, f)$ and let $\varphi_t : \mathcal{M}^n \to \mathcal{M}^n$ be the associated diffeomorphisms. Note that $g = g(0)$ and $f = f(0)$. For any $x \in \mathcal{M}^n$,

$$0 \leq \frac{\partial}{\partial t}f(\varphi_t(x)) = \frac{1}{1-t}|\nabla_g f|^2(\varphi_t(x)) \leq \frac{1}{1-t}f(\varphi_t(x)).$$

Integrating this over $[t, 0]$ for $t \leq 0$ we have

$$\frac{f(x)}{1-t} \leq f(\varphi_t(x)) \leq f(x).$$

Hence, for $t \in (-\infty, 0]$ we have $f(\varphi_t(x)) \leq f(x)$ and for $t \in [-1, 0]$ we have $f(\varphi_t(x)) \geq \frac{1}{2}f(x)$.

Let $x_0 \in \mathcal{M}^n$ and let $x \in \{y \in \mathcal{M}^n : f(y) \leq 2(f(x_0) + 1)\}$. Then $f(\varphi_t(x)) \leq 2(f(x_0) + 1)$ for all $t \leq 0$. Hence for all $t \leq 0$,

$$|\mathrm{Rm}|(x, t) = \frac{1}{1-t}|\mathrm{Rm}|(\varphi_t(x), 0)$$

$$\leq \frac{C_0}{1-t}(f(\varphi_t(x)) + 1)^{\frac{\alpha}{2}}$$

$$\leq 2^{\frac{\alpha}{2}}C_0(f(x_0) + 2)^{\frac{\alpha}{2}},$$

where we used $\alpha \geq 0$ and our curvature hypothesis. Clearly we have the inclusion $B_{c(f(x_0)+1)^{-\alpha/4}}(x_0) \subset \{f \leq 2\,(f(x_0)+1)\}$. Hence, by Shi's local derivative estimates, we have

$$|\nabla^m \operatorname{Rm}|\,(x_0, 0) \leq C_m\,(f(x_0)+1)^{\frac{\alpha}{2}(1+\frac{m}{2})} \leq C_m(d(x, o)+1)^{\alpha(1+\frac{m}{2})}. \qquad \square$$

The following estimates were proved by Perelman in the setting of 3-dimensional ancient κ-solutions [**251**, §1.5].

Conjecture 4.25. *Given a shrinking or steady GRS $(\mathcal{M}^n, g, f)$ with positve scalar curvature, there exist constants C_m such that*

$$|\nabla^m \operatorname{Rm}|\,(x) \leq C_m R^{1+\frac{m}{2}}(x) \quad \text{for } x \in \mathcal{M}^n.$$

4.4.2. Evolution of invariants for GRS in canonical form.

Now assume that we are on a shrinking GRS $(\mathcal{M}^n, g, f)$. Then $\frac{\partial}{\partial t}\varphi_t\,(x) = \frac{1}{1-t}\nabla f(\varphi_t\,(x))$ and

$$0 \leq \frac{\partial}{\partial t}f(\varphi_t\,(x)) = \frac{1}{1-t}\,|\nabla f|^2\,(\varphi_t\,(x)) \leq \frac{1}{1-t}f(\varphi_t\,(x)).$$

Hence for each $t \in (-\infty, 1)$ the value $f(\varphi_t\,(x))$ lies between $f(x)$ and $\frac{f(x)}{1-t}$. Considering $2\sqrt{f}$ as a proxy for the distance function (by (2.63) and by (4.6)), we interpret this as saying that for $t \in (0, 1)$ the pullback expands the metric. Note that $\left|\operatorname{Rm}_{g(t)}\right|(x) = \frac{1}{1-t}\left|\operatorname{Rm}_g\right|(\varphi_t\,(x))$. Therefore for $t \leq 0$ we have

$$f(x)\left|\operatorname{Rm}_{g(t)}\right|(x) \leq (1-t)\left|\operatorname{Rm}_{g(t)}\right|(x)f(\varphi_t\,(x)) = f(\varphi_t\,(x))\left|\operatorname{Rm}_g\right|(\varphi_t\,(x)).$$

Hence, if the curvature decays quadratically, i.e., $f(y)\left|\operatorname{Rm}_g\right|(y) \leq C$ for all $y \in \mathcal{M}^n$ and some C (recall that f is comparable to the distance squared to a fixed point), then

$$f(x)\left|\operatorname{Rm}_{g(t)}\right|(x) \leq C \quad \text{for } x \in \mathcal{M}^n \text{ and } t \leq 0.$$

In particular, for $t \leq 0$, $\sup_{\{f \geq c\}}\left|\operatorname{Rm}_{g(t)}\right| \leq Cc^{-1}$. Hence, by applying Shi's local derivative estimates, we obtain for each $m \geq 0$,

$$(4.77) \qquad f(x)^{\frac{m}{2}+1}\,|\nabla^m \operatorname{Rm}_g|\,(x) \leq C_m \quad \text{for } x \in \mathcal{M}^n.$$

Remark 4.26. Note that, for a shrinking GRS and modulo scaling, going forward/backward in time corresponds to going toward/away from infinity in space. Thus we expect the geometry near infinity of a shrinking GRS to be smoothed out. In general,

$$\sup_{x \in \Omega}|\operatorname{Rm}|\,(x, t) = \frac{1}{1-t}\sup_{x \in \varphi_t(\Omega)}|\operatorname{Rm}|\,(x, 0).$$

Hence for $t \leq 0$ we have

$$\sup_{\{f \geq c\}} \left|\operatorname{Rm}_{g(t)}\right| \leq \frac{1}{1-t} \sup_{\{f \geq \frac{c}{1-t}\}} \left|\operatorname{Rm}_{g(0)}\right| \leq Cc^{-1}.$$

The GRS canonical form reduces the proof of certain formulas for GRS to known formulas for the Ricci flow. For example, from the standard Ricci flow evolution equation

$$(4.78) \qquad \frac{\partial}{\partial t} \operatorname{Ric} = \Delta_L \operatorname{Ric},$$

where the Lichnerowicz Laplacian is defined by

$$(4.79) \qquad \Delta_L v_{ij} := (\Delta_L v)_{ij} := \Delta v_{ij} + 2R_{kij\ell} v^{k\ell} - R_{ik} v_j^k - R_{jk} v_i^k$$

(we use the Einstein summation convention), and from the equation $\frac{\partial}{\partial t} \operatorname{Ric} = \mathcal{L}_{\nabla f} \operatorname{Ric}$ satisfied by the canonical form, for any GRS we have

$$(4.80) \qquad \Delta_L \operatorname{Ric} = \mathcal{L}_{\nabla f} \operatorname{Ric}.$$

Observe that

$$(4.81) \qquad (\mathcal{L}_{\nabla f} \operatorname{Ric})_{ij} = (\nabla_{\nabla f} \operatorname{Ric})_{ij} + \nabla_i \nabla^k f R_{kj} + R_{ik} \nabla_j \nabla^k f.$$

Hence, by (4.80) and (2.5), we have the alternate expression for a shrinking GRS:

$$(4.82) \qquad \Delta_f R_{ij} := (\Delta_f \operatorname{Ric})_{ij} = -2R_{kij\ell} R^{k\ell} + R_{ij}$$
$$= 2R_{kij\ell} \nabla^k \nabla^\ell f,$$

where $\Delta_f = \Delta - \nabla f \cdot \nabla$ is the f-Laplacian.

A tensor $T(g)$ is said to be a **tensor invariant of g of degree** k if $T(cg) = c^k T(g)$ for $c > 0$ and $T(\varphi^* g) = \varphi^* T(g)$ for any diffeomorphism φ. For example, the 4-tensor Rm is a tensor invariant of g of degree 1, the 2-tensor Ric is a tensor invariant of g of degree 0, and the function R is a tensor invariant of g of degree -1. For a dynamical GRS, modulo scaling, by (2.29) the operator $\frac{\partial}{\partial t}$ corresponds to $\mathcal{L}_{\nabla f}$. This is made precise by the following result.

Lemma 4.27. *Let T be a tensor invariant of g of degree k. Then, on a GRS in canonical form, $T(g(t))$ satisfies*

$$(4.83) \qquad \mathcal{L}_{\nabla f(t)} T(g(t)) = \frac{\partial}{\partial t} T(g(t)) + \frac{\lambda k}{1 - \lambda t} T(g(t)).$$

Proof. Since $g(t) = (1 - \lambda t)\varphi_t^* g$ and since T is a tensor invariant of g of degree k, we have

$$T(g(t)) = (1 - \lambda t)^k \varphi_t^* T(g).$$

The lemma follows from computing the time derivative of this expression. $\qquad \square$

Let $\mathrm{Met}(\mathcal{M}^n)$ denote the space of Riemannian metrics on $\mathcal{M}^n$ and let $\mathrm{Diff}(\mathcal{M}^n)$ denote the diffeomorphism group of $\mathcal{M}^n$. Define the action of $\mathrm{Diff}(\mathcal{M}^n)$ on $\mathrm{Met}(\mathcal{M}^n)$ by $(\varphi, g) \mapsto \varphi^*(g)$. We may think of solutions to the Ricci flow as integral curves in $\mathrm{Met}(\mathcal{M}^n)$ to the vector field $g \mapsto -2\,\mathrm{Ric}$ on $\mathrm{Met}(\mathcal{M}^n)$. If $g(t)$ is a solution to the Ricci flow and $\varphi \in \mathrm{Diff}(\mathcal{M}^n)$, then $\varphi^* g(t)$ is also a solution to the Ricci flow. By this diffeomorphism invariance, we may consider solutions as curves in the quotient space $\mathrm{Met}(\mathcal{M}^n)/\mathrm{Diff}(\mathcal{M}^n)$. Additionally, $\mathbb{R}_+$ acts on $\mathrm{Met}(\mathcal{M}^n)$ by scaling: $(c, g) \mapsto cg$. An evolving GRS corresponds to a fixed point in $\mathrm{Met}(\mathcal{M}^n)/\mathrm{Diff}(\mathcal{M}^n) \times \mathbb{R}_+$.

Note that

$$(4.84) \qquad R_{g(t)} + \Delta_{g(t)} f(t) = \frac{n\lambda}{2\tau}.$$

If $(\mathcal{M}^n, g, f)$ is a normalized shrinking GRS, then

$$(4.85) \qquad R_{g(t)} + |\nabla f(t)|_{g(t)}^2 = \frac{f(t)}{1-t}.$$

Hence

$$(4.86) \qquad \frac{\partial f}{\partial t}(t) = \Delta_{g(t)} f(t) + \frac{f(t)}{1-t} - \frac{n}{2(1-t)}.$$

Without assuming completeness, we have the following, which is an extension by Kotschwar [205] of Bando's [31] results in the complete case.

Theorem 4.28. *If $(\mathcal{M}^n, g(t))$, $t \in (\alpha, \omega)$, $-\infty \leq \alpha < \omega \leq \infty$, is a solution to the Ricci flow, then for each fixed $t \in (\alpha, \omega)$ the Riemannian metric $g(t)$ is real analytic.*

Regarding real analyticity in the time variable, Kotschwar [206] proved the following.

Theorem 4.29. *Let $(\mathcal{M}^n, g(t))$, $t \in (0, T)$, be a complete solution to the Ricci flow with uniformly bounded curvature. Let $t_0 \in (0, T)$ and let $(\mathcal{U}, \{x^i\})$ be local coordinates such that $g(t_0)_{ij} = g(t_0)\left(\frac{\partial}{\partial x^i}, \frac{\partial}{\partial x^j}\right) \in C^\omega(\mathcal{U})$, i.e., real analytic. Then for any $x_0 \in \mathcal{U}$ there exists a neighborhood $\mathcal{V}$ of x_0 in $\mathcal{U}$ and $\varepsilon > 0$ such that $g(t)_{ij}(x)$ is in $C^\omega(\mathcal{V} \times (t_0 - \varepsilon, t_0 + \varepsilon))$.*

4.5. Quadratic curvature decay and asymptotically conical shrinkers

In this section, by applications of the elliptic maximum principle, we prove a pair of lower and upper quadratic decay estimates for the curvature of noncompact shrinking GRS. The upper bound requires the assumption that the curvature tends to zero at infinity.

4.5.1. Lower bound for the scalar curvature for nonflat shrinking GRS.

We begin by considering a lower bound for the scalar curvature R which improves Theorem 2.14(a) provided that the shrinking GRS is not Gaussian.

Let $(\mathcal{M}^n, g, f)$ be a complete noncompact shrinking GRS with $R > 0$. This implies $f = R + |\nabla f|^2 > 0$. From (2.60) with $\lambda = 1$ we have

$$(4.87) \qquad \Delta_f R = R - 2\,|\mathrm{Ric}|^2 \le R.$$

As potential barrier functions, we compute that for any $p \in \mathbb{R} - \{0\}$,

$$(4.88) \qquad \Delta_f\left(\frac{1}{p}f^{-p}\right) = f^{-p} - f^{-p-1}\left(\frac{n}{2} - (p+1)\frac{|\nabla f|^2}{f}\right).$$

Observe for $p > 0$ that f^{-p} is f-subharmonic outside of the compact set $\{f \le \frac{n}{2}\}$, and $\lim_{x\to\infty} f^{-p}(x) = 0$ by (4.6).

The two special cases that we shall use are

$$(4.89) \qquad \Delta_f(f^{-1}) = f^{-1} - f^{-2}\left(\frac{n}{2} - 2\frac{|\nabla f|^2}{f}\right),$$

$$(4.90) \qquad \Delta_f(f^{-2}) = 2f^{-2} - f^{-3}\left(n - 6\frac{|\nabla f|^2}{f}\right).$$

Remark 4.30. We may think of f^{-p} as asymptotically approaching an eigenfunction of Δ_f with eigenvalue $-p$. Similarly, assuming that $|\mathrm{Ric}| \le CR$ and $|\mathrm{Ric}|\,(x) \to 0$ as $x \to \infty$, we may consider (4.87) as saying that R is asymptotically approaching an eigenfunction of Δ_f with eigenvalue -1.

Theorem 4.31. *For any non-Gaussian complete noncompact shrinking GRS* $(\mathcal{M}^n, g, f)$ *there exists* $c > 0$ *such that*

$$(4.91) \qquad R(x) \ge c\,(d(x,o) + 1)^{-2}.$$

Proof. Using the functions above to construct a lower barrier for R, we first compute for $c \in \mathbb{R}_+$

$$(4.92) \qquad \Delta_f\left(R - cf^{-1}\right) = R - cf^{-1} + cf^{-2}\left(\frac{n}{2} - 2\frac{|\nabla f|^2}{f}\right) - 2\,|\mathrm{Ric}|^2.$$

To correct for the positive (bad) term $cf^{-2}\frac{n}{2}$ on the right-hand side of (4.92), we consider the quantity $\phi := R - c\left(f^{-1} + nf^{-2}\right)$. We compute that

$$(4.93) \qquad \Delta_f\phi = \phi - cnf^{-3}\left(\frac{f}{2} - n\right) - 2\,|\mathrm{Ric}|^2 - cf^{-4}\,(2f + 6n)\,|\nabla f|^2$$

$$\le \phi - cnf^{-3}\left(\frac{f}{2} - n\right).$$

Remarkably, each term on the right-hand side of (4.93) is good. Since $R > 0$, by choosing $c > 0$ sufficiently small, we may assume that $\phi > 0$ in $B_{8n}(o)$, where o is a minimum point of f.

Suppose that $\inf_{\mathcal{M}^n \setminus B_{8n}(o)} \phi < 0$. Then, since $\liminf_{x \to \infty} \phi(x) \geq 0$ by $R > 0$ and $f(x) \to \infty$ as $x \to \infty$, a negative minimum of ϕ is attained at some point x_0, where $x_0 \in \mathcal{M}^n \setminus B_{8n}(o)$. By the maximum principle, $\Delta_f \phi(x_0) \geq 0$. Hence (4.93) implies

$$(4.94) \qquad\qquad \frac{f(x_0)}{2} - n < 0.$$

On the other hand, since $d(x_0, o) \geq 8n$, (4.7) implies $f(x_0) \geq \frac{9}{4} n^2$, yielding a contradiction to (4.94). We conclude that

$$(4.95) \qquad\qquad R - cf^{-1} \geq \phi \geq 0 \quad \text{on all of } \mathcal{M}^n. \qquad\qquad \square$$

The proof above is based on obtaining a contradiction if ϕ is negative somewhere. Some maximum principle arguments are of this form, while others derive an estimate from the inequality satisfied at a minimum or maximum point.

By applying Theorem 4.31 to the derivative of curvature estimate (4.77), we obtain the following.

Corollary 4.32. *If a complete shrinking GRS $(\mathcal{M}^n, g, f)$ has quadratic curvature decay and is not the Gaussian soliton, then*

$$(4.96) \qquad\qquad |\nabla^m \operatorname{Rm}_g|(x) \leq C_m R^{\frac{m}{2}+1} \quad \text{for } x \in \mathcal{M}^n.$$

Definition 4.33. We say that a Riemannian cone $(C_0 \Sigma, g_{\text{cone}} = dr^2 + r^2 g_\Sigma)$ is an **asymptotic Riemannian cone** of $(\mathcal{M}^n, g)$ if there exists r_0, an end-complementary compact set $\mathcal{K}$, a diffeomorphism $\Phi : C_{r_0} \Sigma \to \mathcal{M}^n \setminus \mathcal{K}$, and $\lambda_i \to \infty$ such that $(C_{\lambda_i^{-1} r_0} \Sigma, \lambda_i^{-2} \rho_{\lambda_i}^* \Phi^* g)$ converges to $(C_0 \Sigma, g_{\text{cone}})$ in C_{loc}^2.

Corollary 4.34. *Under the hypotheses of Theorem 4.31, any asymptotic Riemannian cone of $(\mathcal{M}^n, g)$ has positive scalar curvature and hence is non-flat.*

Remark 4.35. Using a result of X. Cao and Q. Zhang [**72**, Theorem 4.1], Li [**220**, Theorem A.5] proved that any nonflat κ-noncollapsed Type I ancient solution $(\mathcal{M}^n, g(t))$, $t \in (-\infty, 0]$, and $p \in \mathcal{M}^n$ must satisfy

$$\liminf_{x \to \infty} |\operatorname{Rm}|(x, t) d_{g(t)}^2(x, p) > 0$$

for any $t \leq 0$.

Definition 4.36. Let $(\mathcal{M}^n, g)$ be a complete noncompact Riemannian manifold. Its **asymptotic scalar curvature ratio** (ASCR) is

$$(4.97) \qquad \text{ASCR}(g) := \limsup_{d(x,o) \to \infty} R(x)\, d(x,o)^2,$$

where $o \in \mathcal{M}^n$. This definition is independent of the choice of o.

Remark 4.37. Theorem 20.2 in Hamilton [**179**] says that any GRS satisfying the properties that sect > 0, $|\text{Rm}| \leq C$, and R attains its maximum must have ASCR $= \infty$.

4.5.2. Tensorial curvature equations for GRS and Hamilton's matrix Harnack quadratic form.

Following Hamilton [**177**], we derive some fundamental equations satisfied by GRS. The main technique is to differentiate the GRS equation (2.3). To this end, recall the general commutator formula

$$(4.98) \quad \nabla_i \nabla_j \beta_{k_1 \cdots k_r} - \nabla_j \nabla_i \beta_{k_1 \cdots k_r} = -\sum_{h=1}^{r} \sum_{p=1}^{n} R^p_{ijk_h} \beta_{k_1 \cdots k_{h-1}\, p\, k_{h+1} \cdots k_r}$$

for the covariant derivative acting on an r-tensor β. We will use the metric to raise the indices of tensors.

Firstly, by commuting third covariant derivatives acting on the potential function, we obtain

$$\nabla_i \nabla_j \nabla_k f - \nabla_j \nabla_i \nabla_k f = -R_{ijk\ell} \nabla^\ell f.$$

By substituting into this the GRS equation (2.3), while using that $\nabla g = 0$, we have that

$$(4.99) \qquad P_{ijk} := \nabla_i R_{jk} - \nabla_j R_{ik} = R_{ijk\ell} \nabla^\ell f.$$

So, by tracing once the second Bianchi identity, we have that the divergence of the Riemann curvature tensor satisfies

$$(4.100) \qquad \nabla^p R_{pjk\ell} = \nabla_\ell R_{kj} - \nabla_k R_{\ell j} = P_{\ell k j} = R_{\ell k j m} \nabla^m f.$$

Denote $\Delta R_{jk} := (\Delta \text{Ric})_{jk}$. Now, by taking the divergence of (4.99), we have

$$\Delta R_{jk} = \nabla^i \nabla_j R_{ik} + \nabla^i R_{ijk\ell} \nabla^\ell f + R_{ijk\ell} \nabla^i \nabla^\ell f$$

$$= \nabla_j \nabla^i R_{ik} + R^i_{ij\ell} R^\ell_k - R_{ijk\ell} R^{i\ell} + P_{\ell k j} \nabla^\ell f + R_{ijk\ell} \left(-R^{i\ell} + \frac{\lambda}{2} g^{i\ell} \right)$$

$$= \frac{1}{2} \nabla_j \nabla_k R + R_{j\ell} R^\ell_k - 2 R_{ijk\ell} R^{i\ell} + P_{\ell k j} \nabla^\ell f + \frac{\lambda}{2} R_{jk}.$$

Define the symmetric 2-tensor

$$(4.101) \qquad M_{jk} := \Delta R_{jk} - \frac{1}{2} \nabla_j \nabla_k R + 2 R_{ijk\ell} R^{i\ell} - R_{j\ell} R^\ell_k - \frac{\lambda}{2} R_{jk}.$$

Then we have that

$$(4.102) \qquad M_{jk} = P_{\ell kj}\nabla^\ell f.$$

The main identities for GRS we derived are formulas (4.100) and (4.102). We can combine these formulas to obtain **Hamilton's Harnack quadratic form for the Ricci flow**, which vanishes on GRS:

$$(4.103) \qquad H_{jk} := M_{jk} - 2P_{\ell kj}\nabla^\ell f + R_{ijk\ell}\nabla^i f \nabla^\ell f = 0.$$

Observe also that

$$(4.104) \qquad M_{jk} = P_{\ell kj}\nabla^\ell f = R_{\ell kjm}\nabla^\ell f \nabla^m f.$$

Hamilton's matrix Harnack estimate for the Ricci flow is the following [**177**].

Theorem 4.38. *Let $(\mathcal{M}^n, g(t))$, $t \in [0, T)$, be a complete solution to the Ricci flow with bounded nonnegative curvature operator. Then, for any 1-form W and any 2-form U, we have the pointwise monotonicity formula*

$$(4.105) \qquad Q(W \oplus U) := M_{jk}W^j W^k + 2P_{ijk}U^{ij}W^k + R_{ijk\ell}U^{ij}U^{\ell k} \geq 0,$$

where $\lambda = -\frac{1}{t}$ in definition (4.101) of M_{jk} and where we have used the metric to raise the indices on the differential forms W and U.

Note that the choice of λ in the theorem corresponds to the expanding case of GRS. In particular, models for this estimate are expanding GRS defined for $t \in (0, \infty)$ coming out of cones. We also have the formula that

$$(4.106) \qquad H_{jk}W^j W^k = Q(W \oplus (W \wedge df)).$$

In this way we see how the consideration of expanding GRS originally motivated Hamilton's Harnack quadratic.

Now let $\{\omega^j\}_{j=1}^n$ be an orthonormal coframe. Given a tangent vector V, by choosing $W = \omega^j$ and $U = \omega^j \wedge V^\flat$ and summing over j ($V^\flat$ denotes the 1-form dual to V), under the same hypotheses as in the theorem, we obtain **Hamilton's trace Harnack estimate**

$$(4.107) \qquad \partial_t R + \frac{R}{t} - 2\langle \nabla R, V\rangle + 2\mathrm{Ric}(V, V) \geq 0.$$

In particular, we have the pointwise monotonicity formula:

$$(4.108) \qquad \partial_t(tR(x, t)) \geq 0.$$

Now

$$\frac{1}{2}\nabla_j \nabla_k R = \nabla_j(R_{k\ell}\nabla^\ell f) = \nabla_j R_{k\ell}\nabla^\ell f - R_{k\ell}R_j^\ell + \frac{\lambda}{2}R_{jk}.$$

Thus, (4.102) implies that

$$0 = \Delta R_{jk} - \nabla_j R_{k\ell} \nabla^\ell f + 2 R_{ijk\ell} R^{i\ell} - P_{\ell kj} \nabla^\ell f - \lambda R_{jk}$$

$$= \Delta_f R_{jk} + 2 R_{ijk\ell} R^{i\ell} + (\nabla_k R_{j\ell} - \nabla_j R_{k\ell}) \nabla^\ell f - \lambda R_{jk}$$

$$= \Delta_f R_{jk} + 2 R_{ijk\ell} R^{i\ell} - \lambda R_{jk},$$

where $\Delta_f = \Delta - \nabla f \cdot \nabla$ denotes the f-Laplacian operator acting on tensors, $\Delta_f R_{jk} = (\Delta_f \mathrm{Ric})_{jk}$, and since

$$(\nabla_k R_{j\ell} - \nabla_j R_{k\ell}) \nabla^\ell f = R_{kj\ell i} \nabla^i f \nabla^\ell f = 0.$$

Summarizing, we have proved that for any GRS, the following identity holds:

$$(4.109) \qquad \Delta_f R_{jk} + 2 R_{ijk\ell} R^{i\ell} - \lambda R_{jk} = 0.$$

We remark that by (2.27),

$$(\mathcal{L}_{\nabla f} \mathrm{Ric})_{jk} = (\nabla_{\nabla f} \mathrm{Ric})_{jk} + \nabla_j \nabla_\ell f R_k^\ell + \nabla_k \nabla_\ell f R_j^\ell$$

$$= (\nabla_{\nabla f} \mathrm{Ric})_{jk} - 2 R_{j\ell} R_k^\ell + \lambda R_{jk}.$$

Therefore we obtain that

$$(4.110) \qquad (\mathcal{L}_{\nabla f} \mathrm{Ric})_{jk} = \Delta R_{jk} + 2 R_{ijk\ell} R^{i\ell} - 2 R_{j\ell} R_k^\ell.$$

Now, if we have a solution $g(t)$ to the Ricci flow, then

$$(4.111) \qquad \partial_t R_{jk} = \Delta R_{jk} + 2 R_{ijk\ell} R^{i\ell} - 2 R_{j\ell} R_k^\ell.$$

By the dynamical version of GRS, provided we know the evolution equation above for the Ricci tensor, we arrive at the same equation $\partial_t R_{jk} = (\mathcal{L}_{\nabla f} \mathrm{Ric})_{jk}$, since the Ricci tensor is scale-invariant.

Next, we derive an equation for $\Delta \mathrm{Rm}$. By applying ∇_i to equation (4.100) and using the commutator formula (4.98), we obtain

$$\nabla_i R_{\ell kjm} \nabla^m f + R_{\ell kjm} \nabla_i \nabla^m f$$

$$= \nabla_i \nabla^p R_{pjk\ell}$$

$$= \nabla^p \nabla_i R_{pjk\ell} - R_{ippq} R_{qjk\ell} - R_{ipjq} R_{pqk\ell} - R_{ipkq} R_{pjq\ell} - R_{ip\ell q} R_{pjkq},$$

where we did not bother to raise indices, as we usually do, for the quadratic in Rm expressions on the right-hand side. Now, by applying the second Bianchi identity and then commuting covariant derivatives, the first term on the right-hand side above may be rewritten in terms of $\Delta \mathrm{Rm}$ as

$$\nabla^p \nabla_i R_{pjk\ell} = \nabla^p \nabla_p R_{ijk\ell} + \nabla^p \nabla_j R_{pik\ell}$$

$$= \Delta R_{ijk\ell} + \nabla_j \nabla^p R_{pik\ell}$$

$$\quad - R_{pjpq} R_{qik\ell} - R_{pjiq} R_{pqk\ell} - R_{pjkq} R_{piq\ell} - R_{pj\ell q} R_{pikq};$$

here, $\Delta R_{ijk\ell} := (\Delta \, \mathrm{Rm})_{ijk\ell}$ denotes the components of the rough Laplacian of the Riemann curvature tensor. Furthermore, by (4.100),

$$\nabla_j \nabla^p R_{pik\ell} = \nabla_j \left(R_{\ell kim} \nabla^m f \right).$$

Hence, by combining all of the above, we have that

$$
\begin{aligned}
\Delta R_{ijk\ell} &= \nabla_i R_{\ell kjm} \nabla^m f + R_{\ell kjm} \nabla_i \nabla^m f \\
&\quad + R_i^q R_{qjk\ell} + R_{ipjq} R_{pqk\ell} + R_{ipkq} R_{pjq\ell} + R_{ip\ell q} R_{pjkq} \\
&\quad - \nabla_j R_{\ell kim} \nabla^m f - R_{\ell kim} \nabla_j \nabla^m f \\
&\quad - R_j^q R_{qik\ell} + R_{pjiq} R_{pqk\ell} + R_{pjkq} R_{piq\ell} + R_{pj\ell q} R_{pikq} \\
&= \nabla^m f \nabla_m R_{\ell kji} + \lambda R_{ijk\ell} + R_{ipjq} R_{pqk\ell} + R_{ipkq} R_{pjq\ell} + R_{ip\ell q} R_{pjkq} \\
&\quad + R_{pjiq} R_{pqk\ell} + R_{pjkq} R_{piq\ell} + R_{pj\ell q} R_{pikq},
\end{aligned}
$$

where for the last equality we use the second Bianchi identity to obtain the term $\nabla^m f \nabla_m R_{\ell kji}$, and we used the GRS equation (2.3) to obtain the term $\lambda R_{ijk\ell}$. By combining the quadratic in Rm terms, we may rewrite this as

$$(4.112) \qquad \Delta_f R_{ijk\ell} - \lambda R_{ijk\ell} = 2 \left(R_{ipjq} R_{pqk\ell} + R_{ipkq} R_{pjq\ell} + R_{ip\ell q} R_{pjkq} \right),$$

where $\Delta_f R_{ijk\ell} := (\Delta_f \mathrm{Rm})_{ijk\ell}$. Applying the first Bianchi identity to the factor $R_{pqk\ell}$ and denoting

$$(4.113) \qquad\qquad B_{ijk\ell} := -R_{ipjq} R_{kp\ell q},$$

we obtain

$$(4.114) \qquad \Delta_f R_{ijk\ell} - \lambda R_{ijk\ell} = 2(-B_{ijk\ell} + B_{ij\ell k} - B_{ikj\ell} + B_{i\ell jk}).$$

4.5.3. Decay upper estimate for $|\mathrm{Rm}|$ assuming $|\mathrm{Rm}| \to 0$.

Suppose that $\mathcal{M}^n$ is noncompact. Next, we consider an upper bound for $|\mathrm{Rm}|$ assuming that $|\mathrm{Rm}|\,(x) \to 0$ as $x \to \infty$. The idea of the proof is essentially the same as Theorem 4.31, except that some signs are flipped and one has to be more careful with the choice of constants in the barrier function.

We conclude from (4.112) that on a shrinking GRS,

$$(4.115) \qquad\qquad \Delta_f \, \mathrm{Rm} = \mathrm{Rm} + \mathrm{Rm} * \mathrm{Rm}.$$

Here, $A * B$ denotes some product of tensors A and B, including traces. We write formulas in this way when we do not care what the specific form of the product is.

From this we compute that

$$
\begin{aligned}
(4.116) \qquad \Delta_f |\, \mathrm{Rm}\,|^2 &= 2|\nabla \, \mathrm{Rm}\,|^2 + 2\Delta_f \, \mathrm{Rm} \cdot \mathrm{Rm} \\
&= 2|\nabla \, \mathrm{Rm}\,|^2 + 2|\, \mathrm{Rm}\,|^2 + \mathrm{Rm} * \mathrm{Rm} * \mathrm{Rm}.
\end{aligned}
$$

On the other hand,

$$(4.117) \qquad \Delta |\operatorname{Rm}| = \frac{\Delta |\operatorname{Rm}|^2}{2|\operatorname{Rm}|} - \frac{|\nabla |\operatorname{Rm}||^2}{|\operatorname{Rm}|}.$$

Since $|\nabla|\operatorname{Rm}|| \leq |\nabla \operatorname{Rm}|$ by Kato's inequality,

$$(4.118) \qquad |\nabla |T||^2 = \frac{1}{|T|^2} |\langle \nabla T, T\rangle|^2 \leq |\nabla T|^2$$

for any tensor T, we obtain

$$(4.119) \qquad \Delta_f |\operatorname{Rm}| \geq |\operatorname{Rm}| - C|\operatorname{Rm}|^2,$$

where C is a universal constant; for example, the inequality holds for $C = 100$.

The technique used to prove the scalar curvature gap theorem (see Theorem 4.31) can be applied to prove a quadratic curvature decay estimate provided the curvature tends to zero at infinity. In abstract, we have the following result of Munteanu and Wang [**239**, **241**].

Theorem 4.39. *Suppose, on an n-dimensional complete noncompact shrinking GRS with $n \geq 4$, that a function $w \geq 0$ satisfies*

$$(4.120) \qquad \Delta_f w \geq w - C_0 w^2$$

and $w(x) \to 0$ as $x \to \infty$, where C_0 is a positive constant. Then there exists a constant C such that

$$(4.121) \qquad w \leq \frac{C}{f}.$$

Hence

$$(4.122) \qquad w(x) \leq C \left(1 + d(x, o)\right)^{-2}.$$

The proof of this theorem is carried out below.

The **asymptotic (Riemann) curvature ratio** (ACR) of a complete noncompact Riemannian manifold $(\mathcal{M}^n, g)$ is defined by

$$(4.123) \qquad \operatorname{ACR}(g) = \limsup_{x \to \infty} |\operatorname{Rm}|(x) d(x, p)^2.$$

It is easy to see that this value is independent of the choice of $p \in \mathcal{M}^n$. Note that if the $\operatorname{ACR}(g) < \infty$, then $\operatorname{ASCR}(g) < \infty$.

By taking $w = |\operatorname{Rm}|$ in Theorem 4.39, we immediately obtain the following result of Munteanu and Wang [**241**].

Theorem 4.40. *If a complete noncompact n-dimensional shrinking GRS with $n \geq 4$ satisfies $|\operatorname{Rm}|(x) \to 0$ as $x \to \infty$, then the Riemann curvature decays quadratically:*

$$(4.124) \qquad |\operatorname{Rm}|(x) \leq C \left(1 + d(x, o)\right)^{-2}.$$

In other words, the $\operatorname{ACR}(g) < \infty$. *Of course,* $\operatorname{AVR}(g) > 0$.

By (4.124) and Shi's local derivative of curvature estimates (see (4.77)), there exists an asymptotic cone of $(\mathcal{M}^n, g)$ which is a Riemannian cone.

Proof of Theorem 4.39. Since $n \geq 4$, (4.89) implies that

$$(4.125) \qquad \Delta_f(f^{-1}) \leq f^{-1}.$$

From (4.90) we have $\Delta_f(f^{-2}) \geq 2f^{-2} - nf^{-3}$. Hence, if $f \geq 2n$, then

$$(4.126) \qquad \Delta_f(f^{-2}) \geq \frac{3}{2}f^{-2}.$$

Therefore, if $f \geq 2n$ and $a \geq 0$, then

$$(4.127) \qquad \Delta_f(f^{-1} - af^{-2}) \leq f^{-1} - \frac{3}{2}af^{-2}.$$

Note that if $f > \max\{a, 2n\}$, then $f^{-1} - af^{-2} > 0$ and

$$(4.128) \qquad \Delta_f(f^{-1} - af^{-2}) \leq f^{-1} - af^{-2} - \frac{a}{2}(f^{-1} - af^{-2})^2.$$

Let $\psi := b(f^{-1} - af^{-2})$, where $b > 0$. We will choose a and b below. Then $\psi \leq bf^{-1}$ and

$$(4.129) \qquad \Delta_f \psi \leq \psi - \frac{a}{2b}\psi^2.$$

Since $w(x) \to 0$ as $x \to \infty$, there exists $C_1 > 4n$ such that

$$(4.130) \qquad w(x) \leq \frac{1}{10C_0} \quad \text{if } f(x) \geq C_1,$$

where C_0 is the constant in (4.120). Choose a and b so that $\frac{a}{2b} = C_0$. Then (4.120) and (4.129) imply

$$(4.131) \qquad \Delta_f(w - \psi) \geq w - \psi - C_0 w^2 + C_0 \psi^2$$
$$= (1 - C_0(w + \psi)) (w - \psi).$$

Let $a = \frac{1}{2}C_1$. If $f \geq C_1$, then

$$C_0 \psi \leq \frac{C_0 b}{C_1} = \frac{a}{2C_1} = \frac{1}{4}.$$

Thus, if $f \geq C_1$, then

$$(4.132) \qquad C_0(w + \psi) \leq \frac{7}{20}.$$

Now, if $f = C_1 = 2a$, then

$$\psi = b(f^{-1} - af^{-2}) = \frac{b}{4a} = \frac{1}{8C_0}.$$

Therefore, if $f = C_1 = 2a$, then

(4.133) $$w - \psi \le \frac{1}{10C_0} - \frac{1}{8C_0} < 0.$$

On the other hand, by (4.131) and (4.132), on $\{f \ge C_1\}$ we have

(4.134) $$\Delta_f(w - \psi) \ge (1 - C_0(w + \psi))\,(w - \psi),$$

where

(4.135) $$1 - C_0(w + \psi) \ge \frac{13}{20}.$$

Since $w - \psi < 0$ on $\{f = C_1\}$ and since $(w - \psi)(x) \to 0$ as $x \to \infty$, if $w - \psi$ is positive somewhere on $\mathcal{M}^n \setminus \{f < C_1\} = \{f \ge C_1\} := \Omega$, then a positive maximum of $w - \psi$ occurs at some point $x_0 \in \Omega$. Then

(4.136) $$0 \ge \Delta_f(w - \psi)(x_0) > 0$$

by (4.134) and (4.135). Since this is a contradiction, we conclude that

(4.137) $$w(x) \le \psi(x) = b(f^{-1} - af^{-2})(x) \le \frac{b}{f(x)}$$

for all $x \in \{f \ge C_1\}$. This completes the proof of Theorem 4.39. $\qquad\square$

Definition 4.41. Let $(\mathcal{M}^n, g)$ be a (not necessarily complete) Riemannian manifold.

(1) We say that $(\mathcal{M}^n, g)$ is **strongly asymptotic to a Riemannian cone** $(C_0\Sigma^{n-1}, g_{\text{cone}})$ **along an end** E of $\mathcal{M}^n$ if there are r_0 and a diffeomorphism $\Phi : C_{r_0}\Sigma \to E$ such that $(C_{\lambda^{-1}r_0}\Sigma, \lambda^{-2}\rho_\lambda^*\Phi^*g)$ converges as $\lambda \to \infty$ to $(C_0\Sigma, g_{\text{cone}})$ in C^2_{loc}. In this case Σ must be connected since E is a single end.

(2) We say $(\mathcal{M}^n, g)$ is **strongly asymptotic to a Riemannian cone** $(C_0\Sigma^{n-1}, g_{\text{cone}})$ if there are r_0 and a diffeomorphism $\Phi : C_{r_0}\Sigma \to \mathcal{M}^n \setminus \mathcal{K}$, where $\mathcal{K}$ is a compact set, such that $(C_{\lambda^{-1}r_0}\Sigma, \lambda^{-2}\rho_\lambda^*\Phi^*g)$ converges as $\lambda \to \infty$ to $(C_0\Sigma, g_{\text{cone}})$ in C^2_{loc}. In this case Σ may be disconnected. For short we say that $(\mathcal{M}^n, g)$ is **asymptotically conical**.

Generalizing Theorem 4.40, Munteanu and Wang [**241**] proved:

Theorem 4.42. *If a complete noncompact shrinking GRS $(\mathcal{M}^n, g, f)$ satisfies $|\operatorname{Ric}|(x) \to 0$ as $x \to \infty$, then (4.124) holds; i.e., the Riemann curvature decays quadratically and hence $(\mathcal{M}^n, g)$ is asymptotically conical in the sense of Definition 4.41(2).*

In dimension 4 they obtained the following improvement.

Theorem 4.43. *If a complete noncompact shrinking GRS $(\mathcal{M}^4, g, f)$ satisfies $R(x) \to 0$ as $x \to \infty$, then the Riemann curvature decays quadratically and hence $(\mathcal{M}^4, g)$ is asymptotically conical.*

4.6. Compactness of shrinkers with positive sectional curvature

Although shrinking GRS that are not isometric to Euclidean space must have positive scalar curvature, they are not expected to have positive curvature in a much stronger sense. For example, there are shrinkers that do not have nonnegative Ricci curvature.

4.6.1. Non-Gaussian shrinking GRS with $\mathrm{Ric} \geq 0$ have the $\mathrm{AVR} = 0$.

In this subsection we discuss some results of Ni [**246**] regarding shrinking GRS with nonnegative Ricci curvature.

Lemma 4.44. *If $(\mathcal{M}^n, g, f)$, $n \geq 2$, is a non-Gaussian shrinking GRS with bounded nonnegative Ricci curvature, then*

$$\underline{R} := \inf_{x \in \mathcal{M}} R(x) > 0.$$

Proof. Let $\beta(t)$ be a maximal integral curve of ∇f. By Theorem 2.27, $\beta(t)$ is defined for all $t \in \mathbb{R}$. Using (2.41), we compute that

$$(4.138) \qquad \frac{d}{dt} R(\beta(t)) = \langle \nabla R, \nabla f \rangle = 2 \operatorname{Ric}(\nabla f, \nabla f) \geq 0,$$

where the quantities on the right-hand side of this equation are evaluated along $\beta(t)$. Thus $t \mapsto R(\beta(t))$ is nondecreasing. The lemma will follow from showing that for any point outside a sufficiently large ball B, by going backward from the point along an integral curve of ∇f, one eventually intersects B.

Fix $p \in \mathcal{M}^n$. For $x \in \mathcal{M}^n$, let $\gamma : [0, r_x] \to \mathcal{M}^n$ be a minimal geodesic joining p to any x, where $r_x = d(x, p)$. By (2.79) and $|\mathrm{Ric}| \leq C$,

$$(4.139) \qquad \int_0^{r_x} \operatorname{Ric}(\gamma'(r), \gamma'(r)) \, dr \leq C.$$

Let $f(r) = f(\gamma(r))$. Applying (4.139) to (4.1), i.e.,

$$(4.140) \qquad \langle \nabla f, \gamma'(r_x) \rangle = \frac{r_x}{2} - \int_0^{r_x} \operatorname{Ric}(\gamma'(r), \gamma'(r)) dr + f'(0),$$

we obtain $\langle \nabla f, \gamma'(r_x) \rangle \geq \frac{r_x}{2} - C$. Let

$$(4.141) \qquad \frac{d_- h}{dt}(t) := \liminf_{\Delta t \to 0_+} \frac{h(t) - h(t - \Delta t)}{\Delta t}.$$

Hence, as long as $r_{\beta(t)} \geq 4C$, we have

$$(4.142) \qquad \frac{d_-}{dt} r_{\beta(t)} \geq \left\langle \nabla f, \gamma'(r_{\beta(t)}) \right\rangle \geq \frac{r_{\beta(t)}}{2} - C \geq \frac{r_{\beta(t)}}{4}.$$

Let $x \in \mathcal{M}^n \setminus B_{4C}(p)$ and choose $\beta(t)$ so that $\beta(0) = x$. From (4.142), there exists $t_0 < 0$ such that $r_{\beta(t_0)} = 4C$. By (4.138) this implies

$$R(x) \geq \min_{y \in \bar{B}_{4C}(p)} R(y) > 0. \qquad \qquad \square$$

We now remove the bounded Ricci curvature assumption.

Theorem 4.45. *For any complete noncompact non-Gaussian shrinking GRS with nonnegative Ricci curvature there exists $\delta > 0$ such that*

$$(4.143) \qquad \qquad R \geq \delta \quad on \; \mathcal{M}^n.$$

In particular, the $\mathrm{AVR}(g) = 0$ *(in fact,* $\mathrm{Vol}(B_r(p)) \leq C r^{n-2\delta}$ *for* $r \geq 1$*).*

Proof of the theorem (modulo a claim). Let $p \in \mathcal{M}^n$ and let $x_0 \in \mathcal{M}^n \setminus B_{8r_1}(p)$, where $r_1 < \infty$ is to be chosen below. Let $\beta : \mathbb{R} \to \mathcal{M}^n$ be the integral curve of ∇f with $\beta(0) = x_0$. From (4.138) we have

$$(4.144) \qquad R(x_0) \geq R(\beta(t)) \quad \text{for all } t \in (-\infty, 0].$$

Claim. *Either*

$$(4.145) \qquad R(\beta(t_1)) \geq 1 \quad for \; some \; t_1 \in (-\infty, 0]$$

or

$$(4.146) \qquad \beta(t_2) \in \bar{B}_{8r_1}(p) \quad for \; some \; t_2 \in (-\infty, 0).$$

The claim yields the theorem since then (4.144) implies

$$R(x_0) \geq \min \left\{ 1, \; \min_{x \in \bar{B}_{8r_1}(p)} R(x) \right\} > 0. \qquad \qquad \square$$

Proof of the claim. Suppose that there exists $x_0 \in \mathcal{M}^n \setminus B_{8r_1}(p)$ such that $R(\beta(t)) < 1$ for all $t \in (-\infty, 0]$. Let $r(\cdot) = d(\cdot, p)$. The claim follows from showing that there exists $r_1 \geq \frac{n-1}{2}$ such that for any $x \in \mathcal{M}^n \setminus B_{4(n-1)}(p)$ with $R(x) \leq 1$, we have

$$(4.147) \qquad \qquad \left\langle \nabla f, \gamma'(r_x) \right\rangle > \frac{r_x}{4} - r_1,$$

where $\gamma : [0, r_x] \to \mathcal{M}^n$ is a minimal unit speed geodesic joining p to x. Then

$$(4.148) \qquad \frac{d_-}{dt} r_{\beta(t)} \geq \left\langle \nabla f, \gamma'(r_{\beta(t)}) \right\rangle \geq \frac{r_{\beta(t)}}{8} \quad \text{as long as } \beta(t) \notin \bar{B}_{8r_1}(p),$$

which proves (4.146).

We now prove (4.147). Adjust definition (2.76) to be

$$
\zeta(r) = \begin{cases} r & \text{if } 0 \le r \le 1, \\ 1 & \text{if } 1 < r \le r_x - \delta_x, \\ \frac{r_x - r}{\delta_x} & \text{if } r_x - \delta_x < r \le r_x, \end{cases}
$$

where $\delta_x := \frac{4(n-1)}{r_x} \le 1$. Then, analogous to (2.79) while using $\mathrm{Ric} \le Rg$, we obtain

(4.149)
$$
\int_0^{r_x} \mathrm{Ric}\left(\gamma'(r), \gamma'(r)\right) dr \le (n-1)\left(\delta_x^{-1} + 1\right) + \frac{2}{3} S(p) + \int_{r_x - \delta_x}^{r_x} R\left(\gamma(r)\right) dr,
$$

where

$$
S(p) := \max\left\{0, \sup\left\{\mathrm{Ric}(V, V) : V \in \mathcal{S}_y^{n-1}, y \in B_1(p)\right\}\right\}.
$$

To estimate the last term on the right-hand side, in view of $R\left(\gamma(r_x)\right) = R(x) \le 1$, it suffices to derive a gradient estimate for R.

By (2.40), $\mathrm{Ric} \ge 0$, and (2.63), we have for any $y \in \mathcal{M}^n$,

$$
|\nabla R|(y) \le 2\left|\mathrm{Ric}\right| |\nabla f|(y) \le 2R\sqrt{f}(y) \le 2R\left(\sqrt{f}(p) + \frac{r_y}{2}\right).
$$

Therefore, for $r \in [r_x - \delta_x, r_x]$, by our choice of δ_x we have

$$
\ln \frac{R(\gamma(r))}{R(x)} \le \int_r^{r_x} |\nabla \ln R|(\gamma(s)) ds
$$
$$
\le 2 \int_r^{r_x} \left(\sqrt{f}(p) + \frac{r_{\gamma(s)}}{2}\right) ds
$$
$$
\le C.
$$

Hence, for $r \in [r_x - \delta_x, r_x]$ we have $R(\gamma(r)) \le e^C$, which implies

$$
\int_{r_x - \delta_x}^{r_x} R\left(\gamma(r)\right) dr \le C\delta_x \le C.
$$

We obtain from (4.149) that

$$
\int_0^{r_x} \mathrm{Ric}\left(\gamma'(r), \gamma'(r)\right) dr \le (n-1)\delta_x^{-1} + C = \frac{r_x}{4} + C.
$$

Applying this to (4.140), we conclude that

$$
\langle \nabla f, \gamma'(r_x) \rangle \ge \frac{r_x}{4} - C,
$$

which proves (4.147). $\qquad\square$

We may rephrase Theorem 4.45 as:

Corollary 4.46. *For any complete noncompact non-Gaussian shrinking GRS $(\mathcal{M}^n, g, f)$ with the property that there exists a sequence of points $\{x_i\}$ in $\mathcal{M}^n$ with $R(x_i) \to 0$, the Ricci curvature of g cannot be everywhere non-negative.*

4.6.2. Shrinking GRS with sect > 0 are compact.

The equations for f are, in a sense, unreasonably effective. One manifestation of this is the ease with which one can prove that shrinking GRS with positive sectional curvature must be compact in all dimensions, a result which originally took much effort to prove in dimension 3. Namely, by assuming positive sectional curvature, we obtain a Ricci curvature gap theorem, which in the noncompact case implies that the metric has too much positive curvature. The following is due to Munteanu and Wang [**242**]; when $n = 3$ this was originally due to the works of Perelman [**251**], Naber [**245**], Ni and Wallach [**247**], and Cao, Chen, and Zhu [**63**].

Theorem 4.47. *Any shrinking GRS with sect ≥ 0 and Ric > 0 must be compact.*

Proof. Since sect ≥ 0, we have $\sum_{k,\ell} R_{kij\ell} R^{k\ell} \geq 0$ (nonnegative semidefinite). Indeed, at any point $p \in \mathcal{M}^n$ we may write Ric $= \sum_{a=1}^n \lambda_a V^a \otimes V^a$, where $\lambda_a \geq 0$ and where $\{V^a\}_{a=1}^n$ is an orthonormal basis of $T_p^*\mathcal{M}$; hence, for any $W \in T_p\mathcal{M}$ we have

$$R_{kij\ell} R^{k\ell} W^i W^j = \sum_{a=1}^n \lambda_a R_{kij\ell} (V^a)^k (V^a)^\ell W^i W^j \geq 0.$$

Therefore (4.82) yields

(4.150) $\Delta_f \operatorname{Ric} \leq \operatorname{Ric}.$

Suppose that $\mathcal{M}^n$ is noncompact. Seeing the analogy with (4.87), we can apply the same idea as in the proof of Theorem 4.31. Define the symmetric 2-tensor

(4.151) $S := \operatorname{Ric} - c\left(f^{-1} + nf^{-2}\right) g,$

where $c > 0$. Since $\nabla g = 0$, from (4.150), (4.89), and (4.90) we may compute analogously to (4.93) that

(4.152) $\Delta_f S \leq S - cnf^{-3}\left(\frac{f}{2} - n\right) g.$

Since Ric > 0, there exists a constant $c > 0$ such that $S \geq 0$ in the set $\{x \in \mathcal{M}^n : f(x) \leq 2n\}$. Suppose that S has a negative eigenvalue somewhere. Since the symmetric 2-tensor S tends towards nonnegativity at infinity, there exists a point $x_0 \in \{x \in \mathcal{M}^n : f(x) > 2n\}$ and a unit tangent vector U_0 at x_0 such that

$$(4.153) \qquad S(U_0, U_0) = \inf_{U \in T\mathcal{M}, |U|=1} S(U, U).$$

Now, extend U_0 to a vector field U defined in a neighborhood of x_0 by parallel transporting U_0 along geodesics emanating from x_0. We then have at x_0 that

$$\nabla U = 0 \quad \text{and} \quad \Delta U = 0.$$

Thus, by applying the inequality (4.152) to U in a neighborhood of x_0 (this method is known as the tensor maximum principle), we obtain at x_0 that

$$(4.154) \qquad 0 \le S(U_0, U_0) - cnf^{-3}\left(\frac{f}{2} - n\right) < -cnf^{-3}\left(\frac{f}{2} - n\right),$$

which implies $\frac{f(x_0)}{2} - n < 0$; this is a contradiction. Therefore there exists $c > 0$ such that

$$\mathrm{Ric} \ge c\left(f^{-1} + nf^{-2}\right)g \ge cf^{-1}g$$

on all of $\mathcal{M}^n$. That is, the Ricci curvature decays at most quadratically.

Let $\beta : \mathbb{R} \to \mathcal{M}^n$ be an integral curve of ∇f. We have

$$(4.155) \qquad \frac{d}{dt}R(\beta(t)) = 2\,\mathrm{Ric}(\nabla f, \nabla f) \ge 2c\frac{|\nabla f|^2}{f} = 2c\left(1 - \frac{R}{f}\right),$$

where all of the terms above are evaluated at $\beta(t)$. Let $x \in \mathcal{M}^n$ and choose β so that $\beta(0) = x$. If we have $\frac{R}{f}(\beta(t)) \le \frac{1}{2}$ for all $t \in [-\frac{n}{c}, 0]$, then (4.155) implies

$$\frac{d}{dt}R(\beta(t)) \ge c \quad \text{for} \quad -\frac{n}{c} \le t \le 0.$$

Integrating this on $[-\frac{n}{c}, 0]$ yields

$$R(x) \ge n + R\left(\beta\left(-\frac{n}{c}\right)\right) > n.$$

Otherwise, there exists $t \in [-\frac{n}{c}, 0]$ such that $y = \beta(t)$ satisfies $\frac{R}{f}(y) \ge \frac{1}{2}$, so that

$$R(x) \ge R(y) \ge \frac{1}{2}f(y) \ge \frac{1}{2}e^{-\frac{n}{c}}f(x),$$

where the last inequality follows from $\frac{d}{dt}f(\beta(t)) = |\nabla f|^2 \le f(\beta(t))$. We conclude that $R(x) \ge \max\{n, \frac{1}{2}e^{-\frac{n}{c}}f(x)\}$ for all $x \in \mathcal{M}^n$. This contradicts (4.60); i.e., $\int_{\{f<t\}} R\,d\mu \le \frac{n}{2}\mathrm{Vol}\,\{f < t\}$ for t sufficiently large since g has infinite volume and f is a proper function (since f grows quadratically). $\square$

In particular, Theorem 4.47 implies that there do not exist complete noncompact shrinking GRS with positive sectional curvature.

4.6.3. Shrinking GRS with sect ≥ 0.

Munteanu and Wang extended their result above (Theorem 4.47) to the following [**242**].

Theorem 4.48. *If $(\mathcal{M}^n, g, f)$ is a shrinking GRS with sect ≥ 0, then either $\mathcal{M}^n$ is closed, it is the Gaussian shrinking GRS, or its universal covering shrinking GRS is the product of a compact m-dimensional shrinking GRS with sect ≥ 0 and Ric > 0 and the $(n-m)$-dimensional Gaussian soliton, where $2 \leq m \leq n - 1$.*

Proof. Let $(\mathcal{M}^n, g(t), f(t))$, $t \in (-\infty, 1)$, be the canonical form given by Proposition 4.19. By (4.78), we have

$$(4.156) \qquad \frac{\partial}{\partial t} R_{ij} = \Delta R_{ij} + 2 R_{kij\ell} R^{k\ell} - 2 R_{ik} R_j^k.$$

Since sect ≥ 0, the symmetric 2-tensor $R_{kij\ell} R^{k\ell}$ is positive semidefinite. Hence, if $V \in \ker(\mathrm{Ric})$, i.e., $R_{ij} V^j = 0$, then

$$Q_{ij} := 2 R_{kij\ell} R^{k\ell} - 2 R_{ik} R_j^k$$

satisfies $Q_{ij} V^i V^j \geq 0$. By the strong maximum principle (see Theorem 4.49 below), the subbundle

$$\mathcal{K} := \ker(\mathrm{Ric}_{g(t)}) \subset T\mathcal{M}$$

is invariant under parallel transport on $\mathcal{M}^n$ and independent of $t \in (-\infty, 1)$. Thus, (4.156) and $\left(\frac{\partial}{\partial t} R_{ij}\right) V^i V^j = 0$ and $(\Delta \mathrm{Ric})_{ij} V^i V^j = 0$ for $V \in \mathcal{K}$ imply that $R_{kij\ell} R^{k\ell} V^i V^j = 0$ (and hence $R_{kij\ell} R^{k\ell} V^j = 0$) for $V \in \mathcal{K}$. Since Ric > 0 on $\mathcal{K}^\perp$, this implies that

$$\mathrm{Rm}(X, Y) Z = 0 \quad \text{for } Y, Z \in \mathcal{K}^\perp \text{ and } X \in \mathcal{K}.$$

If $\mathcal{K}^\perp = 0$, then $(\mathcal{M}^n, g, f)$ is a shrinking GRS with Ric $= 0$ and hence is the Gaussian shrinker by the equality case of Theorem 2.14. Otherwise, by the de Rham splitting theorem, the (time-dependent) universal covering shrinking GRS $(\widetilde{\mathcal{M}}^n, \tilde{g}(t), \tilde{f}(t))$ splits as the product of an m-dimensional shrinking GRS with sect ≥ 0, Ric > 0, and $2 \leq m \leq n - 1$ (which must be compact by Theorem 4.47) and an $(m - n)$-dimensional shrinking GRS with Ric $= 0$, which must be the $(m - n)$-dimensional Gaussian shrinker. $\qquad\square$

4.7. Splitting theorem for shrinkers with nonnegative curvature

In this section we shall prove the following result of Munteanu and Wang.

Theorem 4.49. *Let $(\mathcal{M}^n, g, f)$ be a shrinking GRS. If the sectional curvature of g is nonnegative everywhere and the Ricci tensor has a zero eigenvalue somewhere, then $(\mathcal{M}^n, g)$ splits locally.*

Theorem 4.49 is a consequence of an application of the strong maximum principle for systems to the nondegenerate elliptic system satisfied by the Ricci tensor. Because of its importance in geometric analysis, we take this opportunity to discuss the more general Brendle and Schoen [**53**] version of the Bony maximum principle for degenerate elliptic equations [**38**].

4.7.1. Introduction to Bony's maximum principle.

In this subsection we let Ω be an open subset in $\mathbb{R}^n$ and we let $F \subset \Omega$ be a relatively closed subset.

Definition 4.50. A vector ξ is said to be **tangential** to F at $x_1 \in F$ if for any $x_0 \in \mathbb{R}^n$ satisfying $|x_0 - x_1| = d(x_0, F)$, the Euclidean distance between the point x_0 and the set F, then we have that $\langle \xi, x_1 - x_0 \rangle = 0$, where the inner product is the Euclidean inner product.

Note that if x_1 lies in the interior of F, then any $\xi \in \mathbb{R}^n$ is tangential to F. If F is a polytope and if x_1 lies in the interior of a k-face of F, then ξ is tangential to F at x_1 if and only if ξ is tangential to the k-face in the usual sense.

Lemma 4.51. *Let $X_1, \ldots, X_m$ be smooth vector fields on Ω that are tangential to F at all points in F. Let $f_1, \ldots, f_m : [0, T] \to \mathbb{R}$ be smooth functions. Suppose $\gamma : [0, T] \to \Omega$ is a smooth path satisfying*

$$(4.157) \qquad \gamma'(s) = \sum_{i=1}^{m} f_i(s) X_i(\gamma(s)).$$

If $\gamma(0) \in F$, then $\gamma(s) \in F$ for all $s \in (0, T]$.

Proof. We define the function $\rho : [0, T] \to \mathbb{R}_{\geq 0}$ by

$$\rho(s) = d^2(\gamma(s), F).$$

Let $\varepsilon > 0$ be such that $\inf_{s \in [0,T]} d(\gamma(s), \partial\Omega) \geq 2\varepsilon$. Since $\rho(0) = 0$, it suffices to show that there exists $L \geq 0$ such that

$$\frac{d^+\rho(s)}{ds} := \limsup_{h \to 0_+} \frac{\rho(s + h) - \rho(s)}{h} \leq L\rho(s)$$

whenever $\rho(s) \leq \varepsilon^2$. (The display implies $\frac{d^+}{ds}(e^{-Ls}\rho(s)) \leq 0$.) To this end, we fix $s \in [0, T]$ and we let $x_1 \in F$ be such that $\rho(s) = d^2(\gamma(s), x_1) \leq \varepsilon^2$. Since X_i is tangential to F at x_1, we have $\langle X_i(x_1), x_1 - \gamma(s) \rangle = 0$ for

$i = 1, \ldots, n$. We compute

$$
\begin{aligned}
\limsup_{h \to 0_+} \frac{\rho(s+h) - \rho(s)}{h} &\leq \limsup_{h \to 0_+} \frac{|x_1 - \gamma(s+h)|^2 - |x_1 - \gamma(s)|^2}{h} \\
&= -2\langle \gamma'(s), x_1 - \gamma(s) \rangle \\
&= -2 \sum_{i=1}^{m} f_i(s) \langle X_i(\gamma(s)), x_1 - \gamma(s) \rangle \\
&= 2 \sum_{i=1}^{m} f_i(s) \Big\langle X_i(x_1) - X_i(\gamma(s)), x_1 - \gamma(s) \Big\rangle \\
&\leq L|x_1 - \gamma(s)|^2 \\
&= L\rho(s)
\end{aligned}
$$

for some constant L. $\square$

Bony's maximum principle says the following.

Theorem 4.52. *Let $X_1, \ldots, X_m$ be smooth vector fields on an open subset Ω of $\mathbb{R}^n$ and let $f_1, \ldots, f_m : [0, T] \to \mathbb{R}$ be smooth functions. Let $u : \Omega \to \mathbb{R}$ be a smooth nonnegative function which is a supersolution to the weakly elliptic equation*

$$(4.158) \qquad \sum_{i=1}^{m} (D^2 u)(X_i, X_i) \leq -K \inf_{|\xi| \leq 1} (D^2 u)(\xi, \xi) + K|Du| + Ku,$$

where D is the Euclidean covariant derivative and where $K > 0$ is a constant. Let $F = \{x \in \Omega : u(x) = 0\}$ and let $\gamma : [0, T] \to \Omega$ be a smooth curve satisfying $\gamma(0) \in F$ and $\gamma'(s) = \sum_{i=1}^{m} f_i(s) X_i(\gamma(s))$. Then $\gamma(s) \in F$ for every $s \in [0, T]$.

Proof. According to Lemma 4.51, it suffices to show that $X_1, \ldots, X_m$ are tangential to F at every point in F. For a contradiction, we assume there exist points $x_1 \in F$ and $x_0 \in \Omega$ such that $|x_1 - x_0| = d(x_0, F)$ but

$$(4.159) \qquad \sum_{i=1}^{m} \langle X_i(x_1), x_1 - x_0 \rangle^2 > 0.$$

By taking x_0 a little closer to x_1, we may also assume that $|x - x_0| > |x_1 - x_0|$ for all $x \in F \setminus \{x_1\}$.

For the barrier argument that follows, we observe that by (4.159) there exists $\alpha > 0$ such that
(4.160)

$$4\alpha^2 \sum_{i=1}^{m} \langle X_i(x_1), x_1 - x_0 \rangle^2 - 2\alpha \sum_{i=1}^{m} |X_i(x_1)|^2 > 2K\alpha + 2K\alpha|x_1 - x_0| + K.$$

By continuity, there exists an open neighborhood U of x_1 such that

$$(4.161) \quad 4\alpha^2 \sum_{i=1}^{m} \langle X_i(x), x - x_0 \rangle^2 - 2\alpha \sum_{i=1}^{m} |X_i(x)|^2 > 2K\alpha + 2K\alpha|x - x_0| + K$$

for every $x \in \bar{U}$.

Now we introduce the barrier function

$$v(x) = \exp(-\alpha|x - x_0|^2) - \exp(-\alpha|x_1 - x_0|^2).$$

Denote $B = B_{|x_1 - x_0|}(x_0)$ so that $\bar{B} \cap F = \{x_1\}$. It then follows that $u(x) > 0$ for all $x \in \partial U \cap \bar{B}$ because F is the zero-set of u. Hence we may find $\lambda > 0$ large enough so that $\lambda u(x) > v(x)$ for every $x \in \bar{B} \cap \partial U$. On the other hand, we have $\lambda u(x) \geq 0 > v(x)$ for every $x \in \partial U - \bar{B}$ by the definition of v. We conclude that

$$(4.162) \quad\quad\quad \lambda u(x) - v(x) > 0 \text{ for all } x \in \partial U.$$

Next, we fix a point $x_2 \in \bar{U}$ such that

$$\lambda u(x_2) - v(x_2) \leq \lambda u(x) - v(x) \text{ for all } x \in \bar{U}.$$

Since $\lambda u(x_1) - v(x_1) = 0$, we have that $\lambda u(x_2) - v(x_2) \leq 0$. It then follows from the boundary condition (4.162) that x_2 lies in the interior of U. Hence, we may compare the first and second derivatives of λu and v at the point x_2 to obtain

$$\lambda u(x_2) \leq v(x_2),$$
$$\lambda Du\Big|_{x_2} = Dv\Big|_{x_2},$$
$$\lambda D^2 u\Big|_{x_2} \geq D^2 v\Big|_{x_2}.$$

Therefore, using (4.158) we compute at the point x_2 that

$$(4.163) \quad \sum_{i=1}^{m} D^2 v(X_i, X_i) \leq \lambda \sum_{i=1}^{m} D^2 u(X_i, X_i)$$
$$\leq -K\lambda \inf_{|\xi| \leq 1} (D^2 u)(\xi, \xi) + K\lambda|Du| + Ku$$
$$\leq -\lambda \inf_{|\xi| \leq 1} (D^2 v)(\xi, \xi) + K|Dv| + Kv.$$

Now we proceed to use the definition of v to show that (4.163) contradicts (4.161). By direct computation, we have at the point x_2 that

(4.164)
$$\sum_{i=1}^{m} (D^2 v)(X_i, X_i)$$

$$= \left(4\alpha^2 \sum_{i=1}^{m} \langle x_2 - x_0, X_i(x_2) \rangle^2 - 2\alpha \sum_{i=1}^{m} |X_i(x_2)|^2 \right) \exp(-\alpha |x_2 - x_0|^2)$$

and

(4.165) $$\inf_{|\xi| \leq 1} (D^2 v)(\xi, \xi) = -2\alpha \exp(-\alpha |x_2 - x_0|^2).$$

By combining (4.163), (4.164), and (4.165), we have

$$4\alpha^2 \sum_{i=1}^{m} \langle X_i(x_2), x_2 - x_0 \rangle^2 - 2\alpha \sum_{i=1}^{m} |X_i(x_2)|^2 \leq 2K\alpha + 2K\alpha |x_2 - x_0| + K.$$

This contradicts (4.161). $\square$

Remark 4.53. In our application in the next section, we shall consider the inequality

$$\sum_{i=1}^{m} (D^2 u)(X_i, X_i) \leq K|Du| + Ku.$$

In this case there is no second derivative term on the right-hand side of (4.158). In fact, in [**53**], the term $\inf_{|\xi| \leq 1} (D^2 u)$ arises from the quadratic curvature term in the evolution equation for Rm. In both [**174**] and our case this term does not occur because the quadratic term in each of these cases has a favorable sign.

4.7.2. Proof of the splitting theorem.

Proof of Theorem 4.49. On the orthonormal frame bundle $\mathcal{O}$ of the manifold $(\mathcal{M}^n, g)$, we have the following equation for 2-tensors (expressed in terms of their components):

$$\sum_{i=1}^{n} \nabla^i \nabla_i R_{ab} - \sum_{i=1}^{n} \nabla^i f \nabla_i R_{ab} = -2 \sum_{i,j=1}^{n} R_{aijb} R^{ij} \leq 0,$$

where the last inequality means negative semidefinite and is a consequence of our curvature assumption sect ≥ 0. We consider the following function on the bundle $\mathcal{O}$:

$$u : \mathcal{O} \to \mathbb{R},$$

$$u : \{e_1, \ldots, e_n\} \mapsto R_{11},$$

where $R_{11} := \mathrm{Ric}(e_1, e_1)$.

According to our assumption that somewhere Ric has a 0 eigenvalue, we have that the set

$$F := \left\{ \{e_1, \ldots, e_n\} \in \mathcal{O} : u(\{e_1, \ldots, e_n\}) = 0 \right\} \subset \mathcal{O}$$

is nonempty.

Let $\gamma : [0, T] \to \mathcal{M}^n$ be a smooth curve and let $\{e_1, \ldots, e_n\}$ be an orthonormal frame at $\gamma(0)$ such that $R_{11} = 0$. Let

$$P(s) = \{P_\gamma e_1(s), \ldots, P_\gamma e_n(s)\}$$

be the horizontal lift of $\gamma(s)$ to $\mathcal{O}$, where $P_\gamma e_i(s)$ stands for parallel transport along γ. Clearly we have $P(0) \in F$. Furthermore, we have

$$\frac{d}{ds} P(s) = \sum_{i=1}^{n} \frac{d\gamma^i(s)}{ds} \nabla_i^H(s)$$

for any $s \in [0, T]$, where $\nabla_i^H(s)$ stands for the horizontal lift to $T\mathcal{O}$ of $P_\gamma e_i(s)$ for each $i = 1, \ldots, n$. It then follows from Theorem 4.52 that $P(s) \in F$ for any $s \in [0, T]$; that is, $\mathrm{Ric}(P_\gamma e_1(s), P_\gamma e_1(s)) = 0$ for any $s \in [0, T]$. Therefore, there exists a nontrivial parallel vector field on $(\mathcal{M}^n, g)$ and it follows that $(\mathcal{M}^n, g)$ splits locally. $\square$

4.8. Notes and commentary

An abbreviated list of works on Ricci solitons not discussed or referenced elsewhere in this book is: Appleton [11], Bamler, Chan, Ma, and Zhang, [23], Bamler, Chow, Deng, Ma, and Zhang, [24], Cabezas-Rivas and Topping [58, 59], H.-D. Cao and Q. Chen [64], H.-D. Cao and Liu [66], X. Cao, Wang, Z. Zhang [71], Carrillo and Ni [73], Catino, Mastrolia, and Monticelli [74], Chan [75, 77], Chan, Ma, and Zhang [78, 79], Chan and Zhu [80], Chen and Wang [95], Chow, Deng, and Ma [105], Chow, Freedman, Shin, and Zhang [106], Chow and Lu [109, 110], Chow, Lu, and Yang [112, 113], Colding and Minicozzi [118, 120], Derdzinski [132], Deruelle [133, 136], Eminenti, La Nave, and Mantegazza [144], Fang, Man, and Zhang [146], Fernández-López and García-Río [148, 149], Ge and Jiang [154], Glickenstein [158], Guan, Lu, and Xu [164], Guenther, Isenberg, and Knopf [165], Guo [168], Hall and Murphy [171], Jablonski [197], Kotschwar and Wang [209], Kröncke [211], Lauret [215], Li, Ni, and Wang [219], Lott and Wilson [224], Munteanu [232], Munteanu and Šešum [234], Munteanu, Sung, and Wang [233], Munteanu and Wang [235–237, 239, 241–243], Munteanu and M.-T. Wang [244], Ni and Wallach [248], Petersen and Wylie [254–256], Šešum [282], Weber [284], Wei and Wylie [285], Wu [288, 289], Wylie [292], Yang [294], Yokota [298, 299] Y. Zhang [302], and Z.-L. Zhang [304–306].

For surveys, see Cao [**61, 62**], Chu [**115**], and Kotschwar [**208**].

Self-similar and ancient solutions to the mean curvature flow is an active area of research. There are works by Andrews, Angenent, Bernstein, Bourni, Brendle, Bryan, K. Choi, Colding, Daskalopoulos, Hamilton, Haslhofer, Hershkovits, Hoffman, Huisken, Ilmanen, Ivaki, Kapouleas, Ketover, Kleene, Langford, Martin, Minicozzi, Møller, Risa, Sáez Trumper, Scheuer, Šešum, W. Sheng, Sinestrari, Tinaglia, L. Wang, X.-J. Wang, White, and others. As a small selection of the vast literature, we refer the reader to [**6**], [**39**], [**48**], [**184**], [**188**], and the references therein, which, along with the book [**5**], include references to papers by the aforementioned authors.

4.9. Exercises

4.9.1. Busemann functions.

Exercise 4.1 (Existence of a ray). Let $(\mathcal{M}^n, g)$ be a complete noncompact Riemannian manifold. Prove that for any point $p \in \mathcal{M}^n$ there exists a (unit speed) minimal geodesic $\gamma : [0, \infty) \to \mathcal{M}^n$ with $\gamma(0) = p$. We call γ a **geodesic ray**.

HINT: Choose any sequence of points x_i with $d(x_i, p) \to \infty$ and join p and x_i by a minimal geodesic segment.

Exercise 4.2 (Busemann function properties). Let $\gamma : [0, \infty) \to \mathcal{M}^n$ be a geodesic ray with $\gamma(0) = p$. Prove that for any point $x \in \mathcal{M}^n$, the function $t \mapsto t - d(\gamma(t), x)$, $t \geq 0$, is nondecreasing and bounded above by $d(x, p)$. Thus, the **Busemann function** associated to γ,

$$(4.166) \quad b_\gamma(x) := \lim_{t \to \infty} \big(t - d(\gamma(t), x) \big) = \lim_{t \to \infty} \big(d(\gamma(t), p) - d(\gamma(t), x) \big),$$

is well-defined.

Show that

$$(4.167) \qquad\qquad |b_\gamma(x) - b_\gamma(y)| \leq d(x, y)$$

for all $x, y \in \mathcal{M}$.

Exercise 4.3 (Busemann functions on $\mathbb{R}^n$). Observe that geodesic rays γ in $\mathbb{R}^n$ (which are "straight") are in one-to-one correspondence with points in $\mathbb{R}^n \times \mathbb{S}^{n-1}$ via the map $\gamma \mapsto (\gamma(0), \gamma'(0))$. Show that the Busemann function associated to γ is given by

$$(4.168) \qquad\qquad b_\gamma(x) = \langle x - \gamma(0), \gamma'(0) \rangle.$$

Exercise 4.4 (Busemann functions are distance functions). Let $(\mathcal{M}^n, g)$ be a complete noncompact Riemannian manifold. Prove that at a point x where a Busemann function b_γ is differentiable,

$$(4.169) \qquad\qquad |\nabla b_\gamma|(x) = 1.$$

We call a function with this property a **distance function**.

HINT: For each $t \geq 0$, let α_t denote the minimal geodesic from x to $\gamma(t)$. Show that for any $\varepsilon > 0$ there exist $t_i \to \infty$ such that $\alpha_{t_i}(\varepsilon)$ converges to some $y \in \mathcal{M}^n$. Prove that $b_\gamma(y) - b_\gamma(x) = \varepsilon$.[1]

Exercise 4.5 (Ric ≥ 0 implies Busemann functions are subharmonic). Let $(\mathcal{M}^n, g)$ be a complete noncompact Riemannian manifold with nonnegative Ricci curvature, and let γ be a geodesic ray. Prove that $\Delta b_\gamma \geq 0$ in the barrier sense.

HINT: Choose any $t_i \to \infty$ such that $\alpha'_{t_i}(0)$ converges to some unit tangent vector V at x. Show that the unique geodesic $\alpha : [0, \infty) \to \mathcal{M}^n$ with $\alpha'(0) = V$ is minimal, i.e., a ray. Show that for all $s \geq 0$ the function $f(y) := s - d(y, \alpha(s)) + b_\gamma(x)$ is a lower barrier for b_γ at x. Apply the Laplacian comparison theorem to f.

4.9.2. Volume comparison.

Exercise 4.6 (Laplacian and volume comparison for Ric ≥ 0). Let $(\mathcal{M}^n, g)$ be a complete noncompact Riemannian manifold with nonnegative Ricci curvature, and let $p \in \mathcal{M}^n$. Using the Riccati inequality (4.27), prove that for any $x \neq p$ and not in the cut locus of p, we have the Laplacian comparison theorem:

$$(4.170) \qquad \Delta r = H \leq \frac{n-1}{r}.$$

Prove that the Jacobian $\mathrm{J}(r)$ along a minimal geodesic emanating from p defined by (4.20) satisfies

$$(4.171) \qquad \frac{d}{dr}\left(\frac{\mathrm{J}(r)}{r^{n-1}}\right) \leq 0$$

and hence $\mathrm{J}(r) \leq r^{n-1}$. Conclude the Bishop volume comparison theorem, which says that

$$(4.172) \qquad r \mapsto \frac{\mathrm{Vol}\, B_r(p)}{r^n}$$

is nonincreasing. In particular, $\mathrm{Vol}\, B_r(p) \leq \omega_n r^n$. Also conclude that the asymptotic volume ratio given by (4.54) is well-defined.

4.9.3. Dynamical GRS.

Exercise 4.7 (The dynamical Gaussian and cigar solitons).

(1) What is the potential function $f(t)$ for the evolving GRS associated to the Gaussian soliton $(\mathbb{R}^n, g_{\mathrm{Euc}}, \frac{\lambda}{4}|x|^2, \lambda)$?

(2) What are the metric and potential functions $(g_\Sigma(t), f_\Sigma(t))$ for the evolving GRS associated to the cigar soliton $(\mathbb{R}^2, g_\Sigma, f_\Sigma)$?

---

[1] For intuition, describe y in the case where $(\mathcal{M}^n, g) = \mathbb{R}^n$.

4.9.4. Yau's Liouville theorem.

Exercise 4.8 (Gradient estimate for harmonic functions). Let $(\mathcal{M}^n, g)$ be a complete noncompact Riemannian manifold with nonnegative Ricci curvature, and let ϕ be a positive harmonic function on $\mathcal{M}^n$. Prove that $u = \log \phi$ satisfies $-\Delta u = |\nabla u|^2 =: Q$ and

$$\frac{1}{2}\Delta Q \geq -\nabla u \cdot \nabla Q + |\nabla \nabla u|^2 \geq -\nabla u \cdot \nabla Q + \frac{1}{n}Q^2.$$

Exercise 4.9 (Localizing the gradient estimate). Continuing with the previous exercise, let $p \in \mathcal{M}^n$, let $r > 0$, and consider a cutoff function $\eta(x) = \phi(d(x,p))$ satisfying $\phi(s) = 1$ for $s \leq r$, $\phi(s) = 0$ for $s \geq 2r$, $\phi'(s) \leq 0$, and

(4.173) $$\frac{(\phi')^2}{\phi}(s) \leq Cr^{-2}, \quad |\phi''|(s) \leq Cr^{-2},$$

for some constant $C < \infty$. Prove that

$$\frac{1}{2}\Delta(\eta Q) \geq \left(\frac{1}{2}\eta^{-1}\Delta\eta - \eta^{-2}|\nabla\eta|^2 + \eta^{-1}\nabla\eta \cdot \nabla u\right)(\eta Q)$$

$$+ \left(\eta^{-1}\nabla\eta - \nabla u\right) \cdot \nabla(\eta Q) + \frac{1}{n\eta}(\eta Q)^2.$$

By applying the elliptic maximum principle and taking $r \to \infty$, prove that $|\nabla u| = 0$ and hence f is constant on $\mathcal{M}^n$. This is Yau's **Liouville theorem**.

4.9.5. Growth estimate for R.

Exercise 4.10 (At most linear growth for R on a net for shrinkers [**99**]). Using the methods of the proof of Theorem 4.3, prove the following estimate: Let $(\mathcal{M}^n, g, f)$ be a complete noncompact shrinking GRS and suppose that f attains its minimum at $o \in \mathcal{M}^n$. For any $\varepsilon > 0$, there exists a constant $C(\varepsilon, n)$ such that for any $x \in \mathcal{M}^n$ there exists $y \in B_\varepsilon(x)$ such that

(4.174) $$R(y) \leq C(\varepsilon, n)(d(y, o) + 1).$$

By an ε-**net** we mean a subset $\mathcal{N}$ of $\mathcal{M}^n$ with the property that for every $x \in \mathcal{M}^n$ there exists $y \in \mathcal{N}$ such that $d(x, y) < \varepsilon$. So this exercise shows that for any $\varepsilon > 0$ there exists an ε-net on which the scalar curvature grows at most linearly.

HINT: For the geodesic stability inequality (2.77), choose $\zeta : [0, r(x)] \to [0, 1]$ so that $\zeta(0) = \zeta(r(x)) = 0$, $\zeta(\varepsilon) = \zeta(r(x)-\varepsilon) = 1$, and ζ is linear in between. Use this to show that there exists $r_1 \in [r(x) - \varepsilon, r(x)]$ such that

(4.175) $$|\nabla f|^2(\gamma(r_1)) \geq \frac{1}{2}(r_1 - C(\varepsilon, n))^2.$$

4.9.6. Riemann curvature tensor for GRS.

Exercise 4.11. Using (4.114) for GRS, derive the formula

(4.176)
$$(\mathcal{L}_{\nabla f}\mathrm{Rm})_{ijk\ell} = \Delta R_{ijk\ell} + 2(B_{ijk\ell} - B_{ij\ell k} + B_{ikj\ell} - B_{i\ell jk})$$
$$- R_i^p R_{pjk\ell} - R_j^p R_{ipk\ell} - R_k^p R_{ijp\ell} - R_\ell^p R_{ijkp} + \lambda R_{ijk\ell}.$$

4.9.7. Bony's maximum principle.

Exercise 4.12. Generalize Theorem 4.52 from Euclidean space to Riemannian manifolds.

HINT: The definition of a vector being tangential to a closed set can be modified as follows: ξ is tangential to F at $x_1 \in F$ if for any $x_0 \in \mathcal{M}^n \setminus \mathrm{Cut}(x_1)$ such that $d(x_0, F) = d(x_0, x_1)$ it holds that $\langle \xi, \gamma'(1) \rangle_{x_1} = 0$, where $\gamma : [0,1] \to \mathcal{M}^n$ is the unique minimizing constant speed geodesic such that $\gamma(0) = x_0$ and $\gamma(1) = x_1$. Furthermore, (4.164) now becomes

$$\sum_{i=1}^m (D^2 v)(X_i, X_i) \geq \left(4\alpha^2 \sum_{i=1}^m \left\langle \frac{\partial}{\partial r}, X_i(x_2) \right\rangle^2 - C\alpha \right) \exp(-\alpha r^2)$$

and (4.165) becomes

$$\inf_{|\xi| \leq 1} (D^2 v)(\xi, \xi) \geq -C\alpha \exp(-\alpha r^2),$$

where $r(x) = d(x_0, x)$ and where C is a constant depending on the local geometry. Inequality (4.161) should also be adjusted accordingly.

Chapter 5

Classification of 3-Dimensional Shrinkers

In this chapter we present the classification of complete 3-dimensional shrinking gradient Ricci solitons. This classification is important in the study of 3-dimensional Ricci flow. The strong maximum principle is used to show that 3-dimensional shrinkers either have positive sectional curvature or their universal coverings split as the product of $\mathbb{R}$ and a 2-dimensional shrinker. By Theorem 4.47, any shrinker with positive sectional curvature must be compact. In this chapter we also present an earlier proof of this last result in dimension 3.

5.1. 3-Dimensional ancient solutions have nonnegative sectional curvature

In view of the canonical form for GRS and the strong maximum principle for the Ricci flow $\partial_t g = -2\mathrm{Ric}$, we recall some facts about the curvature tensor evolution under the Ricci flow.

Let $(\mathcal{M}^n, g)$ be a Riemannian manifold. The Riemann **curvature operator** is the bundle endomorphism

$$\mathrm{Rm} : \Lambda^2 T^*\mathcal{M} \to \Lambda^2 T^*\mathcal{M}$$

defined by

$$(5.1) \qquad \operatorname{Rm}(\alpha)_{ij} := \sum_{k,\ell=1}^{n} R_{ijk\ell} \alpha^{\ell k} := \sum_{k,\ell,p,q=1}^{n} g^{kp} g^{\ell q} R_{ijk\ell} \alpha_{qp}$$

for $\alpha \in \Lambda^2 T^* \mathcal{M}$. As a bilinear form on each fiber $\Lambda^2 T_x^* \mathcal{M}$,

$$(5.2) \qquad \operatorname{Rm}(\alpha, \beta) := \langle \operatorname{Rm}(\alpha), \beta \rangle.$$

Definition 5.1. We say that the Riemannian metric g has **positive curvature operator** if Rm is positive definite, i.e., has positive eigenvalues, denoted by $\operatorname{Rm} > 0$; g has **nonnegative curvature operator** if Rm is nonnegative definite, i.e., has nonnegative eigenvalues, denoted by $\operatorname{Rm} \geq 0$.

Now assume that we are in dimension $n = 3$. In this dimension the Hodge star operator

$$* : \Lambda^2 T^* \mathcal{M} \to T^* \mathcal{M},$$

defined by

$$\alpha \wedge *\beta = \langle \alpha, \beta \rangle \, d\mu,$$

is a bundle isomorphism (and the fibers of each of these vector bundles are isomorphic to $\mathbb{R}^3$). The volume form is equal to $d\mu = \omega^1 \wedge \omega^2 \wedge \omega^3$, where $\omega^1, \omega^2, \omega^3$ is a positively oriented orthonormal basis of $T_x^* \mathcal{M}$, and the inner product on $\Lambda^2 T_x^* \mathcal{M}$ is given by

$$(5.3) \qquad \langle \alpha, \beta \rangle := \frac{1}{2} \sum_{i,j=1}^{3} \alpha_{ij} \beta^{ij}.$$

We have

$$*(\omega^a \wedge \omega^b) = \omega^c,$$

where a, b, c is any cyclic permutation of $1, 2, 3$.[1] Observe that $*$ is a bundle isometry and in particular $\{\omega^1 \wedge \omega^2, \omega^2 \wedge \omega^3, \omega^3 \wedge \omega^1\}$ is an orthonormal basis for $\Lambda^2 T_x^* \mathcal{M}$.

Via this bundle isomorphism we may consider the Riemann curvature operator as a bundle endomorphism

$$\operatorname{Rm} : T^* \mathcal{M} \to T^* \mathcal{M}$$

define by

$$\operatorname{Rm}(\alpha) := *\big(\operatorname{Rm}(*\alpha)\big) \qquad \text{for } \alpha \in T^* \mathcal{M}.$$

By the metric duality isomorphism between $T^* \mathcal{M}$ and $T \mathcal{M}$, we may also consider Rm as a symmetric 2-tensor on $\mathcal{M}^3$: $\operatorname{Rm} \in C^\infty(T^* \mathcal{M} \otimes T^* \mathcal{M})$.

---

[1] The wedge product of two 1-forms α and β is defined by $\alpha \wedge \beta = \alpha \otimes \beta - \beta \otimes \alpha$.

Given $x \in \mathcal{M}^3$, let $\omega^1, \omega^2, \omega^3$ be a positively oriented orthonormal basis of $T_x^*\mathcal{M}$ of eigen-1-forms of Rm $: T^*\mathcal{M} \to T^*\mathcal{M}$ with associated eigenvalues $\lambda_1, \lambda_2, \lambda_3$, respectively. So

$$\mathrm{Rm}(\omega^a) = \lambda_a \omega^a \qquad \text{for } a = 1, 2, 3.$$

Then, considering Rm as an endomorphism of $\Lambda^2 T_x^*\mathcal{M}$, it has the orthonormal basis of eigen-2-forms $\omega^a \wedge \omega^b$ with associated eigenvalues λ_c, where a, b, c is a cyclic permutation of $1, 2, 3$; indeed,

$$\mathrm{Rm}(\omega^a \wedge \omega^b) = *\mathrm{Rm}(\omega^c) = *(\lambda_c \omega^c) = \lambda_c \omega^a \wedge \omega^b.$$

Denote the components of the Riemann curvature 4-tensor with respect to the orthonormal frame $\{e_1, e_2, e_3\}$ dual to $\{\omega^1, \omega^2, \omega^3\}$ by

$$R_{abcd} := \langle \mathrm{Rm}(e_a, e_b)e_c, e_d \rangle$$

for $1 \leq a, b, c, d \leq 3$ and denote the components of Rm, as a symmetric 2-tensor, by

$$\mathrm{Rm}_{ad} := \mathrm{Rm}(e_a, e_d) = g^*(\mathrm{Rm}(\omega^a), \omega^d),$$

where g^* is the inner product on $T^*\mathcal{M}$ induced by g and duality, so that

$$\mathrm{Rm} = \sum_{a,d=1}^{3} \mathrm{Rm}_{ad} \omega^a \otimes \omega^d$$

as a 2-tensor. By (5.2) and (5.3), we have that

$$\mathrm{Rm}_{ad} = \mathrm{Rm}(\omega^b \wedge \omega^c, \omega^e \wedge \omega^f) = \langle \mathrm{Rm}(\omega^b \wedge \omega^c), \omega^e \wedge \omega^f \rangle = 2R_{bcfe},$$

where a, b, c and d, e, f are cyclic permutations of $1, 2, 3$. From this we see that the matrix (Rm_{ad}) is diagonal and

$$\lambda_a = 2\mathrm{Rm}_{aa} = 2R_{bccb}$$

is equal to two times the sectional curvature of the 2-plane spanned by e_b and e_c.

Let φ be a 2-form at a point in $\mathcal{M}^3$. Recall that in dimension 3 every 2-form is the wedge product of two 1-forms. Thus, $\varphi = \alpha \wedge \beta$, where α and β are 1-forms. We calculate that

$$\mathrm{Rm}(\varphi, \varphi) = \langle \mathrm{Rm}(\alpha \wedge \beta), \alpha \wedge \beta \rangle = 2\langle \mathrm{Rm}(\alpha^\natural, \beta^\natural)\beta^\natural, \alpha^\natural \rangle,$$

which is equal to the sectional curvature of the plane spanned by the tangent vectors $\alpha^\natural$ and $\beta^\natural$ divided by the area of the parallelogram spanned by $\alpha^\natural$ and $\beta^\natural$. Thus, when $n = 3$, positive (nonnegative) sectional curvature is equivalent to positive (nonnegative) curvature operator.

Define the components of the Ricci tensor by

$$R_{ab} := \mathrm{Ric}(e_a, e_b).$$

We have that the matrix (R_{ab}) is diagonal and

$$R_{aa} = R_{baab} + R_{caac} = \frac{1}{2}(\lambda_c + \lambda_b),$$

where a, b, c is a cyclic permutation of $1, 2, 3$. Therefore the scalar curvature is equal to

$$R = R_{11} + R_{22} + R_{33} = \lambda_1 + \lambda_2 + \lambda_3.$$

From all of this we conclude that

$$(Rg - 2\mathrm{Ric})_{aa} = \lambda_1 + \lambda_2 + \lambda_3 - (\lambda_b + \lambda_c) = \lambda_a = 2(\mathrm{Rm})_{aa}.$$

That is,

$$(5.4) \qquad\qquad \mathrm{Rm} = \frac{1}{2}Rg - \mathrm{Ric}$$

is equal to minus the **Einstein tensor**.

Now let

$$\lambda(\mathrm{Rm}) \geq \mu(\mathrm{Rm}) \geq \nu(\mathrm{Rm})$$

denote the eigenvalues of the curvature operator Rm in nonincreasing order.

Let $\mathbb{V}$ be a 3-dimensional real vector space and let $\mathbf{M}$ be a symmetric bilinear form on $\mathbb{V}$. The maximum eigenvalue is characterized by

$$\lambda(\mathbf{M}) = \max_{|\mathbf{v}|=1} \mathbf{M}(\mathbf{v}, \mathbf{v}).$$

The minimum eigenvalue is characterized by

$$\nu(\mathbf{M}) = \min_{|\mathbf{v}|=1} \mathbf{M}(\mathbf{v}, \mathbf{v}).$$

We also have

$$\mu(\mathbf{M}) = \mathrm{tr}(\mathbf{M}) - \lambda(\mathbf{M}) - \nu(\mathbf{M}).$$

Let $(\mathcal{M}^3, g)$ be a 3-dimensional Riemannian manifold, and let $p \in \mathcal{M}^3$. Let $V_p \in \Lambda^2 T_p^* \mathcal{M}$ be a unit eigen-2-form for Rm_p with eigenvalue $\nu(\mathrm{Rm}_p)$; that is, $|V_p| = 1$ and $\mathrm{Rm}(V_p) = \nu(\mathrm{Rm}_p)(V_p)$. Within the injectivity radius of g at p, parallel translate V_p along geodesics emanating from p to extend V_p to a unit 2-form field V_x defined for $x \in B_\varepsilon(p)$. Since ν is the minimum eigenvalue, we have that

$$(5.5) \quad \mathrm{Rm}(V_x, V_x) \geq \nu(\mathrm{Rm}_x) \quad \text{for } x \in B_\varepsilon(p), \qquad \mathrm{Rm}(V_p, V_p) - \nu(\mathrm{Rm}_p).$$

That is, we have that $\mathrm{Rm}(V_x, V_x)$ is an upper barrier for $\nu(\mathrm{Rm}_x)$ at $x = p$.

Let $\mathcal{S}$ denote the bundle of symmetric 2-tensors on $\mathcal{M}^3$. We have the map

$$(5.6) \qquad\qquad \vec{\kappa} := (\lambda, \mu, \nu) : \mathcal{S} \to \mathbb{R}^3.$$

The image of $\vec{\kappa}$ is contained in the subset $\{(x, y, z) \in \mathbb{R}^3 : x \geq y \geq z\}$. Since, for any Riemannian metric g, Rm is a section of $\mathcal{S}$, we have

$$\vec{\kappa}(\mathrm{Rm}) := \vec{\kappa} \circ \mathrm{Rm} : \mathcal{M}^3 \to \mathbb{R}^3.$$

Under the Ricci flow, the Riemann curvature operator $\mathrm{Rm} = \mathrm{Rm}_{g(t)}$ satisfies an equation of the form (see Hamilton [**174**, §2 and §5]; this is also Exercise 5.10)

$$(5.7) \qquad \frac{\partial}{\partial t} \mathrm{Rm} = \Delta \mathrm{Rm} + Q(\mathrm{Rm}),$$

where Δ is the rough Laplacian acting on symmetric 2-tensors and where the symmetric 2-tensor $Q(\mathrm{Rm})$ is a quadratic expression of Rm. Since $n = 3$, the quadratic Q satisfies

$$(5.8) \qquad Q \begin{pmatrix} \lambda & 0 & 0 \\ 0 & \mu & 0 \\ 0 & 0 & \nu \end{pmatrix} = \begin{pmatrix} \lambda^2 + \mu\nu & 0 & 0 \\ 0 & \mu^2 + \lambda\nu & 0 \\ 0 & 0 & \nu^2 + \lambda\mu \end{pmatrix}.$$

Since symmetric 2-tensors are diagonalizable (with respect to g) and since Q is invariant under conjugation, Q is determined by this formula. Here, we have also swept under the rug the so-called **Uhlenbeck trick** of pulling back time-dependent metrics on vector bundles to fixed metrics; see §A.3 and the original [**174**, §2]. Since we are employing Uhlenbeck's trick to calculate the evolution of Rm, the metrics on the fibers of $\mathcal{S}$ are independent of time.

The analysis of the corresponding system of ODEs

$$(5.9) \qquad \frac{d}{dt} \mathbf{M}(t) = Q(\mathbf{M}(t))$$

on $\mathbb{R}^3$, where $\mathbf{M} = \mathrm{diag}[\lambda, \mu, \nu] = (\lambda, \mu, \nu)$, was instrumental in Hamilton proving the following seminal result which began the subject of Ricci flow [**173**].

Theorem 5.2 (Hamilton, 1982). *For any initial metric with* Ric > 0 *on a closed 3-manifold, the volume-normalized Ricci flow exists for all time and the metric converges exponential fast in each C^k norm to a constant positive sectional curvature metric as $t \to \infty$.*

Consequently, any closed 3-manifold with Ric > 0 *is diffeomorphic to a spherical space form $\mathbb{S}^3/\Gamma$.*

Chen [**86**] proved a local curvature estimate that implies that any 3-dimensional complete ancient solution to the Ricci flow must have nonnegative sectional curvature. Related to this is the following result of Chen, Xu, and Zhang [**88**, Corollary 1.5], which is a consequence of a localization of the **Hamilton–Ivey estimate** ([**179**], [**192**]). Without loss of generality, we will assume that the eigenvalues of Rm satisfy

(5.10) $$\lambda \geq \mu \geq \nu.$$

Theorem 5.3. *Let $(\mathcal{M}^3, g(t))$ be a 3-dimensional complete solution to the Ricci flow on $[0, T]$ with positive scalar curvature. Then, at any point (x, T) with $\nu(\mathrm{Rm})(x, T) < 0$, we have[2]*

(5.11) $$R \geq |\nu(\mathrm{Rm})| \big(\ln \big(T|\nu(\mathrm{Rm})| \big) - 3 \big).$$

In particular, if $g(t)$, $t \in (-\infty, 0]$, is a 3-dimensional complete ancient solution, then $g(t)$ has nonnegative sectional curvature for all t.

Proof. Firstly, we observe that the last statement in the theorem, for ancient solutions, follows easily from (5.11). Indeed, given any t_0, considering this estimate for the solution $g(t)$ restricted to the time interval $[t_0 - T, t_0]$ and taking $T \to \infty$ yields a contradiction to the assumption that $\nu(\mathrm{Rm})(x, t_0) < 0$ for some point $x \in \mathcal{M}^3$. Thus, $g(t)$ has nonnegative sectional curvature for all $t \in (-\infty, 0]$.

The idea to prove (5.11) is to localize the original Hamilton–Ivey estimate. Let $x_0 \in \mathcal{M}^3$ and $A \geq 1$; later, we will first take A sufficiently large and then let A tend to infinity. Let $r_0 > 0$ be such that

(5.12) $$\mathrm{Ric} \leq r_0^{-2} g \quad \text{in } B(x_0, t, r_0) \quad \text{for all } t \in [0, T].$$

Clearly, given x_0, there exists such an r_0. The key is that the estimate we will prove is independent of r_0. Assume that initially,

(5.13) $$\nu(\mathrm{Rm}) > -N \quad \text{on } B(x_0, 0, 2Ar_0).$$

Let

$$K_- := \big\{ (\lambda, \mu, \nu) \in \mathbb{R}^3 : \lambda \geq \mu \geq \nu,\, \nu < 0 \big\} \subset \mathbb{R}^3.$$

Define the Hamilton–Ivey function $F : K_- \to \mathbb{R}$ by

(5.14) $$F(\lambda, \mu, \nu) := \frac{\lambda + \mu + \nu}{-\nu} - \ln(-\nu).$$

Define

$$\mathcal{K}_- := \vec{\kappa}^{-1}(K_-) \subset \mathcal{S},$$

where $\vec{\kappa} : \mathcal{S} \to \mathbb{R}^3$ is given by (5.6) and $\mathcal{S}$ is the vector bundle of symmetric 2-tensors on $\mathcal{M}^3$. That is, $\mathcal{K}_-$ is the set of symmetric 2-tensors whose smallest eigenvalue is negative. Note by Uhlenbeck's trick that the metric on the fibers of $\mathcal{S}$ is independent of time.

We have $F \circ \vec{\kappa} : \mathcal{K}_- \to \mathbb{R}$. In particular, for our solution to the Ricci flow, we have that the function

(5.15) $$F(x, t) := F\big(\vec{\kappa}(\mathrm{Rm})(x, t) \big)$$

---

[2]The original Hamilton–Ivey estimate implies this when $\mathcal{M}$ is compact.

is defined for all space-time points (x, t) in the set

$$\mathcal{M}_- := \left\{ (x, t) \in \mathcal{M}^3 \times [0, T] : \nu(\mathrm{Rm})(x, t) < 0 \right\} \subset \mathcal{M}^3 \times [0, T].$$

This is the space-time subset of points at which some sectional curvature is negative.

Under the ODE (5.9) on $\mathbb{R}^3$ corresponding to the Ricci flow evolution of Rm, we compute that the quantity F evolves by

$$
\begin{aligned}
(5.16) \quad \frac{d}{dt} F(\mathbf{M}(t)) &= \frac{\lambda^2 + \mu^2 + \nu^2 + \lambda\mu + \lambda\nu + \mu\nu}{-\nu} \\
&\quad + \frac{\lambda + \mu + \nu}{(-\nu)^2}(\nu^2 + \lambda\mu) + \frac{\nu^2 + \lambda\mu}{-\nu} \\
&= -2\nu + \frac{(\lambda + \mu)(\lambda + \mu + \nu)}{-\nu} + \frac{\lambda + \mu + \nu}{(-\nu)^2}(\nu^2 + \lambda\mu) \\
&= -2\nu + \frac{\lambda + \mu + \nu}{(-\nu)^2}(\lambda - \nu)(\mu - \nu) \\
&=: G(\mathbf{M}(t))
\end{aligned}
$$

for $\mathbf{M}(t) \in K_- \subset \mathbb{R}^3$. As we can see, $G > 0$ on K_-. We will use the calculation above for the ODE to estimate the solution Rm to the PDE.

Let $(x_1, t_1) \in \mathcal{M}_-$, and let V_{x_1, t_1} be a unit eigen-2-form for Rm_{x_1, t_1} associated to the eigenvalue $\nu(\mathrm{Rm}_{x_1, t_1})$. Let V_{x, t_1} be the unit 2-form field defined for x in a neighborhood of x_1 by (5.5) via parallel transport. We extend the space-defined 2-form field V_{x, t_1} in the time direction to be independent of time; that is, $V_{x, t} := V_{x, t_1}$. We have that

$$(5.17) \qquad \nabla V(x_1, t_1) = 0, \quad \nabla^2 V(x_1, t_1) = 0, \quad \partial_t V(x_1, t_1) = 0;$$

see Exercise 5.13. Furthermore, since ν is the minimum eigenvalue of Rm, we have that the locally defined function

$$(x, t) \mapsto \mathrm{Rm}(V_{x, t}, V_{x, t})$$

is an upper barrier for $\nu(\mathrm{Rm}_{x, t})$ at the point $(x, t) = (x_1, t_1)$.

We now compute the heat-type equation satisfied by $F = F(\vec{\kappa}(\mathrm{Rm}))$ in the barrier sense. Let (x_1, t_1) and $V_{x, t}$ be as above. It is easy to see that we have the upper barrier property:

$$(5.18) \qquad F(x, t) \leq \frac{R(x, t)}{-\mathrm{Rm}_{x, t}(V_{x, t}, V_{x, t})} - \ln\left(-\mathrm{Rm}_{x, t}(V_{x, t}, V_{x, t})\right),$$

with equality at $(x, t) = (x_1, t_1)$. Define the local functions

$$\tilde{\nu}(x, t) := \mathrm{Rm}_{x, t}(V_{x, t}, V_{x, t})$$

and

$$\tilde{F} := \frac{R}{-\tilde{\nu}} - \ln(-\tilde{\nu}).$$

By (5.18), we have

$$F(x,t) \le \tilde{F}(x,t), \quad F(x_1,t_1) = \tilde{F}(x_1,t_1).$$

That is, $\tilde{F}$ is an upper barrier for F.

At (x_1,t_1), we compute that

$$(5.19) \quad (\partial_t - \Delta)F \ge (\partial_t - \Delta)\tilde{F}$$
$$= \frac{(\partial_t - \Delta)R}{-\nu} + \frac{R}{(-\nu)^2}(\partial_t - \Delta)\tilde{\nu} + \frac{(\partial_t - \Delta)\tilde{\nu}}{-\nu}$$
$$- \frac{2}{(-\nu)^2}\nabla R \cdot \nabla \tilde{\nu} - \frac{2R}{(-\nu)^3}|\nabla \tilde{\nu}|^2 - \frac{1}{(-\nu)^2}|\nabla \tilde{\nu}|^2.$$

Let $\lambda := \lambda(\mathrm{Rm})$, $\mu := \mu(\mathrm{Rm})$, and $\nu := \nu(\mathrm{Rm})$. We proceed to calculate the right-hand side of (5.19) using the evolution equation (5.7) for Rm and the familiar equation

$$(\partial_t - \Delta)R = 2|\mathrm{Ric}|^2 = \lambda^2 + \mu^2 + \nu^2 + \lambda\mu + \lambda\nu + \mu\nu.$$

At (x_1,t_1), using (5.17) we compute that

$$(\partial_t - \Delta)\tilde{\nu}(x_1,t_1) = (\partial_t - \Delta)\big(\mathrm{Rm}(V_{x,t}, V_{x,t})\big)\big|_{(x,t)=(x_1,t_1)}$$
$$= \big((\partial_t - \Delta)\mathrm{Rm}\big)(V_{x,t}, V_{x,t})\big|_{(x,t)=(x_1,t_1)}$$
$$= Q(\mathrm{Rm})(V_{x_1,t_1}, V_{x_1,t_1})$$
$$= (\nu^2 + \lambda\mu)(x_1,t_1).$$

Using the evolution equations above, we calculate that $\tilde{F} = \tilde{F}(\vec{\kappa}(\mathrm{Rm}))$ satisfies at any space-time point (x_1,t_1) with $\nu(x_1,t_1) < 0$,

$$(5.20) \quad (\partial_t - \Delta)\tilde{F}(x_1,t_1) \ge G(\mathrm{Rm}_{x_1,t_1}) + 2\nabla \ln(-\tilde{\nu}) \cdot \nabla\tilde{F} + \frac{1}{(-\nu)^2}|\nabla \tilde{\nu}|^2,$$

where G is the curvature quantity defined by (5.16).

Now we localize this calculation. Let $\eta : \mathcal{M}^3 \times [0,T] \to [0,1]$ be a cutoff function of the form

$$(5.21) \qquad \eta(x,t) := \eta\left(\frac{2d_t(x,x_0)}{Ar_0} - 1\right),$$

where $\eta : \mathbb{R} \to [0,1]$ is smooth, nondecreasing, and satisfies $\eta|_{(-\infty,1]} \equiv 1$, $\eta|_{[2,\infty)} \equiv 0$, and

$$(5.22) \qquad |\eta''| + \frac{|\eta'|^2}{\eta} \le C$$

for some universal constant C. Define on $\mathcal{M}_-$ the function

$$(5.23) \qquad H(x,t) := F(x,t) + 3 + \ln\frac{N}{1+Nt}.$$

Consider the function ηH, which has compact support. Define the upper barrier $\tilde{H} := \tilde{F} + 3 + \ln \frac{N}{1+Nt}$ for H at (x_1, t_1). We compute at (x_1, t_1) that the localized Hamilton–Ivey function ηH satisfies

$$(5.24) \quad (\partial_t - \Delta)(\eta H) \geq (\partial_t - \Delta)(\eta \tilde{H})$$

$$= \eta(\partial_t - \Delta)\tilde{H} + \eta' \tilde{H} \frac{2}{Ar_0}(\partial_t - \Delta)d_t - \left(\frac{2}{Ar_0}\right)^2 \eta'' \tilde{H} - 2\nabla\eta \cdot \nabla\tilde{H},$$

where $\nabla\eta = \frac{2}{Ar_0}\eta'\nabla d_t$ and where we used that $|\nabla d_t| = 1$. Regarding the four terms on the right-hand side of (5.24):

(1) By (5.20) we have that the first term of (5.24) has the lower bound

$$(5.25) \qquad \eta(\partial_t - \Delta)\tilde{H} \geq \eta G(\mathrm{Rm}_{x_1,t_1}) - \frac{\eta N}{1+Nt} + \eta|\nabla\ln(-\tilde{\nu})|^2$$

$$+ 2\nabla\ln(-\tilde{\nu}) \cdot \nabla(\eta\tilde{H}) - 2H\nabla\ln(-\tilde{\nu}) \cdot \nabla\eta.$$

(2) Recall that we chose $r_0 > 0$ so that $\mathrm{Ric} \leq r_0^{-2}g$ in $B(x_0, t, r_0)$ for all $t \in [0, T]$. By Perelman's changing distances formula [**250**, Lemma 8.3(a)], we have in the barrier sense that

$$(\partial_t - \Delta)d_t \geq -\frac{8}{3}r_0^{-1} \quad \text{on } \mathcal{M}^3 \setminus B(x_0, t, r_0).$$

Since $\eta' \leq 0$, at any point where $H \leq 0$ we have that the second term satisfies

$$\eta'\tilde{H}\frac{2}{Ar_0}(\partial_t - \Delta)d_t \geq -\eta'\tilde{H}\frac{2}{Ar_0}\frac{8}{3}r_0^{-1} = -\frac{16}{3Ar_0^2}\eta'\tilde{H}.$$

(3) The third term of (5.24) is $-\left(\frac{2}{Ar_0}\right)^2 \eta'' \tilde{H}$.

(4) We express the fourth term of (5.24) as

$$-2\nabla\eta \cdot \nabla\tilde{H} = -2\nabla\ln\eta \cdot \nabla(\eta\tilde{H}) + 2\left(\frac{2}{Ar_0}\right)^2 \frac{(\eta')^2}{\eta}\tilde{H}.$$

Collecting those right-hand side terms of the heat-type equation (5.24) for ηH involving derivatives of η (but not including the $\nabla(\eta\tilde{H})$ terms, which vanish at a minimum point of ηH), we see that they are bounded from below by

(5.26)

$$\left(-\frac{16}{3Ar_0^2}\eta' - \left(\frac{2}{Ar_0}\right)^2 \eta'' + 2\left(\frac{2}{Ar_0}\right)^2 \frac{(\eta')^2}{\eta}\right)\tilde{H} - \left(\frac{2}{Ar_0}\right)^2 \frac{(\eta')^2}{\eta}\tilde{H}^2,$$

where we have dropped the nonnegative term $\eta \left| \nabla \ln(-\tilde{\nu}) - \frac{\nabla \eta}{\eta} \tilde{H} \right|^2$. The remaining terms are

$$(5.27) \qquad \eta G(\mathrm{Rm}_{x_1,t_1}) - \eta \frac{N}{1+Nt} \geq \eta \left(-2\nu - \frac{N}{1+Nt} \right),$$

where we were able to drop the $\eta \frac{(\lambda-\nu)(\mu-\nu)}{(-\nu)^2} R$ term since $R > 0$.

Now we are in a position to carry out the maximum principle argument. Let (x_1, t_1) be a minimum point of ηH on $\mathcal{M}_-$, so that

$$(5.28) \qquad \eta H(x,t) \geq \eta H(x_1, t_1) \quad \text{for all } (x,t) \in \mathcal{M}_-.$$

Then, by the parabolic maximum principle, i.e., by $(\partial_t - \Delta)(\eta H) \leq 0$ and $\nabla(\eta H) = 0$ at (x_1, t_1), the sum of the terms in (5.26) and (5.27) is nonpositive. By applying to this the estimates in (5.22) for the derivatives of the cutoff function η, we obtain at (x_1, t_1) that

$$(5.29) \qquad 0 \geq \eta \left(-2\nu - \frac{N}{1+Nt_1} \right) + \left(\frac{C}{Ar_0^2} + \frac{C}{A^2 r_0^2} \right) H - \frac{C}{A^2 r_0^2} H^2,$$

since $\tilde{H} = H$ at (x_1, t_1).

Because

$$0 > H(x_1, t_1) = \frac{R}{-\nu}(x_1, t_1) - \ln(-\nu)(x_1, t_1) + 3 + \ln \frac{N}{1+Nt_1}$$

$$> -\ln(-\nu)(x_1, t_1) + \ln \frac{N}{1+Nt_1},$$

we have

$$(5.30) \qquad -\nu(x_1, t_1) > e^{-H(x_1,t_1)} \frac{N}{1+Nt_1} > \frac{N}{1+Nt_1} \geq \frac{N}{1+NT}$$

and at (x_1, t_1),

$$(5.31) \qquad -H = |H| < \ln(-\nu) - \ln \frac{N}{1+Nt_1}.$$

By applying this to (5.29), we have at (x_1, t_1) the inequality

$$(5.32) \qquad 0 \geq -\eta\nu + \left(\frac{C}{Ar_0^2} + \frac{C}{A^2 r_0^2} \right) H - \frac{C}{A^2 r_0^2} H^2.$$

We now compare the identity function of $\mathbb{R}$ to the logarithm function. For $k \geq 1$, define

$$(5.33) \qquad \phi_k(u) := \frac{u}{\left(\ln(u) - \ln\left(\frac{N}{1+Nt_1} \right) \right)^k}$$

for $u > \frac{N}{1+Nt_1}$. Note that ϕ_k attains its minimum at $u = e^k \frac{N}{1+Nt_1}$. Therefore, for all $u > \frac{N}{1+Nt_1}$, we have

$$\phi_k(u) \geq \phi_k\left(e^k \frac{N}{1+Nt_1}\right) = \left(\frac{e}{k}\right)^k \frac{N}{1+Nt_1}. \tag{5.34}$$

Since $-\nu(x_1, t_1) > \frac{N}{1+Nt_1}$ by (5.30), we obtain from (5.34) that at (x_1, t_1),

$$
\begin{aligned}
-\nu &\geq \left(\frac{e}{k}\right)^k \frac{N}{1+Nt_1} \left(\ln(-\nu) - \ln\left(\frac{N}{1+Nt_1}\right)\right)^k \\
&> \left(\frac{e}{k}\right)^k \frac{N}{1+Nt_1} |H|^k,
\end{aligned}
\tag{5.35}
$$

where the last equality follows from (5.31). Taking $k = 2$ and $k = 3$, we obtain

$$-\nu > \frac{e^2}{8} \frac{N}{1+Nt_1} |H|^2 + \frac{e^3}{54} \frac{N}{1+Nt_1} |H|^3. \tag{5.36}$$

Applying this inequality to (5.32) yields

$$0 \geq -\frac{2C}{Ar_0^2}|H| - \frac{C}{A^2 r_0^2}|H|^2 + \eta \frac{e^2}{8} \frac{N}{1+Nt_1}|H|^2 + \eta \frac{e^3}{54} \frac{N}{1+Nt_1}|H|^3,$$

where we also assume that $A \geq 1$. That is, at (x_1, t_1),

$$0 \geq |H|\left(-\frac{2C}{Ar_0^2} + \frac{e^2}{8} \frac{N}{1+Nt_1}\eta|H|\right) + |H|^2\left(-\frac{C}{A^2 r_0^2} + \frac{e^3}{54} \frac{N}{1+Nt_1}\eta|H|\right).$$

Thus,

$$\eta|H|(x_1, t_1) \leq \frac{1+NT}{N} \max\left\{\frac{16C}{e^2 A r_0^2}, \frac{54C}{e^3 A^2 r_0^2}\right\}. \tag{5.37}$$

This implies that on $\mathcal{M}_-$ we have the estimate

$$\eta H(x, t) \geq \eta H(x_1, t_1) \geq -\Psi(A^{-1}|N, r_0, T), \tag{5.38}$$

where $\Psi(A^{-1}|N, r_0, T) \to 0$ for $A^{-1} \to 0$ and N, r_0, T fixed. Thus, by taking $A \to \infty$, we obtain

$$\frac{R}{-\nu} - \ln(-\nu) + 3 + \ln \frac{N}{1+Nt} = H(x, t) \geq 0$$

on $\mathcal{M}_-$. Since $\ln \frac{N}{1+Nt} \leq -\ln t$, this completes the proof of (5.11) and hence the theorem. $\qquad\square$

In view of the fact that the canonical forms of shrinking and steady GRS are ancient, the following consequence is immediate.

Corollary 5.4. *Any 3-dimensional complete shrinking or steady GRS must have nonnegative sectional curvature.*

5.2. Classification of 3-dimensional shrinking GRS

5.2.1. Proof of the main classification theorem.

By applying Hamilton's strong maximum principle for systems to the curvature evolution equation (5.7), we obtain the following (see Hamilton [**174**, §8]).

Proposition 5.5. *Suppose that* $(\mathcal{M}^n, g(t))$, $t \in [0, T]$, *is a solution to the Ricci flow on a connected manifold with nonnegative curvature operator* $\mathrm{Rm}_{g(t)} \geq 0$.[3] *Then:*

(1) *For each* $t \in (0, T]$, $\mathrm{image}(\mathrm{Rm}_{g(t)})$ *is a smooth subbundle of* $\Lambda^2 T^* \mathcal{M}$ *which is invariant under parallel translation.*

(2) *There exist times*

$$0 = t_0 < t_1 < t_2 < \cdots < t_k = T$$

such that $\mathrm{image}(\mathrm{Rm}_{g(t)}(x))$ *is a Lie subalgebra of* $\Lambda^2 T_x^* \mathcal{M} \cong \mathfrak{so}(n)$ *independent of time on* $(t_{i-1}, t_i]$ *for all* $x \in \mathcal{M}^n$ *and* $1 \leq i \leq k$. *Here, the Lie algebra structure on* $\Lambda^2 T_x^* \mathcal{M}$ *is given by* $[\alpha, \beta]_{ij} := g^{k\ell}(\alpha_{ik}\beta_{\ell j} - \beta_{ik}\alpha_{\ell j})$.

(3) *We have*

(5.39) $\mathrm{image}(\mathrm{Rm}_{g(t_i)}) \subset \mathrm{image}(\mathrm{Rm}_{g(t_j)})$ *for all* $0 \leq i < j \leq k$.

For an example of an incomplete solution where the inclusions in (5.39) are proper, see [**111**, Exercise 6.63]; the solution is presented on pp. 563–564 therein.

Applying Proposition 5.5 to GRS, we have:

Corollary 5.6. *If* $\left(\mathcal{M}^n, g(t), f(t), \lambda/\tau(t)\right)$, $\lambda t < 1$, *is the canonical form of a complete GRS* $(\mathcal{M}^n, g, f, \lambda)$ *with* $\mathrm{Rm}_g \geq 0$ *(as given by Proposition 4.19), then* $\mathrm{image}(\mathrm{Rm}_{g(t)})$ *is a smooth subbundle of* $\Lambda^2 T^* \mathcal{M}$ *which is invariant under parallel transport, independent of time, and such that* $\mathrm{image}(\mathrm{Rm}_{g(t)}(x))$ *is a Lie subalgebra of* $\Lambda^2 T_x^* \mathcal{M}$.

Specializing to the $n = 3$ case, we obtain:

Theorem 5.7. *Any 3-dimensional complete GRS* $(\mathcal{M}^3, g, f, \lambda)$ *has one of the following properties:*

(1) *The metric* g *is flat. In the case of a shrinker or an expander, the GRS must be a Gaussian soliton.*

(2) *The metric* g *has positive sectional curvature. Such shrinkers must be (compact) spherical space forms. Such steadies and expanders must be noncompact.*

---

[3] We do not assume that the solution is complete or has bounded curvature.

(3) *The universal covering GRS* $(\tilde{\mathcal{M}}^3, \tilde{g}, \tilde{f}, \lambda)$ *splits as the product of the 1-dimensional Gaussian GRS and a nonflat 2-dimensional GRS. In the case of a shrinker, this is a round* $\mathbb{S}^2$ *or its* $\mathbb{Z}_2$*-quotient* $\mathbb{R}P^2$. *In the case of a steady, this is the cigar soliton.*

Proof. Consider the canonical form of the GRS $(\mathcal{M}^3, g(t), f(t), \lambda/\tau(t))$, $\lambda t < 1$, and fix some t. By Corollary 5.6, image($\mathrm{Rm}_{g(t)}$) is invariant under parallel transport, and image($\mathrm{Rm}_{g(t)}(x)$) is a Lie subalgebra of $\Lambda^2 T_x^* \mathcal{M}$ for each $x \in \mathcal{M}^3$, which is isomorphic to $\mathfrak{so}(3)$. On the other hand, any Lie subalgebra of $\mathfrak{so}(3)$ is either $\{0\}$, $\mathfrak{so}(3)$, or isomorphic to $\mathfrak{so}(2)$. We now consider each of these three cases.

(1) image($\mathrm{Rm}_{g(t)}(x)$) = $\{0\}$ for all $x \in \mathcal{M}^3$. Then $\mathrm{Rm}_{g(t)} \equiv 0$; i.e., $g(t)$ is flat. In the shrinking and expanding cases, the soliton must be Gaussian. This is by the equality case of Theorem 2.14 in the shrinking case. On the other hand, any flat metric is a steady GRS.

(2) image($\mathrm{Rm}_{g(t)}(x)$) $\cong \mathfrak{so}(3)$. Then image($\mathrm{Rm}_{g(t)}(x)$) = $\Lambda^2 T_x^* \mathcal{M}$ for all $x \in \mathcal{M}^3$. Since g has nonnegative curvature operator, this implies that $g(t)$ has positive curvature operator.

In the shrinking case, by Theorem 4.47, $\mathcal{M}^3$ must be compact. Hence, by Theorem 5.2, in this case $g(t)$ must have constant positive sectional curvature. On the other hand, since 3-dimensional compact steady and expanding GRS must have constant nonpositive curvature, which is a contradiction, in these cases $\mathcal{M}^3$ must be noncompact.

(3) image($\mathrm{Rm}_{g(t)}(x)$) $\cong \mathfrak{so}(2)$, which is 1-dimensional. Then there exists a parallel unit 2-form $\alpha \in$ image($\mathrm{Rm}_{g(t)}$) which is independent of time. Let $*$ denote the Hodge star operator with respect to $g(t)$. Thus the metric dual $(*\alpha)^\natural$ of the Hodge star of α is a unit vector field on $\mathcal{M}^3$ parallel with respect to the metric $g(t)$ for each t. Hence, by the de Rham splitting theorem, the universal covering GRS $(\tilde{\mathcal{M}}^3, \tilde{g}(t), \tilde{f}(t))$ is the product of $\mathbb{R}$ and a 2-dimensional GRS with positive curvature.

In the shrinking case, by Theorems 3.12 and 3.5, the 2-dimensional GRS must be a round shrinking $\mathbb{S}^2$. $\qquad\square$

In particular, in the 3-dimensional shrinking case, the theorem says:

Corollary 5.8. *Any 3-dimensional complete shrinking GRS* $(\mathcal{M}^3, g, f)$ *is either a quotient of a round 3-sphere, a quotient of a round* $\mathbb{S}^2 \times \mathbb{R}$, *or the Gaussian shrinker. If* $\mathcal{M}^3$ *is orientable, then the only nontrivial quotient of* $S^2 \times \mathbb{R}$ *is by the* $\mathbb{Z}_2$*-action generated by the antipodal map on each factor.*

The noncollapsed positively curved 3-dimensional steady case will be the subject of Chapter 8. Collapsed nonproduct steady examples, as well as results for expanders, all in dimension 3, are discussed in Chapter 7.

5.2.2. Proof of the 3-dimensional version of Theorem 4.47 via a Bochner formula.

In the rest of this section, out of independent interest, we consider the earlier proof of the 3-dimensional case of Theorem 4.47. Originally, Perelman proved this result in the 3-dimensional bounded curvature case.

Theorem 5.9. *There does not exist a noncompact 3-dimensional shrinking GRS with positive sectional curvature.*

The following proof is due to Ni and Wallach [**247**], which is an illustration of how Bochner formulas can be powerful. Assume that $(\mathcal{M}^3, g)$ has positive sectional curvature. Let

$$F := \frac{|\mathrm{Ric}|^2}{R^2} - \frac{1}{3} \geq 0,$$

which is a scale-invariant quantity measuring the difference from being Einstein (and hence constant curvature since we are in dimension 3). Since $\mathrm{Rm} \geq 0$, which for $n = 3$ is equivalent to $0 \leq \mathrm{Ric} \leq \frac{1}{2}Rg$, we have $|\mathrm{Ric}|^2 \leq \left|\frac{1}{2}Rg\right|^2 = \frac{3}{4}R^2$. Hence $F \leq \frac{5}{12}$.

By taking $v = \mathrm{Ric}$ in the evolution equation (8.171) and formula (8.172) below, we obtain Hamilton's formula that under the Ricci flow on a 3-manifold, F satisfies

$$(5.40) \quad \frac{\partial F}{\partial t} = \Delta F + \frac{2}{R} \langle \nabla R, \nabla F \rangle - \frac{2}{R^4} |R\nabla \mathrm{Ric} - \nabla R \otimes \mathrm{Ric}|^2 - \frac{4}{R^3} J,$$

where

$$(5.41) \quad J := |\mathrm{Ric}|^4 + \frac{R}{2} \left(R^3 - 5R|\mathrm{Ric}|^2 + 4\,\mathrm{tr}\left(\mathrm{Ric}^3\right) \right) \geq 0.$$

For a shrinking GRS, since F is a tensor invariant of g of degree 0, by Proposition 4.19 and Lemma 4.27 we have

$$(5.42) \quad \Delta_f F + \frac{2}{R} \langle \nabla R, \nabla F \rangle - \frac{2}{R^4} |R\nabla \mathrm{Ric} - \nabla R \otimes \mathrm{Ric}|^2 - \frac{4}{R^3} J = 0.$$

We can also compute (5.40) in local coordinates. The evolution of the Ricci tensor, using Uhlenbeck's trick (see §5.4.2), is given by (see Exercise 5.4)

$$(5.43) \quad (\partial_t - \Delta)R_{ij} = 2R_{ik\ell j}R_{k\ell}$$
$$= 3RR_{ij} + (2|\mathrm{Ric}|^2 - R^2)g_{ij} - 4R_{ij}^2,$$

where the second equality holds since being in dimension 3 implies that

$$(5.44) \quad R_{ik\ell j} = R_{ij}g_{k\ell} + R_{k\ell}g_{ij} - R_{i\ell}g_{kj} - R_{kj}g_{i\ell} - \frac{1}{2}R(g_{ij}g_{k\ell} - g_{i\ell}g_{kj}).$$

Thus, keeping in mind that we are using Uhlenbeck's trick, we compute that

$$(5.45) \quad \frac{1}{2}(\partial_t - \Delta)|\mathrm{Ric}|^2 = R_{ij}(\partial_t - \Delta)R_{ij} - |\nabla\mathrm{Ric}|^2$$
$$= -|\nabla\mathrm{Ric}|^2 + 5R|\mathrm{Ric}|^2 - R^3 - 4\,\mathrm{tr}(\mathrm{Ric}^3).$$

By combining this with

$$(5.46) \qquad\qquad \frac{1}{2}(\partial_t - \Delta)R^2 = -|\nabla R|^2 + 2R|\mathrm{Ric}|^2$$

(see Exercise 5.8) and the general formula

$$(5.47) \qquad \frac{1}{2}(\partial_t - \Delta)\frac{h}{k} = \frac{1}{2k}(\partial_t - \Delta)h - \frac{h}{2k^2}(\partial_t - \Delta)k + \frac{1}{k}\left\langle \nabla k, \nabla\frac{h}{k}\right\rangle$$

for any functions h and $k \neq 0$, we obtain

$$(5.48)$$
$$\frac{1}{2}(\partial_t - \Delta)\frac{|\mathrm{Ric}|^2}{R^2} - \frac{1}{R^2}(-|\nabla\mathrm{Ric}|^2 + 5R|\mathrm{Ric}|^2 - R^3 \quad 4\,\mathrm{tr}(\mathrm{Ric}^3))$$
$$- \frac{|\mathrm{Ric}|^2}{R^4}(-|\nabla R|^2 + 2R|\mathrm{Ric}|^2) \mid \frac{1}{R^2}\left\langle \nabla(R^2), \nabla\frac{|\mathrm{Ric}|^2}{R^2}\right\rangle.$$

By combining terms, we obtain (5.40).

Observe that the leading order terms of the Bochner formula (5.42) are $\Delta F + \langle\nabla(2\ln R - f), \nabla F\rangle$ and that $e^{2\ln R - f} = R^2 e^{-f}$. By multiplying equation (5.40) by $(F + \frac{1}{3})R^2 e^{-f} = |\mathrm{Ric}|^2 e^{-f}$ and integrating by parts, which is justified when $\mathcal{M}^3$ is noncompact in the next paragraph, we obtain the integral identity

$$(5.49)$$
$$\int_{\mathcal{M}}\left(|\nabla F|^2 R^2 + \frac{2\,|\mathrm{Ric}|^2}{R^4}|R\nabla\mathrm{Ric} - \nabla R \otimes \mathrm{Ric}|^2 + \frac{4\,|\mathrm{Ric}|^2}{R^3}J\right)e^{-f}d\mu = 0.$$

Since $J \geq 0$, we conclude from this that $\nabla F = 0$ so that F is constant,

$$(5.50) \qquad\qquad\qquad J = 0,$$

and

$$R\nabla\mathrm{Ric} - \nabla R \otimes \mathrm{Ric} = 0.$$

By applying $\mathrm{tr}^{1,2}$, i.e., tracing over the first two components, to the latter equation and using the contracted second Bianchi identity, we obtain

$$\left(\frac{R}{2}g - \mathrm{Ric}\right)(\nabla R) = 0.$$

Since $\frac{R}{2}g - \text{Ric} > 0$, we have $\nabla R = 0$. Hence $\nabla \text{Ric} = 0$. Since this implies that the eigenvalues of Ric are constant (and positive), by Myers's theorem we have that $\mathcal{M}^3$ is compact. Since $(\mathcal{M}^3, g)$ is a compact locally symmetric space with positive sectional curvature, we conclude that g has constant sectional curvature.

Alternatively, J defined by (5.41) may be expressed as

$$(5.51) \qquad 4J = \lambda^2\mu^2 + \lambda^2\nu^2 + \mu^2\nu^2 - \lambda^2\mu\nu - \mu^2\lambda\nu - \nu^2\lambda\mu$$

$$= \frac{1}{2}\left(\lambda^2(\mu - \nu)^2 + \mu^2(\lambda - \nu)^2 + \nu^2(\lambda - \mu)^2\right).$$

Since $J = 0$ by (5.50) and since g has positive sectional curvature, we conclude that at each point,

$$\lambda(\text{Rm}) = \mu(\text{Rm}) = \nu(\text{Rm}).$$

This implies, again, that g has constant sectional curvature.

To prove (5.49) when $\mathcal{M}^3$ is noncompact, it suffices to show that

$$(5.52) \qquad \int_{\mathcal{M}} \text{div}\left(e^{-f} |\text{Ric}|^2 \nabla F\right) d\mu = 0.$$

Note that $\nabla F = \frac{\nabla|\text{Ric}|^2}{R^2} - \frac{2|\text{Ric}|^2\nabla R}{R^3}$. Equation (5.52) is easily seen to be true for the following reasons:

(1) Exhaust $\mathcal{M}^3$ by a sequence of balls $B_{r_i}(p)$, where $r_i \to \infty$ and $\text{Area}(\partial B_{r_i}(p))$ grows polynomially in r_i. For simplicity, assume that each $\partial B_{r_i}(p)$ is a smooth compact hypersurface. Then

$$\int_{B_{r_i}(p)} \text{div}\left(e^{-f} |\text{Ric}|^2 \nabla F\right) d\mu = \int_{\partial B_{r_i}(p)} e^{-f} |\text{Ric}|^2 \langle \nabla F, \nu\rangle d\sigma,$$

where ν is the outward unit normal vector field.

(2) By Theorem 4.3, e^{-f} decays exponentially quadratically.

(3) By (4.51), the Jacobian of the exponential map satisfies $J(r) \leq Cr^{n-1}$ (this controls the area and volume forms).

(4) Since $\text{Rm} \geq 0$, we have $|\text{Rm}| \leq \sqrt{3}R$, which grows at most quadratically by (2.63).

(5) By Shi's local derivative estimates (see Lemma 4.24) and by (3), we have that $|\nabla \text{Rm}|$ grows at most cubically.

(6) By Theorem 4.31 and $R > 0$, we have that R^{-1} grows at most quadratically.

5.3. Notes and commentary

Hamilton classified 2-dimensional shrinking GRS in [**175**]. By the Hamilton–Ivey estimate ([**179**], [**192**]), ancient solutions to the Ricci flow with bounded curvature must have nonnegative sectional curvature. For this result, the hypothesis of bounded curvature was removed by B.-L. Chen [**86**]. Consequently, by the strong maximum principle, the 2-dimensional classification yields a 3-dimensional classification in all cases except where the sectional curvatures are positive everywhere. For this last positive-curvature case, Perelman [**250**] proved nonexistence assuming bounded curvature. The bounded curvature condition was removed by Cao, Chen, and Zhu [**63**] and Ni and Wallach [**247**] and Peterson and Wylie [**256**]. Munteanu and Wang [**242**] extended this result to all dimensions, which includes a new proof of the 3-dimensional result.

5.4. Exercises

5.4.1. Evolution of the curvature, in particular in dimension 3.

Exercise 5.1 (Variation of the Ricci tensor, I). In local coordinates,

$$(5.53) \qquad R_{ijk}^{\ell} = \partial_i \Gamma_{jk}^{\ell} - \partial_j \Gamma_{ik}^{\ell} + \sum_{p=1}^{n} \left(\Gamma_{jk}^{p} \Gamma_{ip}^{\ell} - \Gamma_{ik}^{p} \Gamma_{jp}^{\ell} \right),$$

and the Ricci tensor is $R_{ij} = \sum_{q=1}^{n} R_{qij}^{q}$. Suppose that $y(s)$ is a 1-parameter family of metrics satisfying $\partial_s g_{ij} = -2v_{ij}$, where v is a symmetric 2-tensor.

(a) Prove that

$$(5.54) \qquad \partial_s R_{ij} = \nabla_p \left(\frac{\partial}{\partial s} \Gamma_{ij}^{p} \right) - \nabla_i \left(\frac{\partial}{\partial s} \Gamma_{pj}^{p} \right).$$

(b) Show that if $\partial_s g_{ij} = -2v_{ij}$, then

$$(5.55) \qquad \partial_s R_{ij} = -\nabla_\ell \left(\nabla_i v_{j\ell} + \nabla_j v_{i\ell} - \nabla_\ell v_{ij} \right) + \nabla_i \nabla_j V,$$

where $V = g^{ij} v_{ij}$ is the trace of v.
HINT: Recall that in Exercise 1.7 we calculated the variation of Christoffel symbols formula.

Exercise 5.2. (a) Prove that if $\partial_s g_{ij} = -2v_{ij}$, then

$$(5.56) \qquad \partial_s R = 2\Delta V - 2\operatorname{div}(\operatorname{div} v) + 2\langle v, \operatorname{Ric} \rangle.$$

(b) Show that under the Ricci flow,

$$(5.57) \qquad (\partial_t - \Delta) R = 2|\operatorname{Ric}|^2.$$

(c) Prove formula (5.46) for the evolution of R^2 under the Ricci flow.

Exercise 5.3 (Variation of the Ricci tensor, II). Using Exercise 5.1, prove the Ricci tensor variation formula which says that if $\partial_s g_{ij} = -2v_{ij}$, then

$$(5.58) \qquad\qquad \partial_s \operatorname{Ric} = \Delta_L v + \mathcal{L}_Z g,$$

where the Lichnerowicz Laplacian is defined by (4.79):

$$(\Delta_L v)_{ij} := \Delta_L v_{ij} := \Delta v_{ij} + 2R_{kij\ell} v_{k\ell} - R_{ik} v_{jk} - R_{jk} v_{ik},$$

where $Z := \frac{1}{2}\nabla V - \operatorname{div} v$ and where $V := \operatorname{tr} v$.

As a special case of this formula, we have that under the Ricci flow $\partial_t g_{ij} = -2R_{ij}$,

$$(5.59) \qquad\qquad \partial_t \operatorname{Ric} = \Delta_L \operatorname{Ric}.$$

Exercise 5.4 (Evolution of Ric under Ricci flow in dimension 3). Show when $n = 3$ that, under the Ricci flow,

$$(5.60) \qquad (\partial_t - \Delta)R_{ij} = 3RR_{ij} + (2|\operatorname{Ric}|^2 - R^2)g_{ij} - 6R_{ij}^2.$$

HINT: Use (5.44).

Exercise 5.5. Let $(\mathcal{M}^n, g(t))$ be a solution to the Ricci flow. Suppose that a 2-tensor α satisfies $\partial_t \alpha_{ij} = (\Delta\alpha)_{ij} + \beta_{ij}$. Prove that

$$(5.61) \qquad (\partial_t - \Delta)|\alpha|^2 = -2|\nabla\alpha|^2 + 2\langle\alpha,\beta\rangle + 2R_{ij}\alpha_{ik}\beta_{jk} + 2R_{ij}\alpha_{ki}\beta_{kj},$$

where the norms and inner products are with respect to $g(t)$.

Exercise 5.6. Prove (5.45) for the evolution of $|\operatorname{Ric}|^2$ under the Ricci flow.

5.4.2. Uhlenbeck's trick. The reader may consult §A.3 as necessary.

Exercise 5.7 (Uhlenbeck's trick). We consider Uhlenbeck's trick in the case where $\mathcal{M}^n$ is parallelizable, i.e., the tangent bundle $T\mathcal{M}$ is trivial as a vector bundle. (This holds in particular when $\mathcal{M}^n$ is compact and 3-dimensional, i.e., $n = 3$.) Let $g(t)$, $t \in [0, T)$, be a solution to the Ricci flow on $\mathcal{M}^n$. Since $\mathcal{M}^n$ is parallelizable, there exists a global frame field $\{(e_1)_0, \ldots, (e_n)_0\}$ which is orthonormal with respect to $g(0)$. We evolve this frame field by the system of ODEs

$$(5.62a) \qquad\qquad \frac{d}{dt}e_a(t) = \operatorname{Ric}_t(e_a(t)),$$

$$(5.62b) \qquad\qquad e_a(0) = (e_a)_0,$$

for $1 \le a \le n$.

(a) Show that $\{e_a(t)\}_{a=1}^n$ is orthonormal for all $t \in [0, T]$.

(b) Prove that if α_t, $t \in [0, T]$, is a covariant k-tensor on $\mathcal{M}^n$, then

(5.63)

$$\partial_t\big(\alpha_t(e_{a_1}(t), \ldots, e_{a_k}(t))\big) = (\partial_t \alpha_t)\big(e_{a_1}(t), \ldots, e_{a_k}(t)\big)$$
$$+ \sum_{j=1}^k \alpha_t\big(e_{a_1}(t), \ldots, \operatorname{Ric}_t(e_{a_j}(t)), \ldots, e_{a_k}(t)\big).$$

Exercise 5.8. Assume the same hypotheses as in Exercise 5.7. Define the bundle isomorphisms $\iota_t : E := \mathcal{M}^n \times \mathbb{R}^n \to T\mathcal{M}$, $t \in [0, T]$, by

(5.64) $\qquad \iota_t(x, V^1, \ldots, V^n) = \big(V^1 e_1(x, t) + \cdots + V^n e_n(x, t)\big).$

Let $D := \iota^* \nabla$ be the pullback of the Levi-Civita connection on $T\mathcal{M}$ to E. Let D also denote the associated connections on the tensor product bundles of E and E^*. Let $\Delta_D := \operatorname{tr}_{g(t)} D^2$ be the Laplacian acting on these vector bundles.

(a) Explain why if α_t is a covariant 2-tensor on $\mathcal{M}^n$, then $\iota_t^* \alpha_t$ is equivalent to a time dependent function on $\mathcal{M}^n$ with values in the vector space of symmetric $n \times n$ matrices.

(b) Prove that under the Ricci flow, using Uhlenbeck's trick, the Ricci tensor evolves by

(5.65) $\qquad (\partial_t - \Delta_D)(\iota_t^* \operatorname{Ric}_t)_{ab} = 2(\iota_t^* \operatorname{Rm}_t)_{acdb}(\iota_t^* \operatorname{Ric}_t)_{cd}.$

Exercise 5.9. Derive (5.43):

$$(\partial_t - \Delta)R_{ij} = 2R_{ik\ell j} R_{k\ell}$$

for the Uhlenbeck trick version of the evolution of Ric.

Exercise 5.10 (Evolution of Rm under Ricci flow in dimension 3). Here, we consider Rm as a 2-tensor. Compute, using Uhlenbeck's trick, that

(5.66) $\quad (\partial_t - \Delta)(Rg_{ij} - 2R_{ij}) = -6RR_{ij} + (-2|\operatorname{Ric}|^2 + 2R^2)g_{ij} + 8R_{ij}^2$

(5.67) $\qquad\qquad\qquad\qquad\qquad =: Q_{ij}.$

Show that in local coordinates and at a point where $g_{ij} = \delta_{ij}$ and where Rm, and hence Ric, is diagonal, we have

(5.68) $\qquad\qquad\qquad Q_{ii} = \lambda_i^2 + \lambda_j \lambda_k,$

where i, j, k is a cyclic permutation of $1, 2, 3$. Explain why this proves formula (5.7) for the evolution of Rm.

Exercise 5.11. Prove formula (5.47):

$$\frac{1}{2}(\partial_t - \Delta)\left(\frac{h}{k}\right) = \frac{1}{2k}(\partial_t - \Delta)h - \frac{h}{2k^2}(\partial_t - \Delta)k + \frac{1}{k}\nabla k \cdot \nabla\left(\frac{h}{k}\right)$$

for the heat operator acting on the quotient of two functions h and k.

Exercise 5.12. Let $F := \frac{|\mathrm{Ric}|^2}{R^2} - \frac{1}{3}$. Prove, for the curvature quartic J arising in the monotonicity formula (5.40), formula (5.51) which says that

$$4J = \frac{1}{2}\left(\lambda^2(\mu - \nu)^2 + \mu^2(\lambda - \nu)^2 + \nu^2(\lambda - \mu)^2\right).$$

5.4.3. Maximum principle for systems.

Exercise 5.13 (Local extension of a vector). Let $(\mathcal{M}^n, g)$ be a Riemannian manifold, let $p \in \mathcal{M}^n$, and let $X_p \in T_p\mathcal{M}$. Let $r := \mathrm{inj}_g(p)$ and parallel transport X_p along unit-speed geodesics $\gamma_V : [0, r) \to \mathcal{M}^n$ emanating from p, $V \in T_p\mathcal{M}$, to define X smoothly in $B_p(r)$. Of course,

$$\nabla X(p) = 0.$$

Show that

$$\nabla^2 X(p) = 0,$$

and in particular, $\Delta X(p) = 0$.

Exercise 5.14. Based on the discussion in §5.2.2, give a complete proof of Theorem 5.9 on the classification of 3-dimensional shrinkers.

5.4.4. Perelman's energy monotonicity.

Given a closed Riemannian manifold $(\mathcal{M}^n, g)$ and a function f on $\mathcal{M}^n$, **Perelman's $\mathcal{F}$-energy** is defined by

$$(5.69) \qquad \mathcal{F}(g, f) := \int_{\mathcal{M}} (R + |\nabla f|^2)\, e^{-f} d\mu.$$

Recall from (2.47) that the f-scalar curvature is defined by

$$R_f = R + 2\Delta f - |\nabla f|^2.$$

Observe that

$$(5.70) \qquad \mathcal{F}(g, f) = \int_{\mathcal{M}} R_f\, e^{-f} d\mu.$$

Exercise 5.15 (Perelman's energy monotonicity [**250**, §3 and §9]). Let $\alpha \in \mathbb{R}$. Consider the coupled system of equations

$$(5.71\mathrm{a}) \qquad\qquad \partial_t g_{ij} = -2R_{ij},$$

$$(5.71\mathrm{b}) \qquad\qquad \partial_t f = \Delta f + \alpha R_f.$$

Calculate that ([**101**, Lemma 6.88])

$$(5.72) \qquad (\partial_t - (2\alpha + 1)\Delta + 2\alpha \nabla f \cdot \nabla)R_f = 2\left|R_{ij} + \nabla_i \nabla_j f\right|^2.$$

Use this to show that ([**101**, Corollary 6.89])
(5.73)
$$d_t \mathcal{F}(g(t), f(t)) = 2 \int_{\mathcal{M}} |R_{ij} + \nabla_i \nabla_j f|^2 \, \mathrm{e}^{-f} d\mu - (1 + \alpha) \int_{\mathcal{M}} R_f^2 \, \mathrm{e}^{-f} d\mu.$$
In particular, if $\alpha \leq -1$, then we have monotonicity.

Chapter 6

The Bryant Soliton

In this chapter we present the proof of the existence of the rotationally symmetric Bryant steady GRS [**54**]. As conjectured by Hamilton and by the work of Gu and Zhu [**163**] (while formal matched asymptotics are due to Angenent, Isenberg, and Knopf [**8**, **9**]), this steady GRS, defined in dimensions at least 3, is a singularity model for rotationally symmetric degenerate neckpinches.

Here is a road map to this chapter. In §6.1 we present the statement of the main existence theorem for the Bryant soliton and we set up the system of ODEs for the width w and potential function f. In §6.2 we discuss a criterion for when g and f extend smoothly over the origin. In §6.3 we introduce our first change to new dependent variables x and y and independent variable t, yielding a first-order autonomous system of ODEs for $(x(t), y(t))$. In §6.4 and §6.5 we analyze the behavior of the aforementioned ODE near a stationary point. The Bryant soliton is one of the two trajectories emanating from the stationary point and comprising the unstable manifold. In §6.6 we show that the metric g of the Bryant soliton on the cylinder closes up at one end to extend to a smooth metric on a topological $\mathbb{R}^n$. In §6.7 we show that g is complete and that the width grows like the square root of the radius. In §6.8 we first prove general asymptotics of the curvatures at spatial infinity. After another change of variables, we prove more precise asymptotics.

6.1. The existence statement and the rotational symmetry ansatz

Our ansatz is that $(\mathcal{M}^n, g, f)$, $n \geq 3$, is a C^2 nonflat rotationally symmetric complete noncompact steady GRS. Then either $\mathcal{M}^n$ or $\mathcal{M}^n$ minus a point is diffeomorphic to $I \times \mathbb{S}^{n-1}$, where I is an open interval, and we may write

$$(6.1) \qquad g = dr^2 + w^2(r)g_{\mathbb{S}^{n-1}},$$

where $w : I \to \mathbb{R}_+$ is called the **warping function** and where $g_{\mathbb{S}^{n-1}}$ is the standard metric on $\mathbb{S}^{n-1}$. The choice $w(r) = r$ for $r \in \mathbb{R}_+$ yields the Euclidean metric (i.e., the steady Gaussian soliton) by extending $(\mathcal{M}^n, g)$ smoothly over the origin and by taking f to be a constant.

Let the **orbital sectional curvature** K_{orb} of g denote the sectional curvature of a plane tangent to $\{r\} \times \mathbb{S}^{n-1}$ and let the **radial sectional curvature** K_{rad} of g denote the sectional curvature of a plane containing the radial direction $\frac{\partial}{\partial r}$. The eigenvalues of Rm are equal to two times the values of these sectional curvatures. The following classification theorem for complete rotationally symmetric steady GRS was proved by Bryant [**54**].

Theorem 6.1 (Existence of the Bryant soliton). *For each $n \geq 3$ there exists a complete nonflat rotationally symmetric steady GRS $(\mathcal{M}^n, g, f)$ called the **Bryant soliton**. Up to scaling, this is the unique such GRS. For the Bryant soliton, $\mathcal{M}^n$ is diffeomorphic to $\mathbb{R}^n$, g has positive curvature operator everywhere, and $R_{\max} = R(0^n) = 1$, where 0^n denotes the origin in $\mathbb{R}^n$. As $r \to \infty$, we have that the warping function and its first two derivatives satisfy*

$$w(r) = \mathrm{O}(r^{1/2}), \quad w'(r) = \mathrm{O}(r^{-1/2}), \quad -w''(r) = \mathrm{O}(r^{-3/2}),$$

and the orbital and radial sectional curvatures satisfy

$$K_{\mathrm{orb}} = \mathrm{O}\left(r^{-1}\right) \quad and \quad K_{\mathrm{rad}} = \mathrm{O}\left(r^{-2}\right).$$

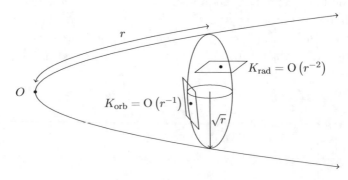

Figure 6.1. The Bryant steady GRS has positive curvature operator and opens up like a paraboloid.

Remark 6.2. More precisely,

$$(6.2a) \qquad K_{\mathrm{orb}} = \frac{1}{2(n-2)r}\left(1 - \frac{5-n}{4}\frac{\ln r}{r}\right) + \mathrm{O}\left(\frac{1}{r^2}\right),$$

$$(6.2b) \qquad K_{\mathrm{rad}} = \frac{(n-2)^2}{4r^2}\left(1 - \frac{5-n}{2}\frac{\ln r}{r}\right) + \mathrm{O}\left(\frac{1}{r^3}\right).$$

In particular, the scalar curvature satisfies

$$(6.3) \qquad R = \frac{n-1}{2r}\left(1 - \frac{5-n}{4}\frac{\ln r}{r}\right) + \mathrm{O}\left(\frac{1}{r^2}\right).$$

The proof of Theorem 6.1 will occupy the rest of this chapter. A standard calculation (e.g., using moving frames) yields under (6.1) that the Ricci tensor is given by

$$(6.4) \qquad \mathrm{Ric}_g = -(n-1)\frac{w''}{w}dr^2 + \left(-ww'' + (n-2)(1-(w')^2)\right)g_{\mathbb{S}^{n-1}}$$

and that its trace, the scalar curvature, is

$$(6.5) \qquad R = -2(n-1)\frac{w''}{w} + (n-1)(n-2)\frac{1-(w')^2}{w^2},$$

where $' := \frac{d}{dr}$.

If f is not a radial function, then $(\mathcal{M}^n, g)$ is flat Euclidean space (see Kotschwar [**204**, Proposition 3]). Hence we may assume that $f = f(r)$ is a radial function. Then

$$(6.6a) \qquad \nabla^2 f = f''dr^2 + ww'f'g_{\mathbb{S}^{n-1}},$$

$$(6.6b) \qquad \Delta f = f'' + (n-1)\frac{w'f'}{w}.$$

Observe that the dependence of the right-hand side of (6.6) on the potential function is linear in f' and f''. The steady GRS equation $\mathrm{Ric}_g + \nabla^2 f = 0$ is hence equivalent to the second-order nonlinear autonomous system of two ODEs for the warping function w and the potential function f, as functions of r:

$$(6.7a) \qquad f'' = (n-1)\frac{w''}{w},$$

$$(6.7b) \qquad ww'f' = ww'' - (n-2)(1-(w')^2).$$

Since we may easily rewrite the system (6.7) in the form $f'' = F(f', w', w)$, $w'' = G(f', w', w)$, where F and G are real analytic (rational) functions defined where $w > 0$, we conclude that:

Lemma 6.3. *The functions w and f are real analytic wherever $w > 0$ (see* [**269**, Theorem 4.2]).

We shall show, by analyzing (6.7), the existence of a rotationally symmetric steady GRS on $(0, \infty) \times \mathbb{S}^{n-1} \cong \mathbb{R}^n \setminus \{0^n\}$, which is complete near ∞, and, with respect to the Euclidean coordinates, that f extends over the origin to a C^3 function while the components of g extend over the origin to a C^2 function. By Theorem B.3, we conclude that g and f are real analytic on all of $\mathbb{R}^n$. This is the Bryant soliton.

6.2. Extending the metric smoothly over the origin

Assuming some results that we prove later for the behavior of the solution $w(r)$ to the ODE near $r = 0$, we first show that the metric g extends as a C^2 metric over the origin 0^n.

Considering $\mathbb{S}^{n-1}$ as the unit sphere in $\mathbb{R}^n$, we have Cartesian coordinates $\{x^i\}_{i=1}^n$ on $\mathbb{R}^n \setminus \{0^n\} \cong (0, \infty) \times \mathbb{S}^{n-1}$ with the relation $r^2 = \sum_{i=1}^n (x^i)^2$. Using $dr = \frac{1}{r} \sum_{i=1}^n x^i dx^i$ and $r^2 g_{\mathbb{S}^{n-1}} = \sum_{i=1}^n (dx^i)^2 - dr^2$, we see that (6.1) becomes

$$g = \frac{r^2 - w^2(r)}{r^4} \left(\sum_{i=1}^n x^i dx^i \right)^2 + \frac{w^2(r)}{r^2} \sum_{i=1}^n (dx^i)^2.$$

Define

$$\rho(r) := \frac{w(r)}{r}$$

for $r > 0$. In terms of $\rho(r)$, the metric is
(6.8)
$$g = \sum_{i=1}^n \left(1 + \left(r^2 - (x^i)^2 \right) \frac{\rho^2(r) - 1}{r^2} \right) (dx^i)^2 - \sum_{i<j} 2x^i x^j \frac{\rho^2(r) - 1}{r^2} dx^i dx^j.$$

By (6.44a) and (6.44b) below (these are the aforementioned results that we prove later) for the behavior of $w(r)$ and $w'(r)$ near $r = 0$, we have

(6.9) $$\rho(r) = 1 - \frac{1}{3(n-1)} r^2 + o\left(r^2\right)$$

and

$$\rho(r) + r\rho'(r) = 1 - \frac{1}{n-1} r^2 + o\left(r^2\right),$$

so that

(6.10) $$\rho'(r) = -\frac{2}{3(n-1)} r + o(r).$$

In particular, $\lim_{r \to 0_+} \rho(r) = 1 := \rho(0)$ and $\lim_{r \to 0_+} \rho'(r) = 0 = \rho'(0)$. Thus, from (6.8) we see that g extends continuously over the origin by defining $g(0^n) := \sum_{i=1}^n (dx^i)^2$.

Given $k \geq 1$, by (6.8) we see that for g to extend to be C^k over the origin, it suffices for

$$(6.11) \qquad \phi_{ij}(x) := x^i x^j \left(\frac{\rho^2(r) - 1}{r^2} \right)$$

to extend to be C^k over the origin for all i, j. Define $\phi_{ij}(0^n) := 0$. Observe that

$$(6.12) \qquad \left(\frac{\rho^2(r) - 1}{r^2} \right)'(r) = \frac{2}{r^3} \left(1 - \rho^2(r) + r\rho(r)\rho'(r) \right).$$

Differentiating (6.11), we have for $x \neq 0^n$ and $V \in \mathbb{R}^n$,

$$(6.13)$$
$$V(\phi_{ij})(x) = (x^i V^j + x^j V^i) \frac{\rho^2(r) - 1}{r^2} + 2x^i x^j \frac{\langle V, \frac{x}{r} \rangle}{r^3} \left(1 - \rho^2(r) + r\rho(r)\rho'(r) \right),$$

since $V(r) = \langle V, \frac{x}{r} \rangle$. By (6.9) and (6.10), we have

$$(6.14) \qquad 1 - \rho^2(r) + r\rho(r)\rho'(r) = \mathrm{o}\left(r^2\right).$$

Applying this to (6.13) yields $V(\phi_{ij})(x) = \mathrm{O}(r)$, so that $d\phi_{ij}(x) = \mathrm{O}(r)$. Hence ϕ_{ij} is differentiable at 0^n and

$$(6.15) \qquad d\phi_{ij}(0^n) = 0$$

by the mean value theorem. In particular, ϕ_{ij} extends to be C^1 over the origin.

Let D denote the Euclidean covariant derivative. Differentiating (6.13) while using (6.12) yields

$$D^2 \phi_{ij}(V, W)(x)$$
$$= W\left(V(\phi_{ij})\right)(x)$$
$$= (W^i V^j + W^j V^i) \frac{\rho^2(r) - 1}{r^2}$$
$$+ 2 \left(\begin{array}{c} (x^i V^j + x^j V^i) \langle W, x \rangle \\ + (x^i W^j + x^j W^i) \langle V, x \rangle \end{array} \right) \frac{1 - \rho^2(r) + r\rho(r)\rho'(r)}{r^4}$$
$$+ 2x^i x^j \left(\frac{\langle V, W \rangle}{r^4} - \frac{4}{r^6} \langle V, x \rangle \langle W, x \rangle \right) \left(1 - \rho^2(r) + r\rho(r)\rho'(r) \right)$$
$$+ 2x^i x^j \frac{\langle V, x \rangle \langle W, x \rangle}{r^5} \left(-\rho(r)\rho'(r) + r\rho(r)\rho''(r) + r\left(\rho'(r)\right)^2 \right),$$

since $W(r) = \langle W, \frac{x}{r} \rangle$ and

$$\left(1 - \rho^2(r) + r\rho(r)\rho'(r) \right)' = -\rho(r)\rho'(r) + r\rho(r)\rho''(r) + r\left(\rho'(r)\right)^2.$$

Using (6.9), (6.10), and (6.14), we see that this implies that

$$(6.16) \qquad D^2 \phi_{ij}(V, W)(x) = -\frac{2}{3(n-1)} \left(W^i V^j + W^j V^i \right) + \mathrm{o}(1),$$

where we have also used that

(6.17)
$$\rho''(r) = \frac{w''(r)}{r} + \frac{2w(r)}{r^3} - 2\frac{w'(r)}{r^2}$$
$$= -\frac{2}{3(n-1)} + o(1),$$

which is true by (6.44). Now (6.16) implies that the ϕ_{ij} are C^2 over the origin with

$$D^2\phi_{ij}(V,W)(0^n) = -\frac{2}{3(n-1)}\left(W^i V^j + W^j V^i\right).$$

Hence g is C^2 in a neighborhood of 0^n (later we shall show that $w(r)$ is defined for all $r > 0$).

6.3. The ODE: Change of variables and its geometric interpretation

We now begin our analysis of the ODE system (6.7) for a rotationally symmetric steady GRS in order to prove the existence of the Bryant soliton. We will consider the ODE for w, f as well as the ODE for various transformed variables.

Observe that (6.6b) and assuming the normalization $\Delta_f f = -1$ from (2.56) imply that

(6.18)
$$f'' + (n-1)\frac{w'f'}{w} - (f')^2 = -1.$$

Substituting this into (6.7) yields

(6.19)
$$2(n-1)ww'f' + (n-1)(n-2)(1-(w')^2) - w^2(f')^2 = -w^2,$$

or equivalently,

(6.20)
$$\left(f' - (n-1)\frac{w'}{w}\right)^2 - (n-1)\frac{n-2+(w')^2}{w^2} = 1.$$

To motivate the variable transformations geometrically, let $\mathrm{II}(r)$ and $H_f(r)$ denote the second fundamental form and f-mean curvature of the distance sphere from the origin $\Sigma_r := \{r\} \times \mathbb{S}^{n-1}$, respectively. Let $g_\Sigma := g|_{\Sigma_r}$. We have

(6.21)
$$\mathrm{II} = -\frac{(\nabla^2 r)|_{\Sigma_r}}{|\nabla r|} = \frac{w'}{w}g_\Sigma := \kappa g_\Sigma,$$

since $\nabla^2 r = \frac{1}{2}\mathcal{L}_{\nabla r}(dr^2 + w^2 g_{\mathbb{S}^{n-1}}) = \frac{w'}{w}g_\Sigma$. That is, the distance spheres are totally umbilic with principal curvature κ. Hence

$$H_f = (n-1)\frac{w'}{w} - f'.$$

Since we are on a steady GRS, the f-Riccati equation (4.33) yields that H_f is a monotone function:

$$(6.22) \qquad H'_f = \left((n-1)\frac{w'}{w} - f' \right)' = -(n-1)\left(\frac{w'}{w}\right)^2 \le 0,$$

which also follows from $f'' = (n-1)\dfrac{w''}{w}$ in (6.7).

Consider the following change of dependent and independent variables from w, f, and r to x, y, and t:

$$(6.23a) \qquad\qquad x := w' = w\kappa,$$

$$(6.23b) \qquad\qquad y := (n-1)w' - wf' = wH_f,$$

$$(6.23c) \qquad\qquad dt := \frac{dr}{w}.$$

These variables are scale invariant in the sense that they are invariant under simultaneously scaling r and w. They are also invariant under adding a constant to f since only f' occurs. As in the study of any ODE, there may not be a best change of variables for all situations; later in our analysis we shall consider other change of variables. Regarding the variable t, we note that the metric can be expressed as

$$(6.24) \qquad\qquad g = w^2\left(dt^2 + g_{\mathbb{S}^{n-1}}\right),$$

which is conformal to the standard cylinder. The variables x and y are the scaled principal curvature and f-mean curvature, respectively.

From (6.7) we obtain the equivalent first-order autonomous system of ODEs for x and y as functions of t:

$$(6.25a) \qquad\qquad \frac{dx}{dt} = x^2 - xy + n - 2,$$

$$(6.25b) \qquad\qquad \frac{dy}{dt} = x\big(y - (n-1)x\big).$$

Note that this system of equations is invariant under time translation and that if $(x(t), y(t))$ is a solution, then $(-x(-t), \; y(-t))$ is also a solution. We may write (6.25) more compactly as $\frac{d}{dt}\mathbf{z} = F(\mathbf{z})$, where $F : \mathbb{R}^2 \to \mathbb{R}^2$ and $\mathbf{z}(t) = (x(t), y(t))$. One of the advantages of (6.25) over the original system (6.7) is that the right-hand side is a polynomial of the dependent variables and in particular the ODE no longer has singular points.

Without assuming a normalization, from (2.56) we have $\Delta_f f = -C$ for some $C \in \mathbb{R}$. Then (6.20) becomes the first integral:

$$(6.26) \qquad\qquad y^2 - (n-1)\left(x^2 + n - 2\right) = Cw^2.$$

For example, as a special case, for $C = 0$, we have the invariant hyperbola:

$$(6.27) \qquad\qquad y^2 = (n-1)\left(x^2 + n - 2\right).$$

The stationary point $(1, n-1)$ lies on this hyperbola. Note that when $C = 0$, we obtain $R + |\nabla f|^2 = 0$. In this case, g is not complete unless it is the Euclidean metric. Indeed, completeness implies that $R \geq 0$, which in turn implies that $R \equiv |\nabla f| \equiv 0$, which with the rotational symmetry of g implies that the metric g is Euclidean. However, we will assume as in (2.56) that $C = 1$.

6.4. The ODE near a stationary point

If (x, y) is a stationary point of the transformed ODE system (6.25), then $y = (n-1)x$ and hence $x^2 = 1$ since $n \geq 3$. Thus the two stationary points are $(1, n-1)$ and $(-1, -n+1)$. The stationary point $(1, n-1)$ corresponds to the solution $w(r) = r$ and $f(r)$ is constant, for $r > 0$. This is the Euclidean steady GRS.

6.4.1. Linearizing about stationary points of the ODE.

We first analyze the behavior of (6.25) near $\mathbf{z}_0 = (x_0, y_0) := (1, n - 1)$ via linearization. We refer the reader to Chapter 9 of Teschl's book [**269**] for general ODE theory. Note that $\mathbf{z}(t) = (x(t), y(t)) \equiv \mathbf{z}_0$ corresponds to the trivial Euclidean GRS. The linearization of (6.25) at $\mathbf{z}_0$ is the following linear system for $\bar{x}, \bar{y}$:

$$(6.28) \qquad \frac{d}{dt} \begin{pmatrix} \bar{x} \\ \bar{y} \end{pmatrix} = \begin{pmatrix} -(n-3) & -1 \\ -(n-1) & 1 \end{pmatrix} \begin{pmatrix} \bar{x} \\ \bar{y} \end{pmatrix}.$$

Abbreviate this as $\frac{d}{dt} \bar{\mathbf{z}} = A\bar{\mathbf{z}}$, where $\bar{\mathbf{z}}(t) = (\bar{x}(t), \bar{y}(t))$. Note that $A = dF_{\mathbf{z}_0}$ and the solutions are $\bar{\mathbf{z}}(t) = e^{tA}\bar{\mathbf{z}}(0)$. The 2×2 matrix in this display has eigenvectors and associated eigenvalues:

$$(6.29) \qquad v_- = \begin{pmatrix} -\frac{1}{n-1} \\ 1 \end{pmatrix}, \ \mu_- = 2 \quad \text{and} \quad v_+ = \begin{pmatrix} 1 \\ 1 \end{pmatrix}, \ \mu_+ = 2 - n.$$

Hence $\mathbf{z}_0$ is a saddle point. The fact that the eigenvalues $\mu_- > 0$ and $\mu_+ < 0$ are nonzero, i.e., the fixed point $\mathbf{z}_0$ is **hyperbolic**, makes the ODE easier to analyze due to the nondegeneracy.

6.4.2. The stable and unstable manifolds.

Consider the behavior of (6.28) near $\mathbf{z}_0 = (1, n - 1)$. The **unstable manifold** of the linearization (6.28) is the line E^- spanned by v_-. Note that a solution to (6.28) stays bounded as $t \to -\infty$ if and only if $\bar{\mathbf{z}}(0) \in E^-$, in which case $\bar{\mathbf{z}}(t) = e^{2t}\bar{\mathbf{z}}(0) \to 0^2$ as $t \to -\infty$. The **stable manifold** of the linearization is the line E^+ spanned by v_+. Given $\mathbf{z} \in \mathbb{R}^2$, we may write it uniquely as $\mathbf{z} = \mathbf{z}_+ + \mathbf{z}_-$, where $\mathbf{z}_\pm \in E^\pm$. Define the projections $P^\pm : \mathbb{R}^2 \to E^\pm$ by $P^\pm(\mathbf{z}) = \mathbf{z}_\pm$.

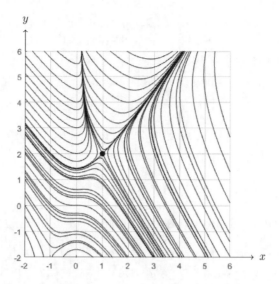

Figure 6.2. Phase portrait for the first-order autonomous system (6.25) for $(x(t), y(t))$ with $n = 3$. The stationary point $(1, 2)$, through which one branch of the invariant hyperbola $y^2 = 2(x^2 + 1)$ passes, corresponds to Euclidean 3-space.

Let $\Phi_t(\mathbf{z})$ denote the **flow** associated to the ODE (6.25). That is,

$$\Phi_t(\mathbf{z}) = \mathbf{z}(t) := (x(t), y(t))$$

is the solution to (6.25) with $\mathbf{z}(0) = \mathbf{z}$ and is defined for t in the **maximal interval** $(T_-(\mathbf{z}), T_+(\mathbf{z}))$ about 0. Define the **unstable set** of the ODE (6.25) by

$$(6.30) \qquad W^-(\mathbf{z}_0) := \{\mathbf{z} \in \mathbb{R}^2 : \lim_{t \to -\infty} \Phi_t(\mathbf{z}) = \mathbf{z}_0\}.$$

Then, in a neighborhood U of $\mathbf{z}_0$, $W^-(\mathbf{z}_0)$ is a smooth curve and is equal to the **unstable manifold** of the ODE (6.25),

$$(6.31) \qquad M^-(\mathbf{z}_0) = \bigcup_{\alpha > 0} M^{-,\alpha}(\mathbf{z}_0)$$

defined by

$$M^{-,\alpha}(\mathbf{z}_0) = \{\mathbf{z} : \gamma_-(\mathbf{z}) \subset U \text{ and } \sup_{t \le 0} e^{-\alpha t} |\Phi_t(\mathbf{z}) - \mathbf{z}_0| < \infty\}.$$

Here, the **backward orbit** of $\mathbf{z}$ is defined by

$$\gamma_-(\mathbf{z}) := \bigcup_{t \in (T_-(\mathbf{z}), 0)} \Phi_t(\mathbf{z}).$$

Moreover, the curve $M^-(\mathbf{z}_0) = W^-(\mathbf{z}_0)$ is tangent to the line E^- at $\mathbf{z}_0$.

The **stable manifold** of (6.25) is the branch containing $(1, n-1)$ of the hyperbola $y^2 = (n-1)\left(x^2 + n - 2\right)$ defined by (6.27) since we easily see that this invariant set is tangent to v_+.

Let $\mathbf{z} = \mathbf{z}_+ + \mathbf{z}_-$, where $\mathbf{z}_\pm \in E^\pm$. Define $g : \mathbb{R}^2 \to \mathbb{R}^2$ by

$$(6.32) \qquad\qquad\qquad F(\mathbf{z}) := A(\mathbf{z} - \mathbf{z}_0) + g(\mathbf{z}).$$

Then $g(\mathbf{z}) = \mathrm{o}(|\mathbf{z} - \mathbf{z}_0|)$. Using (6.32), the ODE (6.25) may be rewritten as the integral equation

$$(6.33) \qquad \mathbf{z}(t) - \mathbf{z}_0 = e^{tA}(\mathbf{z}(0) - \mathbf{z}_0) + \int_0^t e^{(t-s)A} g\left(\mathbf{z}(s)\right) ds.$$

The **(un)stable manifold theorem** says the following.

Lemma 6.4. *There exists a neighborhood U of $\mathbf{z}_0$ and a real analytic function $h^- : E^- \cap (U - \mathbf{z}_0) \to E^+$, where $U - \mathbf{z}_0 := \{\mathbf{z} - \mathbf{z}_0 : \mathbf{z} \in U\}$, satisfying*

$$M^-(\mathbf{z}_0) \cap U = \left\{\mathbf{z}_0 + \mathbf{z}_- + h^-(\mathbf{z}_-) : \mathbf{z}_- \in E^- \cap (U - \mathbf{z}_0)\right\}.$$

Moreover, $h^-(0^2) = 0^2$ and $dh^-_{0^2} = 0$. Hence $M^-(\mathbf{z}_0)$ is tangent to E^- at $\mathbf{z}_0$.

By the Hartman–Grobman theorem (see Theorem 9.9 in [**269**]):

Lemma 6.5. *There exists a homeomorphism $\varphi(\bar{\mathbf{z}}) = \mathbf{z}_0 + \bar{\mathbf{z}} + h(\bar{\mathbf{z}})$, where h is bounded, such that*

$$(6.34) \qquad\qquad\qquad \varphi \circ e^{tA} = \Phi_t \circ \varphi$$

*in some neighborhood of 0^2. Moreover, φ is differentiable at 0^2 and $d\varphi_{0^2}$ is the identity (see [**170**]).*

6.4.3. The ODE trajectory of the Bryant soliton.

Recall that the unstable manifold of the fixed point $(1, n-1)$ of the ODE (6.25), the line E^-, is spanned by the vector $v_- = \begin{pmatrix} -\frac{1}{n-1} \\ 1 \end{pmatrix}$. Of the two solutions $(x(t), y(t))$ of the ODE system (6.25) satisfying

$$(x(t), y(t)) \to (1, n-1) \quad \text{as } t \to -\infty$$

and tangent to the vectors $\pm v_-$ at $(1, n-1)$ (which exist by Lemma 6.4), we will show that the **left-hand trajectory**, which is tangent to v_-, corresponds to the Bryant soliton; see Figure 6.3.

For this solution, to which we now restrict our consideration, $t \mapsto x(t)$ is decreasing and $t \mapsto y(t)$ is increasing for t negative and sufficiently large in magnitude. Recall that, up to a *constant of integration*, using (6.23) we first obtain $w(t)$ from $(x(t), y(t))$ by integrating $\frac{dw}{w} = x dt$, then we obtain $r(t)$

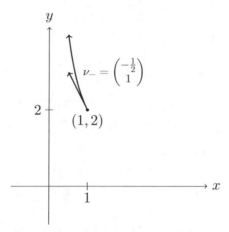

Figure 6.3. Trajectory of the 3-dimensional Bryant soliton, which is tangent to the vector ν_- at $(1, 2)$.

from $\frac{dr}{dt} = w$ which yields $w(r)$, and finally we obtain $f(r)$ from integrating $f' = \frac{(n-1)w'-y}{w}$.

We now show for this solution that w is C^1 in a neighborhood of 0. Later we will show that w is C^3 (before finally proving that g and f are real analytic); see (6.50) below.

Note from (6.25) that $x > 0$ is preserved. Furthermore, as long as $\frac{dx}{dt} < 0$ (i.e., $y > \frac{x^2+n-2}{x}$), which implies

$$(6.35) \qquad\qquad 0 < w' = x < 1,$$

we have $\frac{dy}{dt} > 0$. That is, $y > (n-1)x$, and hence

$$(6.36) \qquad\qquad f' < 0.$$

Therefore, differentiating (6.25) yields $\frac{d}{dt}\left(\frac{dx}{dt}\right) = (2x-y)\frac{dx}{dt} - x\frac{dy}{dt}$, which shows that $\frac{dx}{dt} < 0$ is preserved and hence so is $\frac{dy}{dt} > 0$; equivalently,

$$(6.37) \qquad\qquad w'' = \frac{dx}{dr} < 0.$$

Since $\frac{dw}{dr}(t) = x(t) \to 1$ as $t \to -\infty$ and since $w > 0$, we have that $r(t)$ must limit to a finite number as $t \to -\infty$, which we may translate to be 0. Because $dr = w\,dt$ and $r \to 0_+$ as $t \to -\infty$, we have

$$(6.38) \qquad\qquad \lim_{r\to 0_+} w(r) = 0 := w(0).$$

By the above, $w \in C^1$ and

$$(6.39) \qquad\qquad \lim_{r\to 0_+} w'(r) = 1 = w'(0).$$

Hence $0 < w'(r) < 1$ for $r > 0$, and the curvature operator is positive by (6.67) below (we will see later that is true also at $r = 0$).

6.5. Behavior of the solution near the stationary point

We claim that from the solution to the nonlinear equation emanating from the stationary point approximating the solution to the linearization, we have after a time translation and for t near $-\infty$,

$$(6.40a) \qquad x(t) = 1 - \frac{1}{n-1}e^{2t} + o(e^{2t}),$$

$$(6.40b) \qquad y(t) = n - 1 + e^{2t} + o(e^{2t}).$$

To see this, suppose that $\bar{\mathbf{z}} \in \mathbb{R}^2$ is such that $\varphi(\bar{\mathbf{z}}) \in M^-(\mathbf{z}_0) \cap U$, where φ is as in (6.34). From the definition of $M^-(\mathbf{z}_0)$ and Lemma 6.5, we have

$$\lim_{t \to -\infty} \varphi(e^{tA}\bar{\mathbf{z}}) = \lim_{t \to -\infty} \Phi_t(\varphi(\bar{\mathbf{z}})) = \mathbf{z}_0,$$

$$\varphi \circ e^{tA} = \Phi_t \circ \varphi.$$

So, by the properties of φ near the origin, we have

$$\mathbf{z}_0 = \lim_{t \to -\infty} \varphi(e^{tA}\bar{\mathbf{z}}) = \mathbf{z}_0 + \lim_{t \to -\infty}(e^{tA}\bar{\mathbf{z}} + o(e^{tA}\bar{\mathbf{z}}))$$

and hence

$$\lim_{t \to -\infty} e^{tA}\bar{\mathbf{z}} = 0^2.$$

It follows by definition that $e^{tA}\bar{\mathbf{z}}$ is in the stable manifold of the linearization of the ODE and hence $\bar{\mathbf{z}} \in E^-$.

It is worth noticing that $e^{tA}\bar{\mathbf{z}}$ goes to either 0^2 or ∞ as $t \to -\infty$ unless $\mathbf{z} = 0^2$ and that $e^{tA}\bar{\mathbf{z}}$ must stay in a small neighborhood of 0^2 since its image under the homeomorphism φ is in a small neighborhood of $\mathbf{z}_0$.

Since $d\varphi_{0^2} = \mathrm{id}_{\mathbb{R}^2}$, we have

$$\varphi(e^{tA}\bar{\mathbf{z}}) = \mathbf{z}_0 + e^{tA}\bar{\mathbf{z}} + o(|e^{tA}\bar{\mathbf{z}}|).$$

Hence, for any solution $\mathbf{z}(t)$ to (6.25) with $\mathbf{z}(0) \in M^-(\mathbf{z}_0) \cap U$ and $\mathbf{z}(0) \neq \mathbf{z}_0$, there exists $\bar{\mathbf{z}} \in E^- - \{0^2\}$ such that

$$\mathbf{z}(t) = \Phi_t(\varphi(\bar{\mathbf{z}})) = \mathbf{z}_0 + e^{2t}\bar{\mathbf{z}} + o(e^{2t}).$$

Since the line E^- is spanned by v_-, this proves that (6.40) is true after a suitable time translation.

Now, by substituting (6.40) into (6.25), we obtain

$$(6.41a) \qquad \frac{dx}{dt}(t) = -\frac{2}{n-1}e^{2t} + o(e^{2t}),$$

$$(6.41b) \qquad \frac{dy}{dt}(t) = 2e^{2t} + o(e^{2t}).$$

By (6.38) and (6.39) we have

$$(6.42) \qquad \frac{dr}{dt} = w = r + o(r).$$

Hence $\frac{dt}{dr} = \frac{1}{r} + o(\frac{1}{r})$, so that $t = (1 + o(1)) \ln r$. Since $\ln r = (1 + o(1)) t$, we obtain $r = e^t + o(e^t)$ and hence

$$(6.43) \qquad w(t) = e^t + o(e^t)$$

for t near $-\infty$. We have that (6.41a) and (6.43) imply

$$w'' = \frac{1}{w}\frac{dx}{dt} = -\frac{2}{n-1}e^t + o(e^t).$$

Moreover, $dr = w\,dt = (e^t + o(e^t))dt$, which implies that we can translate r so that $r = e^t + o(e^t)$ and hence $e^t = r + o(r)$. So, for $r > 0$ near 0,

$$w''(r) = -\frac{2}{n-1}r + o(r).$$

This implies $w \in C^2$ and

$$w''(0) = 0.$$

Hence we obtain the following asymptotics for w and its first two derivatives near the origin ($r = 0$):

$$(6.44a) \qquad w(r) = r - \frac{1}{3(n-1)}r^3 + o\left(r^3\right),$$

$$(6.44b) \qquad w'(r) = 1 - \frac{1}{n-1}r^2 + o\left(r^2\right),$$

$$(6.44c) \qquad w''(r) = -\frac{2}{n-1}r + o\left(r\right).$$

Combined with the analysis of the ODE in §6.2, which relied on (6.44), this completes the proof that the metric g is C^2 in a neighborhood of 0^n.

6.6. Real analyticity of g and f at the origin

Let $(w(r), f(r))$ be the solution to (6.7), which we will show yields the Bryant soliton, corresponding to the solution $(x(t), y(t))$ of (6.25) satisfying $(x(t), y(t)) \to (1, n-1)$ as $t \to -\infty$ and tangent to v_-.

So far we have shown, in §6.4.3, that $\lim_{r \to 0_+} w'(r) = 1$, $\lim_{r \to 0_+} w(r) = 0$, $0 < w' < 1$, and $w'' < 0$. In this section we show that the GRS structure extends real analytically over the origin.

By (6.7),

$$(6.45) \qquad w'' = (n-2)\frac{1 - (w')^2}{w} + w'f'.$$

From this and (6.44) we obtain

(6.46) $$\lim_{r \to 0_+} f'(r) = 0 = f'(0).$$

Now, since $f'(r) < 0$ for $r > 0$, we have that $\lim_{r \to 0_+} f(r) := f(0)$ exists. By adding a constant to f if necessary, we may assume that $f(0) = -1$. Furthermore, by (6.7) and (6.44),

$$f''(r) = (n-1)\frac{w''}{w} = -2 + o(1).$$

In particular, $f \in C^2$ and $f''(0) = -2$. Therefore

(6.47a) $$f(r) = -1 - r^2 + o\left(r^2\right),$$

(6.47b) $$f'(r) = -2r + o(r).$$

Now, from (6.25), as functions of r we have

$$\frac{dw}{dr} = x,$$
$$\frac{dx}{dr} = \frac{1}{w}\left(x^2 - xy + n - 2\right),$$
$$\frac{dy}{dr} = \frac{1}{w}x\left(y - (n-1)x\right).$$

We have shown that $w(0) = 0$, $x(0) = 1$, $y(0) = n-1$, $w'(0) = 1$, $x'(0) = 0$, and $y'(0) = 0$, where the last equality follows from

$$y' = (n-1)w'' - w'f' - wf''$$

(which is obtained by differentiating (6.23)).

By further differentiating (6.40) we have

$$\frac{d^2x}{dt^2} = (n+1)x^3 - 4x^2y + xy^2 + 2(n-2)x - (n-2)y,$$
$$\frac{d^2y}{dt^2} = -3(n-1)x^3 + 2nx^2y - xy^2 - 2(n-1)(n-2)x + (n-2)y.$$

Using this, (6.40), and $\frac{dw}{dt} = wx$, we compute that

(6.48) $$w''' = \frac{1}{w}\frac{d}{dt}\left(\frac{1}{w}\frac{dx}{dt}\right)$$
$$= \frac{1}{w^2}\left(\frac{d^2x}{dt^2} - x\frac{dx}{dt}\right)$$
$$= \frac{1}{w^2}\left(nx^3 - 3x^2y + xy^2 + (n-2)x - (n-2)y\right).$$

On the other hand,

$$w(r) = r - \frac{1}{3(n-1)}r^3 + \mathrm{o}\left(r^3\right), \tag{6.49a}$$

$$x(r) = 1 - \frac{1}{n-1}r^2 + \mathrm{o}\left(r^2\right), \tag{6.49b}$$

$$y(r) = n - 1 + r^2 + \mathrm{o}\left(r^2\right), \tag{6.49c}$$

with the last equality following from $y := (n-1)w' - wf'$. Hence

$$w'''(r) = -\frac{2}{n-1} + \mathrm{o}\left(1\right). \tag{6.50}$$

From this we obtain that $w \in C^3$ and

$$w(r) = r - \frac{1}{3(n-1)}r^3 + \mathrm{o}(r^3). \tag{6.51}$$

Moreover, using (6.5) we see that $R \in C^0$ and

$$R - 2n + \mathrm{o}\left(1\right). \tag{6.52}$$

Moreover, one can show that $\mathrm{Ric} \in C^0$. Since

$$\Delta(df) = d(\Delta f) + \mathrm{Ric}\left(\nabla f\right) = -dR + \mathrm{Ric}\left(\nabla f\right) = -\mathrm{Ric}\left(\nabla f\right) \tag{6.53}$$

in a punctured neighborhood of 0^n, $df \in C^1$, and $\mathrm{Ric}\left(\nabla f\right) \in C^0$, we have that $\Delta(df) \in C^0$ and hence $df \in C^{1,\alpha}$ (see [157]). Hence $\mathrm{Ric} = -\nabla^2 f \in C^\alpha$ and thus $\mathrm{Ric}\left(\nabla f\right) \in C^\alpha$. From (6.53) and standard Schauder theory we obtain $df \in C^{2,\alpha}$ and hence $f \in C^{3,\alpha}$ for any $\alpha \in (0,1)$, and so in particular $f \in C^3$. Since $g \in C^2$ and $f \in C^3$, by Theorem B.3, we have that g and f are real analytic in a neighborhood of 0^n.

Remark 6.6. For the original proof that g and f are real analytic near the origin, see Proposition 1 in [54].

6.7. Global geometric properties of the Bryant soliton

Now we proceed to consider the global properties of the steady GRS we discussed in the previous sections to show that it is the Bryant soliton as asserted in Theorem 6.1. The remaining issues are the global completeness of g, which is equivalent to $w(r)$ being defined for all $r \in [0,\infty)$,[1] the positivity of the curvature operator, and the asymptotic behavior of the sectional curvatures. In this section we prove completeness.

---

[1] Since the exponential map at 0^n is then defined on all of $T_{0^n}\mathbb{R}^n$.

Define the change of independent variables:

$$s(r) := (n-1)\ln w(r) - f(r) = \ln(\mathrm{J}_f(r)),$$

where $\mathrm{J}_f(r) := w^{n-1}(r)\mathrm{e}^{-f(r)}$ is the f-Jacobian. Then s is defined on an interval extending to $-\infty$ and

$$(6.54) \qquad ds = \left((n-1)\frac{w'}{w} - f'\right)dr = H_f dr$$

and $\mathrm{e}^s = w^{n-1}\mathrm{e}^{-f}$. Note that $\mathrm{e}^s dr \wedge d\mu_{\mathbb{S}^{n-1}} = \mathrm{e}^{-f}d\mu_g$ is the f-volume form and $\lim_{r\to 0} s(r) = -\infty$. In any case, the geometric significance of the variable s is that it is the logarithm of the f-volume form density.

We also change dependent variables by defining

$$(6.55)$$

$$X := \frac{\frac{w'}{w}}{(n-1)\frac{w'}{w} - f'} = \frac{\kappa}{H_f} = \frac{x}{y}, \qquad Y := \frac{1}{w\left((n-1)\frac{w'}{w} - f'\right)} = \frac{1}{wH_f} = \frac{1}{y}.$$

So Y is the reciprocal of the scaled f-mean curvature of the distance sphere and X is the ratio of the principal curvature over the f-mean curvature. We compute that (6.7) is equivalent to the following nonlinear autonomous system of ODEs for X, Y:

$$(6.56\mathrm{a}) \qquad \frac{dX}{ds} = (n-1)X^3 - X + (n-2)Y^2,$$

$$(6.56\mathrm{b}) \qquad \frac{dY}{ds} = Y\left((n-1)X^2 - X\right),$$

which is an autonomous first-order system of ODEs whose right-hand side is a polynomial of the dependent variables. Since $w(r) = r + o(r)$ for the Bryant soliton solution from (6.42) and since $f'(r) = o(1)$ as $r \to 0_+$ from (6.47b), we have

$$\lim_{s\to -\infty} X(s) = \lim_{s\to -\infty} Y(s) = \frac{1}{n-1}.$$

The linearization of (6.56) at the stationary point $(X_0, Y_0) := \left(\frac{1}{n-1}, \frac{1}{n-1}\right)$ is the following linear system for $\bar{X}, \bar{Y}$:

$$\frac{d}{ds}\begin{pmatrix}\bar{X}\\ \bar{Y}\end{pmatrix} = \begin{pmatrix} -\frac{n-4}{n-1} & \frac{2(n-2)}{n-1} \\ \frac{1}{n-1} & 0 \end{pmatrix}\begin{pmatrix}\bar{X}\\ \bar{Y}\end{pmatrix}.$$

The 2×2 matrix in the display has eigenvectors and associated eigenvalues

$$V_1 = \begin{pmatrix}2\\1\end{pmatrix}, \ \lambda_1 = \frac{2}{n-1} \quad \text{and} \quad V_2 = \begin{pmatrix}2-n\\1\end{pmatrix}, \ \lambda_2 = -\frac{n-2}{n-1},$$

where $\lambda_1 > 0$ and $\lambda_2 < 0$. Hence (X_0, Y_0) is a saddle point.

By Lemma 6.4, there are exactly two solutions $(X(s), Y(s))$ to (6.56) with $(X(s), Y(s)) \to (X_0, Y_0)$ as $s \to -\infty$ and tangent to $\pm V_1$. The Bryant soliton solution is the left-hand solution tangent to $-V_1$, since it corresponds to the solution $(x(t), y(t))$ limiting to $(x_0, y_0) = (1, n-1)$ as $t \to -\infty$ and tangent to $v_- = (-\frac{1}{n-1}, 1)$. For this solution, $\frac{dX}{ds}(s) < 0$ and $\frac{dY}{ds}(s) < 0$ for $-s$ sufficiently large.

By (6.20) and (6.55),

$$(6.57) \qquad L := (n-1)\left(X^2 + (n-2)Y^2\right) = 1 - \frac{1}{H_f^2} < 1.$$

We have that L is a positive monotone function satisfying

$$(6.58) \qquad \frac{dL}{ds} = -2(n-1)X^2(1-L) < 0.$$

Observe that (6.57) implies that X and Y are uniformly bounded. We now show that equation (6.58) and $\left|\frac{dX}{ds}\right|$ being bounded (by (6.56)) imply the following.

Lemma 6.7. *The solution is defined for all* $s \in (-\infty, \infty)$ *and we have* $(X(s), Y(s)) \to (0, 0)$ *as* $s \to \infty$.

Proof. Suppose $X(s)$ does not limit to zero. Then there exists a sequence $s_i \to \infty$ such that $s_{i+1} \geq s_i + 1$ and $X(s_i) \geq c$ for all i, where c is a positive constant. Since $\left|\frac{dX}{ds}\right|(s) \leq C$ for all s, we have $X(s) \geq c/2$ on $[s_i, s_i + \delta]$ for $\delta \in (0, 1]$ sufficiently small independent of i. Moreover, (6.58) implies $1 - L \geq c'$ for some constant $c' > 0$. Thus (6.58) further implies that $\frac{dL}{ds}(s) \leq -c''$ on $[s_i, s_i + \delta]$ for some $c'' > 0$. From this and $\frac{dL}{ds} < 0$ we conclude that $L(s)$ is negative for s sufficiently large, a contradiction. Thus $X(s) \to 0$ as $s \to \infty$.

Since $x > 0$, $\frac{dx}{dt} < 0$, and $\frac{dy}{dt} > 0$ are preserved and since $\frac{ds}{dt} = H_f w > 0$ by (6.23) and (6.54), we have that $X(s) = \frac{x}{y}$ and $Y(s) = \frac{1}{y}$ are positive and decreasing functions of s. Since $\frac{dX}{ds} < 0$ and $\left|\frac{d^2X}{ds^2}\right| \leq C$ from differentiating (6.56), by the same argument as in Lemma 6.7 we obtain that $\frac{dX}{ds}(s) \to 0$ as $s \to \infty$. By (6.56a), this implies that $Y(s) \to 0$ as $s \to \infty$. $\square$

We compute

$$\frac{d}{ds}\left(\frac{X}{Y^2}\right) = n - 2 - \frac{X}{Y^2}\left(1 - 2X + (n-1)X^2\right),$$

so that

$$\frac{d}{ds}\left(\frac{X}{Y^2} - (n-2)\right) = -\left(\frac{X}{Y^2} - (n-2)\right)\left(1 - 2X + (n-1)X^2\right)$$
$$+ (n-2)(2 - (n-1)X)X.$$

Since $X(s) = o(1)$ as $s \to \infty$, we have

(6.59) $\dfrac{d}{ds}\left(\dfrac{X}{Y^2} - (n-2)\right) = -(1+o(1))\left(\dfrac{X}{Y^2} - (n-2)\right) + o(1).$

It follows that (see Exercise 6.8)

(6.60) $\displaystyle\lim_{s\to\infty}\dfrac{X}{Y^2} = n-2.$

Next we compute that

$$\dfrac{d}{ds}\left(\dfrac{X-(n-2)Y^2}{Y^4}\right) = \dfrac{1}{Y^2}\dfrac{d}{ds}\left(\dfrac{X}{Y^2} - (n-2)\right) - \dfrac{2}{Y^3}\left(\dfrac{X}{Y^2} - (n-2)\right)\dfrac{dY}{ds}$$

$$= -\dfrac{X-(n-2)Y^2}{Y^4}\left(1 - 4X + 3(n-1)X^2\right)$$

$$+ \dfrac{X}{Y^2}(n-2)\left(2 - (n-1)X\right).$$

Thus

$$\dfrac{d}{ds}\left(\dfrac{X-(n-2)Y^2}{Y^4}\right) = -(1+o(1))\dfrac{X-(n-2)Y^2}{Y^4} + (1+o(1))2(n-2)^2.$$

It follows that

(6.61) $\displaystyle\lim_{s\to\infty}\dfrac{X-(n-2)Y^2}{Y^4} = 2(n-2)^2.$

Applying (6.60) to (6.58), we obtain

$$\dfrac{dL}{ds} = -\dfrac{2}{n-1}(1+o(1))L^2.$$

Hence $L = (1+o(1))\frac{n-1}{2s}$, so that

(6.62) $X = (1+o(1))\dfrac{1}{2s}$ and $Y^2 = (1+o(1))\dfrac{1}{2(n-2)s}.$

Since $1 - \frac{1}{H_f^2(s)} = L(s) \to 0$ as $s \to \infty$ by (6.57) and Lemma 6.7, we have

(6.63) $H_f(s) \to 1$ as $s \to \infty.$

Hence, by (6.54),

$$r = (1+o(1))s$$

and r is defined on all of $[0,\infty)$. Moreover, by this, (6.55), (6.62), and (6.63), we obtain

(6.64) $w(r) = \dfrac{1}{YH_f} = \left(\sqrt{2(n-2)} + o(1)\right)r^{1/2}.$

This implies that the rotationally symmetric Bryant steady GRS is defined on all of $\mathbb{R}^n$ and is complete and opens up like a paraboloid.

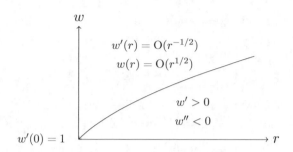

Figure 6.4. The warping function of the Bryant soliton.

6.8. Asymptotic behavior of the curvatures

In the first subsection of this section, we discuss the rough asymptotic behavior of the sectional curvatures at spatial infinity using the behavior of $(X(s), Y(s))$ as $s \to \infty$. In the second subsection, we we discuss a more precise asymptotic behavior of the sectional curvatures using yet another change of variables and computing power series expansions. From now on, we shall use the notation $O(r^p)$ to mean bounded above and below by a positive constant times r^p.

6.8.1. Rough asymptotic behavior of the curvatures.

Since $w' = \frac{X}{Y}$, by (6.62) we have

$$(6.65) \qquad w'(r) = O(r^{-1/2}).$$

By (6.23) and (6.25), we have

$$w'' = \frac{1}{w}\frac{dx}{dt} = \frac{1}{w}\left(x^2 - xy + n - 2\right) = \frac{X^2 - X + (n-2)Y^2}{wY^2},$$

since $x = \frac{X}{Y}$ and $y = \frac{1}{Y}$. Hence, by (6.61), we have

$$(6.66) \qquad -w''(r) = O(r^{-3/2}).$$

Note that the second fundamental form satisfies

$$|\mathrm{II}| = \sqrt{n-1}\left(\frac{w'}{w}\right) = O(r^{-1}).$$

The orbital and radial sectional curvatures are

$$(6.67) \qquad K_{\mathrm{orb}} = \frac{1 - (w')^2}{w^2} \quad \text{and} \quad K_{\mathrm{rad}} = -\frac{w''}{w}.$$

Since $0 < w'(r) < 1$ and $w''(r) < 0$ for $r \in (0, \infty)$ and since $\lim_{r \to 0+} \frac{1-(w')^2}{w^2} = \frac{2}{n-1}$ (by (6.44)), the curvature operator is positive on all of $\mathbb{R}^n$; note that the eigenforms of the curvature operator are primitive 2-forms, i.e., the wedge product of two 1-forms.

In the following, the constant C may change from line to line. By applying the asymptotics for w, w', and w'' proved above to (6.67), we first see that there exists $C > 0$ such that

(6.68) $C^{-1}r^{-1} \leq K_{\mathrm{orb}} \leq Cr^{-1}$ and $C^{-1}r^{-2} \leq K_{\mathrm{rad}} \leq Cr^{-2}$ for $r \geq 1$.

Secondly, since $R + |\nabla f|^2 = 1$, f is radial, and $C^{-1}r^{-1} \leq R \leq Cr^{-1}$ for $r \geq 1$, it follows that there exists C such that $f_{\mathrm{Bry}} := f$ satisfies

(6.69) $-r + C^{-1}\ln r - C \leq f_{\mathrm{Bry}}(r) \leq -r + C\ln r + C$ for $r \geq 1$.

6.8.2. Precise asymptotic behavior of the curvatures.

In this subsection we consider yet another change of variables which not only allows us to prove the existence of the Bryant soliton again but provides us with more precise curvature asymptotics.

Define the change of variables

(6.70) $\hat{x}(r) := w^2(r), \quad \hat{y}(r) := (w')^2(r), \quad \hat{z}(r) := w(r)w''(r),$

which depend only on w and its first and second derivatives. From (6.67), these variables are related to the sectional curvatures by

(6.71) $K_{\mathrm{orb}} = \dfrac{1 - \hat{y}}{\hat{x}}$ and $K_{\mathrm{rad}} = -\dfrac{\hat{z}}{\hat{x}}.$

We have

$$\hat{x}' = 2ww', \quad \hat{y}' = 2w'w'', \quad \text{and} \quad \hat{z}' = ww''' + w'w''.$$

So the curve $\hat{\mathbf{v}}(r) := (\hat{x}(r), \hat{y}(r), \hat{z}(r))$ is an immersion, i.e., $\hat{\mathbf{v}}'(r) \neq 0$, except at r where

   (1) $w = w'' = 0$ or
   (2) $w' = w''' = 0$, since $w = w' = 0$ does not correspond to a smooth point of g.

We have

(6.72) $\hat{x}\,d\hat{y} - \hat{z}\,d\hat{x} = 0.$

We can eliminate f' from the system of ODEs (6.7) by dividing the second equation therein by ww', differentiating, and then substituting in the first equation of (6.7). This yields the third-order ODE for w:

(6.73) $w\left(w'\right)^2 w'' = w^2 w' w''' - w^2(w'')^2 + (n-2)ww''$
$$+ (n-2)(1 - (w')^2)(w')^2.$$

The singular points of this ODE are the values of r for which $w^2(r)w'(r) = 0$. We multiply (6.73) by $d\hat{x} = 2ww'dr$ to obtain, in terms of $\hat{x}$, $\hat{y}$, and $\hat{z}$, the ordinary differential relation

(6.74) $0 = 2\hat{x}\hat{y}\,d\hat{z} - \left(\hat{z}^2 + 2\hat{y}\hat{z} + (n-2)\hat{y}^2 - (n-2)\hat{z} - (n-2)\hat{y}\right)d\hat{x}.$

We have from multiplying (6.26) by $(w')^2$ and substituting in the second equation of (6.7) that

(6.75)
$$\left((w')^2 - ww'' + n - 2\right)^2 - (n-1)\left((w')^2 + n - 2\right)(w')^2 = Cw^2 (w')^2;$$

that is, a first integral of (6.74) is

(6.76)
$$(\hat{y} - \hat{z} + n - 2)^2 - (n-1)(\hat{y} + n - 2)\hat{y} = C\hat{x}\hat{y},$$

or equivalently,

$$\frac{(2-n)\hat{y}^2 - 2\hat{y}\hat{z} + \hat{z}^2 - (n-3)(n-2)\hat{y} - 2(n-2)\hat{z} + (n-2)^2}{\hat{x}\hat{y}} = C.$$

For example, if $n = 3$, then

$$\frac{-\hat{y}^2 - 2\hat{y}\hat{z} + \hat{z}^2 - 2\hat{z} + 1}{\hat{x}\hat{y}} = C.$$

In particular, if $n = 3$ and $C = 0$, then we obtain the hyperbola

(6.77)
$$-\hat{y}^2 - 2\hat{y}\hat{z} + \hat{z}^2 - 2\hat{z} + 1 = 0.$$

The center of this conic section is the point $\left(-\frac{1}{2}, \frac{1}{2}\right)$. The change of variables $\hat{y} =: y - \frac{1}{2}$, $\hat{z} =: z + \frac{1}{2}$ results in the equation

(6.78)
$$-y^2 - 2yz + z^2 + \frac{1}{2} = 0.$$

The major axis of the hyperbola makes the angle $\frac{\pi}{8}$ with the positive x-axis.

For general $n \geq 3$, denoting $m := n - 2$, we have for $C = 0$ the invariant hyperbola

(6.79)
$$-m\hat{y}^2 - (m-1)m\hat{y} - 2\hat{y}\hat{z} + \hat{z}^2 - 2m\hat{z} + m^2 = 0.$$

In general dimensions, by substituting (6.72) into (6.74), we can eliminate $\hat{x}$ to obtain the ordinary differential relation

(6.80)
$$2\hat{y}\hat{z}d\hat{z} - \left(\hat{z}^2 + 2\hat{y}\hat{z} + (n-2)\hat{y}^2 - (n-2)\hat{z} - (n-2)\hat{y}\right)d\hat{y} = 0.$$

The *singular points* of the 1-form on the left-hand side (by definition, where it vanishes) are $(0,0)$, $(0, n-2)$, and $(1,0)$. We have the first-order ODE for $\hat{z}$ as a function of $\hat{y}$:

(6.81)
$$2\hat{y}\hat{z}\frac{d\hat{z}}{d\hat{y}} = \hat{z}^2 + 2\hat{y}\hat{z} + (n-2)\hat{y}^2 - (n-2)\hat{z} - (n-2)\hat{y}.$$

Let $n = 3$. Then we have

(6.82)
$$2\hat{y}\hat{z}\frac{d\hat{z}}{d\hat{y}} = \hat{z}^2 + 2\hat{y}\hat{z} + \hat{y}^2 - \hat{z} - \hat{y}$$
$$= (\hat{y} + \hat{z} - 1)(\hat{y} + \hat{z}).$$

Consider the following nonlinear autonomous system of ODEs associated to (6.82):

(6.83a) $$\frac{d\hat{y}}{dt} = 2\hat{y}\hat{z},$$

(6.83b) $$\frac{d\hat{z}}{dt} = (\hat{y} + \hat{z} - 1)(\hat{y} + \hat{z}).$$

Then, wherever $\hat{y} \neq 0$ and $\hat{z} \neq 0$, we may locally write $\hat{z} = \hat{z}(\hat{y})$ solving (6.81). The fixed points of the system (6.83) are $(0,0)$, $(1,0)$, and $(0,1)$. The integral curves of (6.83) are displayed in Figure 6.5. Note that the fixed points $(1,0)$ and $(0,1)$ lie on the right branch of the hyperbola defined by (6.77).

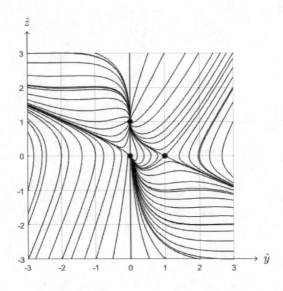

Figure 6.5. Phase portrait for the ODE system (6.83) for $(\hat{y}(t), \hat{z}(t))$, where $n = 3$. The stationary points are $(0,0)$, $(1,0)$, and $(0,1)$, where $(1,0)$ corresponds to the Euclidean metric.

Remark 6.8 (Linearization of (6.83))**.**

(1) The linearization of the ODE system (6.83) for $(\hat{y}, \hat{z})$ at $(0,0)$ is the following system for $(\tilde{y}, \tilde{z})$:

$$\frac{d\tilde{y}}{dt} = 0,$$
$$\frac{d\tilde{z}}{dt} = -\tilde{y} - \tilde{z}.$$

The corresponding 2×2 matrix $\begin{pmatrix} 0 & 0 \\ -1 & -1 \end{pmatrix}$ has eigenvector-eigenvalue pairs

$$\begin{pmatrix} -1 \\ 1 \end{pmatrix} \leftrightarrow 0, \quad \begin{pmatrix} 0 \\ 1 \end{pmatrix} \leftrightarrow -1.$$

(2) The linearization of (6.83) at $(1,0)$ is

$$\frac{d\tilde{y}}{dt} = 2\tilde{z},$$
$$\frac{d\tilde{z}}{dt} = \tilde{y} + \tilde{z}.$$

The corresponding 2×2 matrix $\begin{pmatrix} 0 & 2 \\ 1 & 1 \end{pmatrix}$ has eigenvector-eigenvalue pairs

$$\begin{pmatrix} -2 \\ 1 \end{pmatrix} \leftrightarrow -1, \quad \begin{pmatrix} 1 \\ 1 \end{pmatrix} \leftrightarrow 2.$$

(3) The linearization of (6.83) at $(0,1)$ is

$$\frac{d\tilde{y}}{dt} = 2\tilde{y},$$
$$\frac{d\tilde{z}}{dt} = \tilde{y} + \tilde{z}.$$

The corresponding 2×2 matrix $\begin{pmatrix} 2 & 0 \\ 1 & 1 \end{pmatrix}$ has eigenvector-eigenvalue pairs

$$\begin{pmatrix} 0 \\ 1 \end{pmatrix} \leftrightarrow 1, \quad \begin{pmatrix} 1 \\ 1 \end{pmatrix} \leftrightarrow 2.$$

Define the wedge

$$(6.84) \qquad W = \{(\hat{y}, \hat{z}) : \hat{z} \leq -\hat{y} \leq 0\}.$$

In W we have $\frac{d\hat{y}}{dt} \leq 0$ and $\frac{d\hat{z}}{dt} \geq 0$. Moreover:

(1) If $\hat{y} = 0$ and $\hat{z} \leq 0$, then $\frac{d\hat{y}}{dt} = 0$ and $\frac{d}{dt}(\hat{y} + \hat{z}) = (\hat{z} - 1)\hat{z}$, which is zero if $\hat{z} = 0$.

(2) If $\hat{z} = -\hat{y} \leq 0$, then $\frac{d\hat{y}}{dt} = -2\hat{y}^2$, which is zero if $\hat{y} = 0$, and $\frac{d}{dt}(\hat{y}+\hat{z}) = -2\hat{y}^2 \leq 0$.

This shows that the set W is preserved under the ODE (6.83).

We shall prove the following.

Lemma 6.9. *There exists a unique solution* $\hat{z} = \hat{z}(\hat{y})$, $\hat{y} \in (0,1)$, *with* $\lim_{\hat{y} \to 0} \hat{z}(\hat{y}) = \lim_{\hat{y} \to 1} \hat{z}(\hat{y}) = 0$. *This solution is real analytic and corresponds to the Bryant soliton. We call the corresponding curve in the $\hat{y}\hat{z}$-plane from $(1,0)$ to $(0,0)$ the* **Bryant curve**.

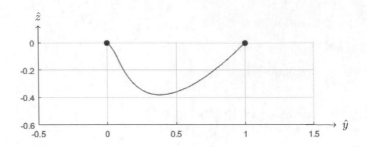

Figure 6.6. The Bryant curve in the $\hat{y}\hat{z}$-plane. Cf. Figure 6.5.

The right branch of the hyperbola given by (6.77) intersects the $\hat{z}$-axis at $\hat{z} = 1$ and the $\hat{y}$-axis at $\hat{y} = 1$. For each initial data in the interior $\mathcal{R}$ (see Figure 6.7) of the region bounded by the half-line $\{(0, \hat{z}) : \hat{z} \leq 1\}$ and the part of the right branch of the above hyperbola with $\hat{z} \leq 1$, under the ODE (6.83) the solution limits to $(0, 0)$ as $t \to \infty$. The reason is that the three stationary points are on the boundary of $\mathcal{R}$, where by Remark 6.8 on the linearization of (6.83):

(1) $(0, 1)$ is a nodal source,

(2) $(1, 0)$ is a saddle with the stable manifold composed of the part of the right branch of the hyperbola given by (6.77) with $\hat{z} \leq 1$,

(3) $(0, 0)$ is a (semistable) node where nearby solutions with $\hat{y} > 0$ converge to $(0, 0)$.

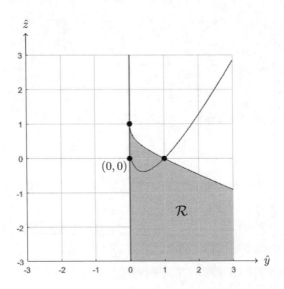

Figure 6.7. In the $\hat{y}\hat{z}$-plane, solutions to the ODE (6.83) with initial data inside $\mathcal{R}$ (shaded region) tend to $(0, 0)$. Compare with Figure 6.5.

Note also that for each initial data in $\mathcal{R}$, the solution eventually enters the wedge W defined by (6.84).

Consider the unique left-trajectory $(\hat{y}(t), \hat{z}(t))$, $t \in (-\infty, \infty)$, emanating from $(1, 0)$. By the above remarks, $\lim_{t \to \infty}(\hat{y}(t), \hat{z}(t)) = (0, 0)$. In particular, there exists a solution $\hat{z}_0 = \hat{z}_0(\hat{y})$, $\hat{y} \in (0, 1)$, with $\lim_{\hat{y} \searrow 0} \hat{z}_0(\hat{y}) = 0$ and $\lim_{\hat{y} \nearrow 1} \hat{z}_0(\hat{y}) = 0$. Translate r so that $r = 0$ at $\hat{y} = 1$. Then $w'(0) = 1$. By the results of §6.6, both g and f extend smoothly over the origin.

Claim 6.10. *The above solution, which is the curve joining $(1, 0)$ to $(0, 0)$, indeed corresponds to the Bryant soliton.*

Proof of Claim 6.10. Recall by the analysis in §6.4.3 that for our solution $(w(r), f(r))$, we have:

(i) $0 < w'(r) < 1$ and $w''(r) < 0$ for $r > 0$,

(ii) $w(r) \to 0$ and $w'(r) \to 1$ as $r \to 0_+$.

Thus, $1 > \hat{y}(t) \to 1$ and $0 > \hat{z}(t) \to 0$ as $t \to -\infty$. This implies that the curve $(\hat{y}(t), \hat{z}(t))$ corresponding to our solution $(w(r), f(r))$ is a curve joining $(1, 0)$ to $(0, 0)$ and hence is the Bryant curve. We can also see that the solution $(\hat{y}(r), \hat{z}(r))$ limits to $(0, 0)$ as $r \to \infty$ from the formula that $(\hat{y}(r), \hat{z}(r)) = ((w')^2(r), w(r)w''(r))$. □

Although $\hat{z}_0(\hat{y})$ is not real analytic at $\hat{y} = 0$ for $n = 3$, its Taylor series $\sum_{k=0}^{\infty} \frac{d^k \hat{z}_0}{d\hat{y}^k}(0)\hat{y}^k$ is a formal power series solution to the ODE (6.81), where $a_0 = 0$.

Now let $\hat{z}(\hat{y})$, $\hat{y} \in (0, \varepsilon)$, be any solution to (6.81) satisfying the condition $\lim_{\hat{y} \searrow 0} \hat{z}(\hat{y}) = 0$, denoted as $\hat{z}(0) = 0$. Let $a_k := \frac{d^k \hat{z}}{d\hat{y}^k}(0)$ for $k \geq 0$. Substituting $\hat{z} = \sum_{k=1}^{\infty} a_k \hat{y}^k$ into the ODE (6.81) yields the recursion formula

$$(6.85) \qquad (n-2)a_j = 2a_{j-1} - \sum_{k-1}^{j-1}(2k-1)a_k a_{j-k} \quad \text{for } j \geq 3.$$

We calculate that for the first three positive powers of $\hat{y}$:

$$n - 2 + a_1(n-2) = 0,$$
$$a_1^2 - 2a_1 - (n-2) + a_2(n-2) = 0,$$
$$2a_2(2a_1 - 1) + a_3(n-2) = 0.$$

Thus
(6.86)
$$a_1 = -1, \quad a_2 = \frac{n-5}{n-2}, \quad a_3 = \frac{6(n-5)}{(n-2)^2}, \quad a_4 = -\frac{3(n-5)(n-21)}{(n-2)^3}, \quad \text{etc.}$$

That is, for $\hat{y}$ near 0, the solution $\hat{z}(\hat{y})$ satisfies

$$(6.87) \qquad \hat{z}(\hat{y}) = -\hat{y} + \frac{n-5}{n-2}\hat{y}^2 + \frac{6(n-5)}{(n-2)^2}\hat{y}^3 + \mathrm{O}(\hat{y}^4).$$

Remark 6.11. Note that if $n = 5$, then (6.81) becomes

$$2\hat{y}\hat{z}\frac{d\hat{z}}{d\hat{y}} = \hat{z}^2 + 2\hat{y}\hat{z} + 3\hat{y}^2 - 3\hat{z} - 3\hat{y}$$

and $\hat{z}(\hat{y}) = -\hat{y}$ is an explicit solution, with $\hat{x} = \frac{9}{\hat{y}}$. However, $\hat{z}(1) = -1$, so this solution does not correspond to the Bryant soliton.

If $n = 3$, then the formal power series $\hat{z} = \sum_{k=1}^{\infty} a_k \hat{y}^k$ diverges. This follows from:

Claim 6.12. *If $n = 3$, then*

$$(6.88) \qquad a_j \le -2^{j-2}j! \quad for \ j \ge 1.$$

Proof of Claim 6.12. Since $a_1 = -1$ and $a_2 = -2$, (6.88) is true for $j \le 2$. Suppose that (6.88) is true for $1 \le j \le \bar{j} - 1$, where $\bar{j} \ge 3$. Then (6.85) implies that

$$a_{\bar{j}} \le 2a_{\bar{j}-1} - 2(\bar{j} - 1)a_{\bar{j}-1}a_1 = 2\bar{j}a_{\bar{j}-1} \le -2^{\bar{j}-2}\bar{j}!.$$

We conclude by induction that (6.88) is true for all $j \ge 1$. $\qquad\square$

We may write the rotationally symmetric metric g as

$$(6.89) \qquad g = \frac{dw^2}{h(w^2)} + w^2 g_{\mathbb{S}^{n-1}},$$

where h is a positive function defined on an open interval. Since
(6.90a)

$$w' = \sqrt{h(w^2)},$$

(6.90b)

$$w'' = \frac{h'(w^2)}{\sqrt{h(w^2)}}ww' = wh'(w^2),$$

(6.90c)

$$w''' = w'h'(w^2) + 2w^2h''(w^2)w' = \sqrt{h(w^2)}\left(h'(w^2) + 2w^2h''(w^2)\right),$$

we may rewrite (6.73) as the second-order ODE for h:
(6.91)

$$2v^2h(v)h''(v) - v^2(h')^2(v) + (n-2)vh'(v) + (n-2)(1 - h(v))h(v) = 0,$$

where $v = w^2$.

By using $df = f'(r)\frac{dw}{w'(r)}$ and substituting (6.90a) and (6.90b) into the second equation in (6.7), we obtain

$$(6.92) \qquad df = \frac{w^2 h'\left(w^2\right) - (n-2)(1 - h\left(w^2\right))}{wh\left(w^2\right)} dw$$

$$= \frac{vh'\left(v\right) + (n-2)(h\left(v\right) - 1)}{2vh\left(v\right)} dv.$$

Applying (6.87) to this, we obtain for $\hat{y}$ near 0

$$(6.93)$$

$$\hat{x}(\hat{y}) = \frac{\left(-(n-2) - 2\hat{y} + \frac{n-5}{n-2}\hat{y}^2 + \frac{6(n-5)}{(n-2)^2}\hat{y}^3\right)^2 - (n-1)\left(n-2+\hat{y}\right)\hat{y}}{\hat{y}}$$

$$= \frac{(n-2)^2}{\hat{y}} + (5-n)\left(n-2\right) - (n+3)\left(n-4\right)\hat{y} + \mathrm{O}(\hat{y}^2)$$

and

$$(6.94) \qquad \frac{d\hat{x}}{d\hat{y}}(\hat{y}) = -\frac{(n-2)^2}{\hat{y}^2} - (n+3)\left(n-4\right) + \mathrm{O}(\hat{y}).$$

We have

$$(6.95) \qquad r(\hat{y}) = -\int_{\hat{y}}^1 \frac{dr}{d\hat{y}}(y)dy$$

$$= -\int_{\hat{y}}^1 \frac{d\hat{x}}{d\hat{y}}(y) \left(\frac{d\hat{x}}{dr}(r(y))\right)^{-1} dy$$

$$= -\int_{\hat{y}}^1 \frac{\frac{d\hat{x}}{dy}(y)}{2\sqrt{y\hat{x}(y)}}dy.$$

Using this, (6.93), and (6.94), we compute the asymptotics

$$r(\hat{y}) = -\int_{\hat{y}}^1 \frac{-\frac{(n-2)^2}{y^2} - (n+3)\left(n-4\right) + \mathrm{O}(y)}{2\sqrt{(n-2)^2 + (5-n)\left(n-2\right)y - (n+3)\left(n-4\right)y^2 + \mathrm{O}(y^3)}}dy$$

$$= \int_{\hat{y}}^1 \frac{\frac{(n-2)^2}{y^2} + (n+3)\left(n-4\right) + \mathrm{O}(y)}{2(n-2) + (5-n)\,y + \mathrm{O}(y^2)}dy$$

$$= \int_{\hat{y}}^1 \left(\frac{n-2}{2y^2} - \frac{5-n}{4y} + \mathrm{O}(1)\right) dy$$

$$= \frac{n-2}{2\hat{y}} + \frac{5-n}{4}\ln\hat{y} + \mathrm{O}(1).$$

In particular, $\lim_{\hat{y}\to 0} r(\hat{y}) = \infty$, which again implies the completeness of g. Assuming the uniqueness part of Theorem 6.1, which will be proved in the next subsection, we obtain the Bryant soliton.

Remark 6.13. If $n = 5$, then for the solution with $\hat{z}(\hat{y}) = -\hat{y}$ and $\hat{x} = \frac{9}{\hat{y}}$, we have $r = \frac{3}{2}\left(\frac{1}{\hat{y}} - 1\right)$ by (6.95), so that $\hat{y} = \frac{1}{\frac{2}{3}r+1}$ and $\hat{x} = 6\left(r + \frac{3}{2}\right)$. Hence, in this case the metric, after translating r so that $\hat{x} := 6r$, is $g = dr^2 + 6r g_{\mathbb{S}^4}$. This metric is singular at the origin.

For r large, we have

$$(6.96) \qquad r = \frac{n-2}{2x^2} + \frac{5-n}{2}\ln x + \beta(x^2),$$

where β is a C^∞ function near $x = 0$ (since $\hat{y} = (w')^2 = x^2$). This implies

$$(6.97) \qquad x = \sqrt{\frac{n-2}{2r}}\left(1 - \frac{5-n}{8}\frac{\ln r}{r} + \frac{\gamma(r^{-1})}{r}\right),$$

where γ is a C^∞ function near $x = 0$.

By (6.67), we have

$$K_{\text{orb}} = \frac{1-\hat{y}}{\hat{x}} \quad \text{and} \quad K_{\text{rad}} = -\frac{\hat{z}}{\hat{x}}.$$

Hence, by (6.93), the orbital sectional curvatures satisfy

$$K_{\text{orb}} = \frac{1}{(n-2)^2}\hat{y}\left(1 - \frac{3}{n-2}\hat{y} + \mathrm{O}(\hat{y}^2)\right);$$

that is,

$$K_{\text{orb}} = \frac{1}{(n-2)^2}\left(x^2 - \frac{3}{n-2}x^4 + \mathrm{O}(x^6)\right).$$

By (6.97), in terms of the distance to the origin, we obtain the asymptotics

$$(6.98) \qquad K_{\text{orb}} = \frac{1}{2(n-2)r}\left(1 - \frac{5-n}{4}\frac{\ln r}{r} + \frac{\delta(r^{-1})}{r}\right),$$

where δ is a C^∞ function near $x = 0$.

In general, the radial sectional curvatures satisfy

$$K_{\text{rad}} = x^4 + 6x^8 + \mathrm{O}(x^{10}).$$

Hence we have the asymptotics

$$(6.99) \qquad K_{\text{rad}} = \frac{(n-2)^2}{4r^2}\left(1 - \frac{5-n}{2}\frac{\ln r}{r} + \frac{\epsilon(r^{-1})}{r}\right),$$

where ϵ is a C^∞ function near $x = 0$. The Ricci tensor is

$$\mathrm{Ric} = (n-1)K_{\text{rad}}dr^2 + ((n-2)K_{\text{orb}} + K_{\text{rad}})g_{\mathbb{S}^{n-1}}$$

and the scalar curvature is

$$R = 2(n-1)K_{\text{rad}} + (n-1)(n-2)K_{\text{orb}}.$$

Hence

$$R = \frac{n-1}{2r}\left(1 - \frac{5-n}{4}\frac{\ln r}{r} + \frac{\delta(r^{-1})}{r}\right).$$

This determines the asymptotics of the sectional curvatures and, modulo the uniqueness statement, completes the proof of Theorem 6.1.

Remark 6.14. As we have seen above and in Remark 6.11, the case where $n = 5$ is special. In fact, when $n = 5$, Betancourt de la Parra, Dancer, and Wang [**36**] and Alexakis, Chen, and Fournodvalos [**2**] showed that the system of ODEs for the Bryant soliton is completely integrable and the solution can be written explicitly.

Remark 6.15. From the fact that the Bryant soliton $(\mathbb{R}^3, g_{\mathrm{Bry}}, f_{\mathrm{Bry}})$ is asymptotically cylindrical (in the sense that the blow-down limits at spatial infinity are round cylinders), we can anticipate the rate of its scalar curvature decay as follows. Consider the time-dependent canonical form. Since g_{Bry} is asymptotically cylindrical, we have $R^2 \sim \frac{\partial R}{\partial t}$. On the other hand, $\frac{\partial R}{\partial t} = \langle \nabla f, \nabla R \rangle$ and $\nabla f \sim -\frac{\partial}{\partial r}$. Assuming that R decays like a power of R, we then have $R^2 \sim \langle \nabla f, \nabla R \rangle \sim \frac{R}{r}$, which implies that $R \sim \frac{1}{r}$.

6.9. Nonexistence of complete rotationally symmetric shrinking GRS

Using a similar analysis as for the Bryant soliton, we can classify rotationally symmetric shrinking GRS.

Sketch of a proof of Theorem 2.11(2) on classifying rotationally symmetric shrinkers. Let I be an open interval and on $I \times \mathbb{S}^{n-1}$ let $g = dr^2 + w^2(r)g_{\mathbb{S}^{n-1}}$, where $w : I \to \mathbb{R}_+$. By (6.4) and (6.6a), the shrinking GRS equation $\mathrm{Ric} + \nabla^2 f = \frac{1}{2}g$ is equivalent to the second-order system of ODEs:

(6.100) $\quad f'' = (n-1)\dfrac{w''}{w} + \dfrac{1}{2}, \qquad ww'f' = ww'' - (n-2)(1-(w')^2) + \dfrac{1}{2}w^2.$

As in (6.23), make the change of variables

$$x := w' = w\kappa, \quad y := (n-1)w' - wf' = wH_f, \quad dt := \frac{dr}{w}.$$

Let $\mathbf{z}(t) = (w(t), x(t), y(t))$ be defined on a maximal interval $-\infty \le T_- < t < T_+ \le \infty$. Then we obtain the equivalent system for $\mathbf{z}(t)$:

(6.101a) $\qquad\qquad \dfrac{dw}{dt} = xw,$

(6.101b) $\qquad\qquad \dfrac{dx}{dt} = x^2 - xy + n - 2 - \dfrac{1}{2}w^2,$

(6.101c) $\qquad\qquad \dfrac{dy}{dt} = x(y - (n-1)x) - \dfrac{1}{2}w^2.$

The two stationary points of (6.101) are $(0, 1, n - 1)$ and $(0, -1, -n + 1)$. Let $\mathbf{z}_0 := (0, 1, n - 1)$. The linearization of (6.101) at $\mathbf{z}_0$ is

$$(6.102) \qquad \frac{d}{dt} \begin{pmatrix} w \\ x \\ y \end{pmatrix} = \begin{pmatrix} 1 & 0 & 0 \\ 0 & -(n-3) & -1 \\ 0 & -(n-1) & 1 \end{pmatrix} \begin{pmatrix} w \\ x \\ y \end{pmatrix}.$$

The 3×3 matrix in this display has eigenvalues $\mu_1 = 1$, $\mu_2 = 2$, and $\mu_3 = 2 - n$ and associated eigenvectors

$$v_1 = \begin{pmatrix} 1 \\ 0 \\ 0 \end{pmatrix}, \quad v_2 = \begin{pmatrix} 0 \\ -\frac{1}{n-1} \\ 1 \end{pmatrix}, \quad \text{and} \quad v_3 = \begin{pmatrix} 0 \\ 1 \\ 1 \end{pmatrix},$$

respectively. Hence $\mathbf{z}_0$ is a saddle point.

Example 6.16. We exhibit the three cases of the conclusion of Theorem 2.11(2) in terms of the ODE system (6.101).

(1) If $f \equiv 0$, then $y = (n - 1)x$. The *round n-sphere* is given by

$$w(r) = \sqrt{2(n-1)} \sin\left(\frac{r}{\sqrt{2(n-1)}} \right) \qquad \text{for } 0 \leq r \leq \sqrt{2(n-1)}\pi.$$

This solution joins $\mathbf{z}_0 = (0, 1, n - 1)$ to $-\mathbf{z}_0 = (0, -1, -(n - 1))$, has $\frac{d\mathbf{z}}{dr}(0) = (1, 0, 0)$, and lies on the ellipse $x^2 + \frac{w^2}{2(n-1)} = 1$ in the plane $y = (n - 1)x$.

(2) The *Gaussian soliton* has $w(r) = r$ and $f(r) = \frac{1}{4}r^2$. Here, $x(r) \equiv 1$ and $y(r) = n - 1 - \frac{1}{2}r^2$. This solution starts at $\mathbf{z}_0$, has $\frac{d\mathbf{z}}{dr}(0) = (1, 0, 0)$, and lies on the parabola $y = n - 1 - \frac{1}{2}w^2$ in the plane $x = 1$.

(3) The *cylinder* has $w(r) \equiv \sqrt{2(n-2)}$ and $f(r) = \frac{1}{4}r^2$. We have $x(r) \equiv 0$ and $y(r) = -\sqrt{\frac{n-2}{2}} r$. This solution has constant velocity and lies on the line $\{(\sqrt{2(n-2)}, 0)\} \times \mathbb{R}$.

We have the following invariant sets of our ODE system.

Lemma 6.17. *Under the ODE system* (6.101), *each of the following regions are preserved as t increases:*

$$\left\{ x \geq 1, \ \frac{dx}{dt} \geq 0 \right\}, \quad \left\{ x \leq -1, \ \frac{dx}{dt} \leq 0 \right\}, \quad \text{and} \quad \{y \leq 0\},$$

and each of the following regions are preserved as t decreases:

$$\left\{ x \geq 1, \ \frac{dx}{dt} \leq 0 \right\}, \quad \left\{ x \leq -1, \ \frac{dx}{dt} \leq 0 \right\}, \quad \text{and} \quad \{y \geq 0\}.$$

Lemma 6.18. *If for some t_+ we have $x(t_+) = 0$, $\frac{dx}{dt}(t_+) > 0$, and $y(t_+) \leq 0$, then $x(t)$ and $\frac{dx}{dt}(t)$ increase as t increases until $x(t) > 1$. On the flip side, if for some t_- we have $x(t_-) = 0$, $\frac{dx}{dt}(t_-) > 0$, and $y(t_-) \geq 0$, then $x(t)$ and $\frac{dx}{dt}(t)$ decrease as t decreases until $x(t) < -1$. Hence, if $x(t_0) < 0$ for some t_0, then either $x(t) < 0$ for all $t \geq t_0$ or $x(t_1) > 1$ for some $t_1 > t_0$.*

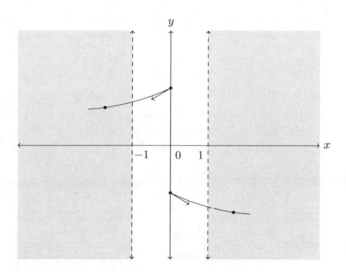

Figure 6.8. Illustration of the statement of Lemma 6.18.

If $\mathcal{M}^n$ is diffeomorphic to $\mathbb{S}^n$, then one shows that $-1 < x(t) < 1$ and $\frac{dx}{dt}(t) < 0$ for all $t \in \mathbb{R}$, $\mathbf{z}(t) \to \mathbf{z}_0$ as $t \to -\infty$, and $\mathbf{z}(t) \to -\mathbf{z}_0$ as $t \to \infty$. From this it follows that $\mathrm{Rm}_g > 0$. By Böhm and Wilking's theorem that the Ricci flow deforms any closed Riemannian manifold with $\mathrm{Rm} > 0$ to a spherical space form (see [**37**]), we conclude that g has constant positive sectional curvature (and hence f is constant).

The condition that g is complete as $t \to T_\pm$ is that

$$(6.103) \qquad \lim_{t \to T} r(t) - r(\bar{t}) = \int_{\bar{t}}^{T_\pm} w(t) dt = \pm\infty.$$

Let K_{orb} denote the sectional curvature of a plane tangent to $\mathbb{S}^{n-1}$ and let K_{rad} denote the sectional curvature of a plane containing the radial direction. Then

$$(6.104) \qquad K_{\mathrm{orb}} = \frac{1 - x^2}{w^2} \quad \text{and} \quad K_{\mathrm{rad}} = -\frac{1}{w^2}\frac{dx}{dt}.$$

Lemma 6.19. *If $x(\bar{t}) > 1$ and $\frac{dx}{dt}(\bar{t}) > 0$ for some $\bar{t} \in (T_-, T_+)$, then $\int_{\bar{t}}^{T_+} w(t) dt < \infty$. In this case, g is not complete and has negative curvature operator.*

One shows that if $\mathcal{M}^n$ is diffeomorphic to $\mathbb{R}^n$, then the unique trajectory $\mathbf{z}(t)$ emanating from $\mathbf{z}_0$ has $x(t) \equiv 1$, yielding the Gaussian soliton. If $\mathcal{M}^n$ is diffeomorphic to $\mathbb{S}^{n-1} \times \mathbb{R}$, then the unique trajectory $\mathbf{z}(t)$ satisfying (6.103) has $x(t) \equiv 0$ and $w(t) \equiv \sqrt{2(n-2)}$, yielding the round cylinder. This completes a very rough sketch of the proof of Theorem 2.11(2). See [**204**] for more details. $\square$

Rotationally symmetric expanding GRS with either positive sectional curvature or negative sectional curvature have been constructed by Bryant. In the positive sectional curvature case, the curvature decays quadratically and the expanding GRS are asymptotically cylindrical.

We now discuss the incomplete rotationally symmetric shrinking GRS asymptotic to rotationally symmetric Riemannian cones. Let $(\mathcal{N}^{n-1}, g_{\mathcal{N}})$ be a closed Riemannian manifold and define the set (thought of as a truncated cone)

$$C_s \mathcal{N} := (s, \infty) \times \mathcal{N}^{n-1} \quad \text{for } s \geq 0.$$

Definition 6.20. The **Riemannian cone** over $(\mathcal{N}^{n-1}, g_{\mathcal{N}})$ is $C_0 \mathcal{N}$ with the metric

$$g_{\text{cone}} := dr^2 + r^2 g_{\mathcal{N}}.$$

Let C denote the topological cone obtained by adding a vertex c_0 to $C_0 \mathcal{N}$ at $s = 0$. Then the metric distance $d_{g_{\text{cone}}}$ extends to a complete metric d_{C} on C and so we have the associated **pointed Riemannian cone** $(\mathrm{C}, d_{\mathrm{C}}, c_0)$ with smooth Riemannian metric g_{cone} on $\mathrm{C} \backslash \{c_0\} = C_0 \mathcal{N}$ inducing $d_{\mathrm{C}}|_{\mathrm{C} \backslash \{c_0\}}$.

Let $(\mathcal{M}^n, g)$ be a complete noncompact Riemannian manifold and let $p \in \mathcal{M}^n$. The **aperture** of g is defined by

$$(6.105) \qquad \alpha(g) := \lim_{r \to \infty} \frac{\text{diam}(\partial B_r(p))}{2r},$$

where $\text{diam}(S) := \sup_{x,y \in S} d(x, y)$ denotes the extrinsic diameter of a subset $S \subset \mathcal{M}^n$. It is easy to show that $\alpha(g)$ is independent of the choice of p. For example, $\alpha(g_{\text{Euc}}) = 1$ and the aperture of hyperbolic space is 1 as well.

We have the following result of Hamilton [**179**, Theorem 18.1].

Proposition 6.21. *If $(\mathcal{M}^n, g(t))$ is a complete noncompact solution to the Ricci flow with bounded curvature and $\text{Ric} \geq 0$, then $\alpha(g(t))$ is independent of t.*

Consider the rotationally symmetric Riemannian cone

$$(6.106) \qquad g_\beta^{\text{cone}} = dr^2 + \beta r^2 g_{\mathbb{S}^{n-1}}$$

on $C_0 \mathbb{S}^{n-1} = (0, \infty) \times \mathbb{S}^{n-1}$, where $\beta > 0$. The following result is proven in Kotschwar and Wang [**210**, Proposition B.3].

Theorem 6.22. *Let $n \geq 2$. For each $\beta > 0$ there exists a rotationally symmetric shrinking GRS $((0,\infty) \times \mathbb{S}^{n-1}, g_\beta, f_\beta)$ strongly asymptotic to the Riemannian cone $(C_0 \mathbb{S}^{n-1}, g_\beta^{\text{cone}})$ along the end $(1,\infty) \times \mathbb{S}^{n-1}$ in the sense of Definition 4.41. Moreover, if $\beta \neq 1$, then $((0,\infty) \times \mathbb{S}^{n-1}, g_\beta, f_\beta)$ cannot extend to a complete shrinking GRS. When $\beta = 1$, we have the Gaussian shrinking GRS.*

6.10. Notes and commentary

The exposition in this chapter follows Bryant [**54**] and Ivey [**195**], [**101**, Chapter 1]. The proof of the existence and uniqueness of a nonflat rotationally symmetric steady GRS (on $\mathbb{R}^n$) was given by Bryant [**54**].

Hamilton conjectured the existence of degenerate neckpinches in §3 of [**179**]. A proof of existence in the rotationally symmetric case is given by Gu and Zhu [**163**]. Also in the rotationally symmetric case, matched asymptotics are provided by Angenent, Isenberg, and Knopf [**8,9**].

Ivey [**193**] proved the existence of steady GRS where the metric is a doubly warped product of a ray with a sphere and an Einstein manifold with positive Ricci curvature.

§6.4. Appleton [**10**] proved the existence of cohomogeneity-1 noncollapsed non-Kähler steady GRS, which in the 4-dimensional case are on complex line bundles over $\mathbb{C}P^1$ with first Chern class $k \geq 3$. They are asymptotic at infinity to the $\mathbb{Z}_k$-quotient of the Bryant steady 4-soliton.

6.11. Exercises

Exercise 6.1. Describe the incomplete metrics (i.e., not the round sphere or the Gaussian soliton) in the 1-parameter family of trajectories of (6.101) in the half-space $\{(w,x,y) : w > 0\}$ which are initially tangent to the vector $(1,0,0)$. Are any of these incomplete metrics smooth at their vertices?

Exercise 6.2. (See [**204**, Proposition 3].) Let $(\mathcal{M}^n, g, f, \lambda)$, where $g = dr^2 + w^2(r)g_{\mathbb{S}^{n-1}}$ and $n \geq 3$, be a GRS and denote the covariant derivative of $(\mathbb{S}^{n-1}, g_{\mathbb{S}^{n-1}})$ by $\bar{\nabla}$. Let $x^0 = r$ and let $\{x^i\}_{i=1}^{n-1}$ be local coordinates on $\mathbb{S}^{n-1}$. Let $f_i = \frac{\partial^2 f}{\partial x^i}$, $f_{ij} = \frac{\partial^2 f}{\partial x^i \partial x^j}$, $w' = \frac{dw}{dr}$, and $w'' = \frac{d^2w}{dr^2}$.

(1) Show that for $i, j \geq 1$,

$$\nabla_0 \nabla_0 f = f_{00},$$

$$\nabla_0 \nabla_i f = f_{0i} - \frac{\omega'}{\omega} f_i,$$

$$\nabla_i \nabla_j f = \bar{\nabla}_i \bar{\nabla}_j f + \omega\omega' f_0 (g_{\mathbb{S}^{n-1}})_{ij}.$$

(2) Show that the GRS equation is equivalent to the following system: for $i, j \geq 1$,

(6.107a)
$$0 = f_{00} - (n-1)\frac{\omega''}{\omega} - \frac{\lambda}{2},$$

(6.107b)
$$0 = f_{0i} - \frac{\omega'}{\omega}f_i,$$

(6.107c)
$$0 = \bar{\nabla}_i \bar{\nabla}_j f + \left((n-2)\left(1 - (\omega')^2\right) - \omega\omega'' + \omega\omega' f_0 - \frac{\lambda}{2}\omega^2\right)(g_{\mathbb{S}^{n-1}})_{ij}.$$

(3) Show that for $i, j, k \geq 1$,
$$f_i \bar{g}_{jk} - f_k \bar{g}_{ij} = \bar{\nabla}_k \bar{\nabla}_i \bar{\nabla}_j f - \bar{\nabla}_i \bar{\nabla}_k \bar{\nabla}_j f$$
$$= \omega\omega'\left(f_{0i}\bar{g}_{jk} - f_{0k}\bar{g}_{ij}\right),$$

so that
$$(n-2)\left(\omega\omega' f_{0i} - f_i\right) = 0$$

and which combined with (6.107b) yields
$$(n-2)|\bar{\nabla}f|_{\bar{g}}^2(1 - (\omega')^2) = 0.$$

(4) Prove that if g is not flat, then f is rotationally symmetric.
HINT: Show that if f is not rotationally symmetric, then $|\omega'| \equiv 1$ and $\omega'' \equiv 0$, which by (6.67) implies that g is flat.

Exercise 6.3. Prove, regarding the smooth extension over the origin in the construction of the Bryant soliton, (6.46): $\lim_{r \to 0_+} f'(r) = 0 = f'(0)$.

Exercise 6.4. Verify, regarding the third derivative of the width function near the origin, (6.50): $w'''(r) = -\frac{2}{n-1} + o(1)$.

Exercise 6.5. Prove, regarding the expansion of the width function near the origin, (6.51).

Exercise 6.6. Verify, regarding the scalar curvature near the origin, (6.52).

Exercise 6.7. Prove, in the construction of the Bryant soliton, that Ric $\in C^0$ (at the origin).

Exercise 6.8. Prove (6.60): $\lim_{s \to \infty} \frac{X}{Y^2} = n - 2$.

Exercise 6.9. Determine if the incomplete GRS metrics in Lemma 6.19 extend to smooth metrics over $t = T_-$.

Exercise 6.10 (Volume growth of the Bryant soliton). Show that the Bryant soliton satisfies

$$C^{-1}r^{(n+1)/2} \leq \operatorname{Vol} B_r(0^n) \leq Cr^{(n+1)/2} \quad \text{for } r \geq 1,$$

where 0^n denotes the origin.

Chapter 7

Expanding and Steady GRS and the Flying Wing

In this chapter we survey some results on expanding GRS, we discuss the classification by Deng and Zhu of 3-dimensional steady solitons with curvature decaying at least linearly without a noncollapsing assumption, and then we discuss Yi Lai's [214] solution to Hamilton's conjecture on the existence of flying wing steady GRS in dimension 3.

7.1. Expanding GRS

The expanding GRS equation for a triple $(\mathcal{M}^n, g, f)$ is (2.3) with $\lambda = -1$; that is,

$$(7.1) \qquad \mathrm{Ric} + \nabla^2 f = -\frac{1}{2}g.$$

Tracing this equation yields

$$(7.2) \qquad R + \Delta f = -\frac{n}{2}.$$

Furthermore, by normalizing f via adding a suitable constant, we may assume that (see (2.45))

$$(7.3) \qquad R + |\nabla f|^2 = -f.$$

Suppose that our expander satisfies $\mathrm{Ric} \geq 0$. Then $\nabla^2 f \leq -\frac{1}{2}g < 0$, so that f is a strictly concave function. In particular, if f has a critical point, then the critical point is unique and $\mathcal{M}^n$ is diffeomorphic to $\mathbb{R}^n$.

Now suppose the stronger assumptions that Ric > 0 and that g is asymptotically conical, so that $|\mathrm{Rm}|$ decays quadratically, and in particular $R \to 0$ at infinity. Thus R attains its maximum over $\mathcal{M}^n$, say at a point o. Hence $0 = \nabla R(o) = 2\mathrm{Ric}(\nabla f)(o)$. Since Ric > 0, this implies that $\nabla f(o) = 0$. Therefore, o is the unique critical point of f, as well as the unique critical point of R.

Recall that the asymptotic volume ratio (AVR) is defined by (4.54). We have the following result of Hamilton (for an exposition, see Proposition 9.46 of [**111**]).

Theorem 7.1. *If* $(\mathcal{M}^n, g, f)$ *is a complete noncompact expanding GRS with* Ric > 0, *then the* AVR$(g) > 0$.

From this theorem one may hope that for expanding GRS, being asymptotically conical is in some sense prototypical.

7.1.1. Basic concepts of metric geometry.

In this subsection we recall some of the basic concepts of metric geometry that we will use in this chapter. Our discussion is terse and is not meant to be an introduction to metric geometry. A wonderful book on the subject, which includes plenty of motivation, intuition, and examples, is by Burago, Burago, and Ivanov [**56**].

Given two closed and bounded subsets A and B of a metric space (Z, d_Z), the **Hausdorff distance** in Z between A and B is defined by

$$(7.4) \qquad d_{\mathrm{H}}(A, B) := \inf\{\varepsilon > 0 : A \subset \mathcal{N}(B, \varepsilon) \text{ and } B \subset \mathcal{N}(A, \varepsilon)\},$$

where $\mathcal{N}(A, \varepsilon) := \{x \in Z : d_Z(x, A) < \varepsilon\}$ is the ε-neighborhood of A.

The **Gromov–Hausdorff distance** $d_{\mathrm{GH}}((X, d), (X', d'))$ between two compact metric spaces (X, d) and (X', d') is the infimum, over all metric spaces (Z, d_Z) and isometric embeddings $f : (X, d) \to (Z, d_Z)$ and $f' : (X', d') \to (Z, d_Z)$, of the Hausdorff distance in Z between $f(X)$ and $f'(X')$.

We say that a sequence of compact metric spaces $\{(X_i, d_i)\}_{i \in \mathbb{N}}$ **converges in the Gromov–Hausdorff distance** to a compact metric space (X_∞, d_∞) if

$$\lim_{i \to \infty} d_{\mathrm{GH}}((X_i, d_i), (X_\infty, d_\infty)) = 0.$$

The **length** of a continuous path $\gamma : [a, b] \to (X, d)$ is defined by

$$(7.5) \qquad \mathrm{L}(\gamma) := \sup \sum_{i=1}^{k} d(\gamma(t_{i-1}), \gamma(t_i)),$$

where the supremum is taken over all integers $k \geq 2$ and all partitions $a \leq t_0 \leq t_1 \leq \cdots \leq t_k \leq b$. We say that γ is **rectifiable** if $\mathrm{L}(\gamma) < \infty$.

We say that a metric space (X, d) is a **length metric space** if for all $x, y \in X$, the distance $d(x, y)$ is equal to the infimum of the lengths of all (rectifiable) paths from x to y.

We say that a rectifiable path $\gamma : [a, b] \to (X, d)$ is a **shortest path** if $L(\gamma) = d(\gamma(a), \gamma(b))$.

We say that a length metric space (X, d) is **complete** if for all pairs of points x and y in X there exists a shortest path from x to y.

A **triangle** $\triangle pqr$ in a complete length metric space (X, d) consists of three points $p, q, r \in X$, called **vertices**, and three shortest paths $\overline{qr}, \overline{rp}, \overline{pq}$, called **sides**, joining these points.

Given a triangle $\triangle pqr$, it is easy to see that there exists a Euclidean triangle $\triangle \tilde{p}\tilde{q}\tilde{r}$, called the **comparison triangle**, with sides of the same lengths as the corresponding sides of $\triangle pqr$.

The Toponogov comparison theorem (see [**84**, Chapter 2]) motivates the following definition. We say that a length metric space (X, d) is an **Alexandrov space with nonnegative sectional curvature** if there exists an open cover $\{\mathcal{U}_\alpha\}$ of X such for any $\mathcal{U}_\alpha$ and any triangle $\triangle pqr$ contained in $\mathcal{U}_\alpha$, we have the comparison inequality

$$(7.6) \qquad\qquad d(q, s) \geq |\tilde{q} - \tilde{s}|$$

for all $s \in \overline{rp}$ and where $\tilde{s} \in \overline{\tilde{r}\tilde{p}}$ is the unique point with $|\tilde{p} - \tilde{s}| = d(p, s)$.

7.1.2. Asymptotically conical expanders.

Recall from Theorem 3.4 that there exists a 1-parameter family of rotationally symmetric complete expanding GRS $(\mathbb{R}^2, g_\alpha, f_\alpha)$ with $R > 0$ and which is asymptotic (at its infinity) to a 2-dimensional Riemannian cone with cone angle $\alpha \in (0, 2\pi)$. As $\alpha \to 2\pi$, $(\mathbb{R}^2, g_\alpha, 0^2)$ converges in the pointed C^∞ Cheeger–Gromov sense to the Euclidean plane.

For all $n \geq 3$, Bryant [**54**] proved the following existence theorem for rotationally symmetric asymptotically conical expanders.

Theorem 7.2. *There exists a 1-parameter family of rotationally symmetric complete expanding GRS $(\mathbb{R}^n, g_c, f_c)$, $c \in (0, 1)$, with $\mathrm{Rm}_{g_c} > 0$ and which is asymptotic to the Riemannian cone $(C_0 \mathbb{S}^{n-1}, dr^2 + (cr)^2 g_{\mathbb{S}^{n-1}})$, i.e., to the Riemannian cone over $(\mathbb{S}^{n-1}, c^2 g_{\mathbb{S}^{n-1}})$. As $c \to 1$, $(\mathbb{R}^n, g_c, 0^n)$ converges in the pointed C^∞ Cheeger–Gromov sense to Euclidean n-space.*

Recall that the asymptotic curvature ratio $\mathrm{ACR}(g)$ is defined by (4.123). Chen and Deruelle [**91**, Theorem 1.2] proved that expanders with quadratic curvature decay are asymptotic to unique metric cones.

Theorem 7.3. *If $(\mathcal{M}^n, g, f)$, $n \geq 3$, is a complete expanding GRS with asymptotic curvature ratio $\mathrm{ACR}(g) < \infty$, then the metric g is asymptotic to a unique metric cone over a Riemannian manifold (Σ^{n-1}, g_Σ), where the metric is $C^{1,\alpha}$, $\alpha \in (0,1)$.*

We remark that if a complete expanding GRS $(\mathcal{M}^n, g, f)$ is asymptotic to a Riemannian cone over (Σ^{n-1}, g_Σ), then the asymptotic volume ratio satisfies

$$(7.7) \qquad\qquad \mathrm{AVR}(g) = \frac{1}{n}\,\mathrm{Vol}(g_\Sigma).$$

Let $\mathfrak{M}$ denote the class of pointed complete expanding GRS $(\mathcal{M}^n, g, f, p)$ satisfying the following conditions:

(1) $\mathrm{Ric} \geq 0$.

(2) The unique critical point of f is p.

(3) $\sup_{\mathcal{M}} |\mathrm{Rm}| \leq \lambda_0$.

(4) $\limsup_{r_p \to \infty} r_p^{k+2} |\nabla^k \mathrm{Rm}| \leq \Lambda_k$ for all $k \geq 0$.

Deruelle [**135**] proved the following compactness theorem for expanders.

Theorem 7.4. *Given any sequence of constants λ_0, Λ_k, $k \geq 0$, the space $\mathfrak{M}$ is compact in the pointed C^∞ topology. Moreover, the asymptotic cones of a convergent sequence subconverge in the Gromov–Hausdorff distance.*

Deruelle [**134**, Theorem 1.3] has also proved the following theorem regarding the existence of a unique asymptotically conical expander with a prescribed asymptotic cone.

Theorem 7.5. *Let (Σ^{n-1}, g_Σ) be a simply connected closed Riemannian manifold with curvature operator $\mathrm{Rm} > \mathrm{id}$. Then there exists a unique (up to isometry) complete expanding GRS $(\mathcal{M}^n, g, f)$ with positive curvature operator such that the metric g is asymptotic to the Riemannian cone $(C_0\Sigma, dr^2 + r^2 g_\Sigma)$. For this unique expander, the gradient of the potential function ∇f is asymptotic to the radial vector field $r\partial_r/2$.*

In [**134**, Theorem 1.4], Deruelle proved the following result regarding the existence of 1-parameter families of asymptotically conical expanders connecting to a Bryant expander.

Theorem 7.6. *Let $(\mathbb{R}^n, g, f)$ be any nonflat complete expanding GRS with $\mathrm{Rm} \geq 0$ and which is asymptotic to the Riemannian cone over a manifold*

(Σ^{n-1}, g_Σ). *Then there exists* $c = c(\mathrm{Vol}(g_\Sigma), n) \in (0, 1)$ *and a 1-parameter family of expanding GRS* $(\mathbb{R}^n, g(s), f(s))$, $s \in [0, 1]$, *satisfying:*

(1) $g(s)$ *is asymptotically conical for* $s \in [0, 1]$.

(2) $(g(0), f(0)) = (g, f)$ *and* $(g(1), f(1)) = (g_c, f_c)$ *is a Bryant expanding GRS.*

(3) $\mathrm{Rm}_{g(s)} > 0$ *for* $s \in (0, 1]$.

Note that in Theorem 7.6, the curvature of the asymptotic slice of g satisfies $\mathrm{Rm}_{g_\Sigma} \geq \mathrm{id}$. Let $g_\Sigma(s)$ denote the asymptotic slice metric of the asymptotically conical metric $g(s)$. We may define $g(s)$ by choosing $g_\Sigma(s)$, $s \in [0, 1]$, to be a time- and space-rescaled solution to the Ricci flow satisfying $\mathrm{Rm}_{g_\Sigma(s)} > \mathrm{id}$ for $s \in (0, 1]$, $g_\Sigma(0) = g_\Sigma$, $g_\Sigma(1) = (g_c)_\Sigma = c^2 g_{\mathbb{S}^{n-1}}$, and applying Theorem 7.5, so that $g(0) = g$ and $g(1) = g_c$.

Chodosh [**97**, Theorem 1.2] has proven the following uniqueness theorem for expanders asymptotic to a rotationally symmetric cone.

Theorem 7.7. *Let* $(\mathcal{M}^n, g, f)$, $n \geq 3$, *be a complete expanding GRS with positive sectional curvature and which is asymptotic to the Riemannian cone over* $(\mathbb{S}^{n-1}, c^2 g_{\mathbb{S}^{n-1}})$, *where* $c \in (0, 1)$. *Then* $(\mathcal{M}^n, f, g)$ *is rotationally symmetric.*

7.2. Steady GRS with nonnegative and linear decaying curvature

In this section we discuss one of the results in the works of Deng and Zhu [**128–130**]. The sign convention for the potential function f that we use in this section, which is *opposite* to that in the rest of the book, is given by the steady GRS equation on $\mathcal{M}^n$ as

(7.8) $$\mathrm{Ric} = \nabla^2 f.$$

We have the following result of Deng and Zhu [**129**, Corollary 1.2], which builds on Brendle's classification of noncollapsed 3-dimensional steady GRS as the Bryant soliton or Euclidean 3-space (which is the topic of the next chapter).

Theorem 7.8. *Let* $(\mathcal{M}^3, g, f)$ *be a simply connected complete steady GRS with at least linear curvature decay in the sense that*

(7.9) $$R(x) \leq \frac{C}{d(x, p)} \quad \text{for all } x \neq p$$

and for some constant C. *Then one of the following is true:*

(1) $(\mathcal{M}^3, g)$ *is Euclidean 3-space.*

(2) $(\mathcal{M}^3, g)$ *is the 3-dimensional Bryant steady soliton.*

If $\mathcal{M}^3$ is not simply connected, then $(\mathcal{M}^3, g)$ is isometric to either:

(1) *a quotient of Euclidean 3-space, i.e., a 3-dimensional Euclidean space form,*

(2) *the quotient of the product of $\mathbb{R}$ with the cigar soliton by an isometric $\mathbb{Z}$-action generated by a g-isometry of the form*

(7.10) $$(x, y) \mapsto (x + \alpha, \phi(y)),$$

where $\phi : \mathbb{S}^1 \to \mathbb{S}^1$ is an isometry.

The remainder of this section is devoted to the proof of this theorem.

Let $(\mathcal{M}^3, g, f)$ be a complete steady GRS with at least linear curvature decay. By Corollary 5.4 (a consequence of the localized Hamilton–Ivey estimate), $(\mathcal{M}^3, g)$ must have nonnegative sectional curvature. By Theorem 5.7 (a consequence of the strong maximum principle for systems), either $(\mathcal{M}^3, g)$ has positive sectional curvature or the universal covering $(\widetilde{\mathcal{M}}^3, \widetilde{g})$ is Euclidean 3-space or the product of $\mathbb{R}$ and the cigar soliton $(\mathbb{R}^2, g_\Sigma)$.

If $(\widetilde{\mathcal{M}}^3, \widetilde{g})$ is Euclidean 3-space, then we are done. On the other hand, we will classify the quotients of $\mathbb{R} \times (\mathbb{R}^2, g_\Sigma)$ at the end of the proof of the theorem (note that $\mathbb{R} \times (\mathbb{R}^2, g_\Sigma)$ itself does not have linear curvature decay). So we assume that $(\mathcal{M}^3, g)$ has positive sectional curvature. In this case it remains only to prove that $(\mathcal{M}^3, g)$ is the 3-dimensional Bryant steady soliton. By Brendle's theorem (see Theorem 8.1 below), it suffices to show that g is noncollapsed, which in turn will follow from showing that g is asymptotically cylindrical.

Since $\mathrm{Ric} = \nabla^2 f$, we have that f is a strictly convex function. Since $R > 0$ and since hypothesis (7.9) implies that $R(x) \to 0$ as $x \to \infty$, we have that R achieves its maximum at some point p. We normalize g so that

(7.11) $$R(p) = \sup_\mathcal{M} R = 1.$$

This point p is a critical point of f since $\nabla R = 2\mathrm{Ric}(\nabla f)$ and $\mathrm{Ric} > 0$. Since f is strictly convex, p must be the unique minimum point of f. We normalize f (by adding a constant to it if necessary) so that

(7.12) $$f(p) = 1.$$

Since $R + |\nabla f|^2$ is constant by (2.44) and by using (7.11), we have

(7.13) $$R + |\nabla f|^2 \equiv 1.$$

By using the curvature hypothesis (7.9), which guarantees that the curvature decays at least linearly, we can show that the potential function f is comparable to the distance-to-a-point function in the following sense.

Lemma 7.9. *There exists a positive constant C such that*

(7.14) $$d(x,p) - C\sqrt{d(x,p)} \leq f(x) \leq d(x,p) + 1$$

for all $x \in \mathcal{M}^3$ such that $d(x,p) \geq 1$.

Proof. Since $R \geq 0$, we obtain $|\nabla f| \leq 1$. This and $f(p) = 1$ imply the upper bound $f(x) \leq 1 + d(x,p)$ for all $x \in \mathcal{M}^3$.

On the other hand, since f is strictly convex, for any unit speed geodesic $\gamma : \mathbb{R} \to \mathcal{M}^3$ we have that $f \circ \gamma$ is strictly convex. Indeed,

$$(f \circ \gamma)''(t) = \frac{d}{dt} df(\gamma'(t)) = \nabla df\left(\gamma'(t), \gamma'(t)\right) > 0 \quad \text{for all } t \in \mathbb{R}.$$

Since p is (the unique) critical point of f and by the compactness of the unit sphere in $T_p\mathcal{M}$, there exists a positive constant c such that $\langle \nabla f, \gamma'(1) \rangle = (f \circ \gamma)'(1) \geq c$ for all γ such that $\gamma(0) = p$. (Note that $(f \circ \gamma)'(t) \geq 0$ for $t \geq 0$.) For any such geodesic γ, we have $\langle \nabla f, \gamma'(t) \rangle \geq c$ for $t \geq 1$, so that

$$f(\gamma(r)) = \int_0^r \langle \nabla f, \gamma'(t) \rangle \, dt \geq c(r-1) \quad \text{for } r \geq 1.$$

Hence, by considering minimal geodesic segments, we see that for any point $x \in \mathcal{M}^3 \setminus B_1(p)$,

(7.15) $$f(x) \geq c(d(x,p) - 1).$$

In particular, f is a proper function (see Definition 4.2). Let

$$\Sigma_s := f^{-1}(s),$$

which is a closed hypersurface diffeomorphic to $\mathbb{S}^2$. Since $R(x) \to 0$ as $x \to \infty$, there exists $s_0 > 1$ such that

(7.16) $$|\nabla f|(x) \geq \frac{1}{2} \quad \text{for all } x \in \mathcal{M}^3 \text{ such that } f(x) \geq s_0.$$

Let $\Omega_s := \{x \in \mathcal{M}^3 : f(x) \leq s\}$, which is a compact subset diffeomorphic to a closed ball for any $s > 1$. Let $r_0 := 1 + \max_{x \in \Sigma_{s_0}} d(x,p)$. Then $\Omega_{s_0} \subset B_{r_0}(p)$. Let $x \in \mathcal{M}^3 \setminus B_{r_0}(p)$ and let $\phi_t : \mathcal{M}^3 \to \mathcal{M}^3$, $t \in \mathbb{R}$, be the 1-parameter group of diffeomorphisms generated by the vector field $-\nabla f$. For $t \geq 0$, as long as $\phi_t(x) \in \mathcal{M}^3 \setminus B_{r_0}(p)$, we have

(7.17) $$\frac{d}{dt}\left(f(\phi_t(x))\right) = \langle \nabla f, d_t\phi_t(x) \rangle = -|\nabla f|^2(\phi_t(x)) \leq -\frac{1}{2}.$$

Let $r \geq r_0$. By (7.17), for each $x \in \mathcal{M}^3 \setminus B_r(p)$ there exists $t_x > 0$ such that

$$\phi_{t_x}(x) \in \partial B_r(p).$$

Observe that

$$\frac{d}{dt}\big(|\nabla f|^2(\phi_t(x))\big) = -2\nabla^2 f(\nabla f, \nabla f)(\phi_t(x)) = -2\mathrm{Ric}(\nabla f, \nabla f)(\phi_t(x)) < 0,$$

provided that $\phi_t(x) \neq p$. Hence

$$f(x) - f(\phi_{t_x}(x)) = \int_0^{t_x} |\nabla f|^2(\phi_t(x))\,dt$$

$$\geq |\nabla f|(\phi_{t_x}(x)) \int_0^{t_x} |\nabla f|(\phi_t(x))\,dt$$

$$\geq |\nabla f|(\phi_{t_x}(x))\,d\big(x, \phi_{t_x}(x)\big),$$

since

$$d\big(x, \phi_{t_x}(x)\big) \leq \mathrm{L}(\phi_t(x)|_{t\in[0,t_x]}) = \int_0^{t_x} |\nabla f|(\phi_t(x))\,dt,$$

where L denotes length.

On the other hand, since $\phi_{t_x}(x) \in \partial B_r(p)$ and by the triangle inequality,

$$d\big(x, \phi_{t_x}(x)\big) \geq d(x,p) - r.$$

Thus,

$$f(x) - f(\phi_{t_x}(x)) \geq |\nabla f|(\phi_{t_x}(x))\,(d(x,p) - r).$$

Now, for any $y \in \partial B_r(p)$, we have by hypothesis (7.9) that

$$(7.18) \qquad |\nabla f|(y) = \sqrt{1 - R(y)} \geq \sqrt{1 - Cr^{-1}} \geq 1 - Cr^{-1},$$

where we from now on assume that $r > C$. So, with the aid of (7.15), we have that

$$f(x) \geq (1 - Cr^{-1})(d(x,p) - r) + c(r - 1).$$

This estimate is qualitatively optimized by taking $r = \sqrt{d(x,p)}$, in which case we obtain

$$f(x) \geq (1 - Cr^{-1})d(x,p) - C\sqrt{d(x,p)}.$$

This completes the proof of the lemma. $\square$

In particular, the hypothesis (7.9) implies that

$$(7.19) \qquad |\mathrm{Rm}(x)|\,f(x) \leq CR(x)f(x) \leq C^2 \quad \text{for all } x \in \mathcal{M}^3.$$

We have the following weighted derivative of curvature estimates.

Lemma 7.10. *For each $k \geq 0$, there exists a constant C_k such that*

$$(7.20) \qquad |\nabla^k \mathrm{Rm}|\,f^{\frac{k}{2}+1} \leq C_k \quad \text{on } \mathcal{M}^3.$$

Proof. By (7.19) we have the estimate for $k = 0$ with $C_0 = C^2$. Let $(\mathcal{M}^3, g(t), f(t))$, $t \in \mathbb{R}$, be the canonical form, which is a Ricci flow extension of our steady GRS. We now proceed to show that we can invoke Shi's local derivative of curvature estimates for Ricci flow (Theorem 1.6). To this end we will show that there exists a positive constant c with the following property:

For any $x_0 \in \mathcal{M}^3 \setminus B_1(p)$, we have the following curvature bound in a parabolic neighborhood:

$$(7.21) \qquad |\mathrm{Rm}|(x, t) \leq \frac{1}{c r_0} \quad \text{for all} \ (x, t) \in B_{c\sqrt{r_0}}^{g(0)}(x_0) \times [-c r_0, 0],$$

where $r_0 := d(x_0, p)$ and $g(0) = g$.

By Theorem 1.6, we then have

$$(7.22) \qquad |\nabla^k \mathrm{Rm}|(x_0, 0) \leq \frac{1}{c_k r_0^{\frac{k}{2}+1}}.$$

Since f is comparable to d_p, this shows that the lemma follows from the curvature estimate (7.21), which we now prove.

Let $(x, t) \in B_{c\sqrt{r_0}}^{g(0)}(x_0) \times [-c r_0, 0]$. By Proposition 4.19, $g(t) = \varphi_t^* g$, where $\partial_t \varphi_t(x) = -(\nabla_g f)(\varphi_t(x))$ and the minus sign on the right-hand side appears because we are using the opposite sign convention for the potential function in this section. Since $|\partial_t \varphi_t(x)| = -|\nabla_g f|(\varphi_t(x)) \leq 1$, we have

$$d(\varphi_t(x), p) \geq r_0 - c\sqrt{r_0} - c r_0 \geq \frac{1}{2} r_0$$

for $c > 0$ sufficiently small. Hence

$$|\mathrm{Rm}|(x, t) = |\mathrm{Rm}|(\varphi_t(x), 0) \leq \frac{C}{d(\varphi_t(x), p)} \leq \frac{2C}{r_0}$$

for some constant C, which proves (7.21). This completes the proof of the lemma. $\qquad \square$

Remark 7.11. If one makes the assumption that $(\mathcal{M}^3, g)$ is noncollapsed in addition to assuming linear curvature decay, then Lemma 7.10 follows from Perelman's derivatives of curvature estimates [**251**, §1.5]. On the other hand, Perelman's derivatives of curvature estimates do not require any curvature decay.

Now we investigate the geometry of the level sets of the potential function. Let

$$\Sigma_s = f^{-1}(s)$$

for $s \geq 1$, which is diffeomorphic to $\mathbb{S}^2$ for $s > 1$. Note that $\Sigma_1 = \{p\}$ by our potential function normalization (7.12). Fix a diffeomorphism $\Phi_2 : \mathbb{S}^2 \to \Sigma_2$, and extend this choice by defining diffeomorphisms $\Phi_s : \mathbb{S}^2 \to \Sigma_s$, $s > 1$, to satisfy the ODE:

$$(7.23) \qquad \partial_s \Phi_s(z) = \frac{\nabla f}{|\nabla f|^2}(\Phi_s(z)) \quad \text{for all } z \in \mathbb{S}^2, \ s > 1.$$

Note that $\partial_s f(\Phi_s(z)) = 1$ for $z \in \mathbb{S}^2$, so that $f(\Phi_s(z)) = s$, and hence Φ_s is indeed a diffeomorphism from $\mathbb{S}^2$ to Σ_s. Let

$$(7.24) \qquad g_s := \Phi_s^*(g)$$

be the pulled back induced metrics on $\mathbb{S}^2$, and let d_s denote the intrinsic distance of $(\mathbb{S}^2, g_s)$. Of course, $(\mathbb{S}^2, g_s)$ is isometric to $(\Sigma_s, g|_{\Sigma_s})$.

The following "changing distances formula", inspired by Perelman [**250**, Lemma 8.3(a)], is Lemma 3.2 in [**129**]. The idea is that, under the hypothesis of curvature decay, the level sets of the potential function asymptotically satisfy the Ricci flow equation; see (7.49) below.

Lemma 7.12. *For all $z_1, z_2 \in \mathbb{S}^2$, $s \geq 2$, and $0 < r_* \leq d_s(z_1, z_2)/2$, we have*

$$(7.25) \qquad \frac{\partial}{\partial s} d_s(z_1, z_2) \leq C \left(\frac{1}{r_*} + \frac{r_*}{s} + \frac{d_s(z_1, z_2)}{s^2} \right).$$

In particular, if $d_s(z_1, z_2) \geq 2\sqrt{s} \geq 2\sqrt{2}$, then we may take $r_* = \sqrt{s}$ to obtain

$$(7.26) \qquad \frac{\partial}{\partial s} d_s(z_1, z_2) \leq C \left(\frac{2}{\sqrt{s}} + \frac{d_s(z_1, z_2)}{s^2} \right).$$

Proof. Denote $r_s := d_s(z_1, z_2)$ and let $\gamma : [0, r_s] \to \mathbb{S}^2$ be a minimal geodesic joining z_1 to z_2 with respect to g_s. Let Ric_{g_s} denote the Ricci tensor of g_s. Since $n - 1 = 2$, we have $\mathrm{Ric}_{g_s} = \frac{1}{2} R_s g_s$, where R_s is the scalar curvature of g_s, but we compute in general dimensions. By the same proof as for Lemma 2.20, except now we define ζ by

$$\zeta(r) = \begin{cases} \frac{r}{r_*} & \text{if } 0 \leq r \leq r_*, \\ 1 & \text{if } r_* < r \leq r_s - r_*, \\ \frac{r_s - r}{r_*} & \text{if } r_s - r_* < r \leq r_s, \end{cases}$$

we obtain from the second variation of arc length formula for γ that

$$(7.27) \qquad \int_0^{r_s} \mathrm{Ric}_{g_s}(T, T) \, dr \leq \frac{2(n-1)}{r_*} + \frac{2}{3} r_* \left(\mathrm{S}_s(z_1) + \mathrm{S}_s(z_2) \right),$$

where $T := \gamma'(r)$ and

$$\mathrm{S}_s(z) := \sup_{V \in \mathcal{S}_y^{n-2}, \, y \in B_{r_*}^{g_s}(z)} \mathrm{Ric}_{g_s}(V, V)_+.$$

Let $N := \frac{\nabla f}{|\nabla f|}$, which is a unit vector field on $\mathcal{M}^3 \backslash \{p\}$ normal to the level sets Σ_s of f. We write the ODE (7.23) as the normal flow of hypersurfaces:

$$(7.28) \qquad \partial_s \Phi_s = \frac{1}{|\nabla f|} N.$$

A standard calculation yields the evolution of the metrics g_s as given by the first equality below (see e.g. [**5**, §5.3] for a formula for Euclidean ambient spaces that is easily generalized to Riemannian ambient spaces):

$$(7.29) \qquad \partial_s g_s = \frac{2}{|\nabla f|} \mathrm{II} = \frac{2}{|\nabla f|^2} \nabla^2 f = \frac{2}{|\nabla f|^2} \mathrm{Ric},$$

where II denotes the second fundamental form of Σ_s and where the symmetric 2-tensors II, $\nabla^2 f$, and Ric are all pulled back to $\mathbb{S}^2$ by the maps Φ_s. In particular, $\mathrm{Ric} := \Phi_s^*(\mathrm{Ric})$. We will show that (7.29) approximates the backward Ricci flow on surfaces for s large.

In any case, by (7.29), for any minimal geodesic γ joining z_1 to z_2 as above, we have

$$(7.30) \qquad \frac{\partial}{\partial s} d_s(z_1, z_2) \leq \int_0^{r_s} \frac{1}{|\nabla f|^2} \mathrm{Ric}(T, T) \, dr,$$

where $T := \gamma'(r)$. On the other hand, by a trace of the Gauss equation (2.109), we have

$$(7.31) \quad \mathrm{Ric}(T, T) = \mathrm{Ric}_{g_s}(T, T) + \mathrm{Rm}(N, T, T, N) - H \, \mathrm{II}(T, T) + \mathrm{II}^2(T, T),$$

where H denotes the mean curvature of Σ_s, Ric_{g_s} is the Ricci tensor of g_s, $\mathrm{Rm}(N, \cdot, \cdot, N) := \Phi_s^*(\mathrm{Rm}(N, \cdot, \cdot, N))$, and $H := H \circ \Phi_s$.

Now, by (4.104), we have that

$$R_{\ell k j m} \nabla_\ell f \nabla_m f = \Delta R_{jk} - \frac{1}{2} \nabla_j \nabla_k R + 2 R_{ijk\ell} R_{i\ell} - R_{j\ell} R_{\ell k}.$$

Thus,

$$|\nabla f|^2 \mathrm{Rm}(N, T, T, N) = \mathrm{Rm}(\nabla f, T, T, \nabla f)$$
$$= \Delta \mathrm{Ric}(T, T) - \frac{1}{2} \nabla^2 R(T, T)$$
$$+ 2 R_{ijk\ell} R_{i\ell} T_j T_k - \mathrm{Ric}^2(T, T).$$

By applying to this the curvature and its derivatives estimates of Lemma 7.10, on Σ_s, $s \geq 2$, we obtain that

$$(7.32) \qquad 0 < \mathrm{Rm}(N, T, T, N) \leq \frac{C}{s^2}.$$

Furthermore, by letting $\{e_1, e_2\}$ be an orthonormal frame field on Σ_s, we have

$$-H\,\mathrm{II}(T,T) + \mathrm{II}^2(T,T) = \frac{1}{|\nabla f|^2}\left(\mathrm{Ric}(e_1, e_1) + \mathrm{Ric}(e_2, e_2)\right)\mathrm{Ric}(T,T)$$
$$- \frac{1}{|\nabla f|^2}\mathrm{Ric}(e_1, T)^2 - \mathrm{Ric}(e_2, T)^2.$$

So by applying the at least linear curvature decay assumption, we have on Σ_s, $s \geq 2$, that

$$(7.33) \qquad\qquad \left|\mathrm{Ric}(T,T) - \mathrm{Ric}_{g_s}(T,T)\right| \leq \frac{C}{s^2}.$$

An easy consequence of this is that

$$(7.34) \qquad\qquad \mathrm{Ric}_{g_s}(T,T) \leq \frac{C}{s}.$$

Hence, the second variation argument used to obtain (7.27) yields

$$\int_0^{r_s} \mathrm{Ric}_{g_s}(T,T)\, dr \leq \frac{2(n-1)}{r_*} + \frac{Cr_*}{s}.$$

Therefore, we conclude by (7.30) and (7.33) that

$$\frac{\partial}{\partial s} d_s(z_1, z_2) \leq \int_0^{r_s} \frac{1}{|\nabla f|^2}\left(\mathrm{Ric}_{g_s}(T,T) + \left|\mathrm{Ric}(T,T) - \mathrm{Ric}_{g_s}(T,T)\right|\right) dr$$
$$\leq \frac{C}{r_*} + \frac{Cr_*}{s} + \frac{C}{s^2} d_s(z_1, z_2).$$

This completes the proof of the changing distances lemma. $\square$

The following level set diameter bound is Proposition 3.3 in [**129**].

Proposition 7.13. *There exists a constant C such that*

$$(7.35) \qquad\qquad \mathrm{diam}(\Sigma_s) \leq C\sqrt{s} \quad \textit{for all } s \geq 2.$$

Proof. Let $s \geq 2$. By Lemma 7.12, if $d_s(z_1, z_2) \geq 2\sqrt{s}$, then

$$\frac{\partial}{\partial s} d_s(z_1, z_2) \leq C\left(\frac{2}{\sqrt{s}} + \frac{d_s(z_1, z_2)}{s^2}\right).$$

On the other hand, suppose that $d_s(z_1, z_2) < 2\sqrt{s}$. By (7.30), we have

$$\frac{\partial}{\partial s} d_s(z_1, z_2) \leq \int_0^{d_s(z_1, z_2)} \frac{1}{|\nabla f|^2}\mathrm{Ric}(T,T)\, dr.$$

On Σ_s we have $|\nabla f| \geq c$, where $c > 0$ is independent of $s \geq 2$. On Σ_s we also have $\mathrm{Ric}(T,T) \leq R \leq \frac{C}{f} = \frac{C}{s}$. Thus, if $d_s(z_1, z_2) < 2\sqrt{s}$, then

$$\frac{\partial}{\partial s} d_s(z_1, z_2) \leq 2\sqrt{s}\,\frac{1}{c^2}\frac{C}{s}.$$

We conclude that there exists a constant C_1 such that for all $s \geq 2$,

$$\frac{\partial}{\partial s} d_s(z_1, z_2) \leq C_1 \left(\frac{2}{\sqrt{s}} + \frac{d_s(z_1, z_2)}{s^2} \right). \tag{7.36}$$

Thus,

$$\frac{\partial}{\partial s} \left(e^{\frac{C_1}{s}} d_s(z_1, z_2) \right) \leq e^{\frac{C_1}{s}} \frac{2C_1}{\sqrt{s}} \leq \frac{C_2}{2\sqrt{s}}$$

for some constant C_2. Integrating this from 2 to s, we obtain

$$e^{\frac{C_1}{s}} d_s(z_1, z_2) - e^{\frac{C_1}{2}} d_2(z_1, z_2) \leq C_2 (\sqrt{s} - \sqrt{2}).$$

We conclude for all $z_1, z_2 \in \mathbb{S}^2$ and $s \geq 2$ that

$$d_s(z_1, z_2) \leq C_3 (\sqrt{s} + 1) \tag{7.37}$$

for some constant C_3. The proposition follows. $\qquad \square$

The following is Proposition 4.3 in [**129**].

Proposition 7.14. *In addition to the inverse linear upper bound, the scalar curvature has an inverse linear lower bound:*

$$\frac{C^{-1}}{d(x,p)} \leq R(x) \leq \frac{C}{d(x,p)} \quad \text{for all } x \text{ such that } d(x,p) \geq 1. \tag{7.38}$$

Proof. We argue by contradiction. Suppose that there exists a sequence of points $p_i \to \infty$ such that $R(p_i) s_i \to 0$, where $s_i := f(p_i)$.

Claim 7.15. *The Riemann curvature tensor of $(\mathcal{M}^3, g)$ has the property that, after passing to a subsequence, we have*

$$s_i \max_{\Sigma_{s_i}} |\mathrm{Rm}| \leq C s_i \max_{\Sigma_{s_i}} R \to 0. \tag{7.39}$$

Note that the inequality part of the statement is simply a consequence of the nonnegativity of the curvature.

Before we prove Claim 7.15, we show that it implies the proposition. Let $R_{\Sigma_{s_i}}$ denote the scalar curvature of the surface $(\Sigma_{s_i}, g_{\Sigma_{s_i}})$, where $g_{\Sigma_{s_i}}$ denotes the metric on Σ_{s_i} induced from $(\mathcal{M}^3, g)$, a.k.a. the first fundamental form which we also denote by $\mathrm{I} := g_{\Sigma_{s_i}}$. By the Gauss equation, we have that

$$R_{\Sigma_{s_i}} = R - 2\mathrm{Ric}(N, N) + 2\frac{\det \mathrm{II}}{\det \mathrm{I}}, \tag{7.40}$$

where $N := \frac{\nabla f}{|\nabla f|}$. By Claim 7.15 and (7.40), we have that

$$R_{\Sigma_{s_i}} = o(s_i^{-1}) + 2\frac{\det \mathrm{II}}{\det \mathrm{I}}. \tag{7.41}$$

On the other hand,

$$(7.42) \qquad \frac{\det \mathrm{II}}{\det \mathrm{I}} \leq |\mathrm{II}|_{\mathrm{I}}^2 = \frac{|\nabla^2 f|_g^2}{|\nabla f|_g^2} = \frac{|\mathrm{Ric}|_g^2}{|\nabla f|_g^2} = \mathrm{o}(s_i^{-2}).$$

So we have that the intrinsic scalar curvature of Σ_{s_i} satisfies

$$(7.43) \qquad \max_{\Sigma_{s_i}} |R_{\Sigma_{s_i}}| = \mathrm{o}(s_i^{-1}).$$

By combining this with the bound $\mathrm{diam}(\Sigma_{s_i}) \leq C\sqrt{s_i}$ given by Proposition 7.13, we obtain

$$(7.44) \qquad \mathrm{diam}(\Sigma_{s_i})^2 \max_{\Sigma_{s_i}} |R_{\Sigma_{s_i}}| = \mathrm{o}(1).$$

Since Σ_{s_i} is diffeomorphic to $\mathbb{S}^2$, this is a contradiction. Indeed, by the Gauss–Bonnet formula and volume comparison, we have that

$$(7.45) \qquad 8\pi = 4\pi\chi(\Sigma_{s_i}) = \int_{\Sigma_{s_i}} R_{\Sigma_{s_i}}\, d\mu_{\Sigma_{s_i}} \leq C\, \mathrm{diam}(\Sigma_{s_i})^2 \max_{\Sigma_{s_i}} |R_{\Sigma_{s_i}}|,$$

which contradicts (7.44). Thus the proposition follows from the claim.

Proof of Claim 7.15. We argue by contradiction again. Suppose that there exist points $q_i \in \Sigma_{s_i}$ such that

$$s_i R(q_i) \geq c$$

for all i and some positive constant c. Consider the sequence of Riemannian surfaces (diffeomorphic to $\mathbb{S}^2$)

$$\{(\Sigma_{s_i}, s_i^{-1} g_{\Sigma_{s_i}})\}.$$

By (7.20), we have that $|\nabla^k \mathrm{Rm}_g|(x) \leq C_k s_i^{-\left(\frac{k}{2}+1\right)}$, so that

$$|\nabla^k \mathrm{Rm}_{s_i^{-1}g}|(x) \leq C_k \quad \text{for all } x \in \Sigma_{s_i}.$$

On the other hand, we have the second fundamental form derivative estimate

$$|\nabla^k \mathrm{II}| \leq \frac{C_k}{s_i^{\frac{k}{2}+1}}.$$

Thus, by the Gauss equation (and its covariant derivatives), we obtain that

$$|\nabla^k \mathrm{Rm}_{s_i^{-1}g_{\Sigma_{s_i}}}| \leq C_k.$$

Moreover, by Proposition 7.13, there exists a constant C such that

$$\mathrm{diam}(\Sigma_{s_i}, s_i^{-1} g_{\Sigma_{s_i}}) \leq C$$

for all i.

Now, given a positive constant C, there exists a positive constant c such that for any closed orientable Riemannian surface $(\mathcal{M}^2, g)$ with $|R| \leq C$, diam $\leq C$, and not diffeomorphic to a 2-torus, we have inj $\geq c$. Indeed, by the Gauss–Bonnet formula, we have

$$(7.46) \qquad 8\pi \leq 4\pi |\chi(\mathcal{M}^2)| \leq \int_{\mathcal{M}} |R| \, d\mu \leq C \operatorname{Area}(g),$$

so that $\operatorname{Area}(g) \geq v/C$. By a result of Cheeger, Gromov, and Taylor [85], this implies that $\operatorname{inj}(g) \geq c$ for some positive constant c depending only on C.

Hence, we have uniform injectivity radius estimates for $s_i^{-1} g_{\Sigma_{s_i}}$ and we may apply the pointed C^∞ Cheeger–Gromov convergence theorem (Theorem 1.2) to the sequence of Riemannian 2-spheres $(\Sigma_{s_i}, s_i^{-1} g_{\Sigma_{s_i}}, p_i)$ to obtain subconvergence to a Riemannian 2-sphere $(\mathbb{S}^2, g_\infty, p_\infty)$ satisfying $R_\infty(p_\infty) = 0$.

The formula (7.31) and the inequality (7.32) hold for any $T \in T\mathbb{S}^2$. This and the fact that $|\operatorname{II}|_{\operatorname{I}} = O(s^{-1})$ imply that for any $T \in T\mathbb{S}^2$,

$$(7.47) \qquad \operatorname{Ric}(T, T) = \operatorname{Ric}_{g_s}(T, T) + O(s^{-2}) = \frac{1}{2} R_{g_s} |T|^2 + O(s^{-2}),$$

where $g_s := \Phi_s^*(g)$. Furthermore, we have that

$$(7.48) \qquad |\nabla f|^2 = 1 - R = 1 + O(s^{-1}).$$

Therefore (7.29) yields an "asymptotically" Ricci flow equation:

$$(7.49) \qquad \frac{\partial}{\partial s} g_s = (-R_{g_s} + O(s^{-2})) g_s.$$

Indeed, $R_{g_s} = O(s^{-1})$ and the Ricci flow equation on surfaces is $\partial_t g_t = -R_{g_t} g_t$. As a consequence of equation (7.49), the derivative estimates for f, and our previous estimates, we have that the (sub)sequence of families of pointed metrics $(g_i(t), p_i)$ on $\mathbb{S}^2$ defined by

$$(7.50) \qquad g_i(t) = s_i^{-1} g_{s_i(1-t)}, \quad -\infty < t < 1,$$

converges in the C^∞ pointed Cheeger–Gromov sense to an ancient solution $(\mathbb{S}^2, g_\infty, p_\infty)$ to the Ricci flow with $g_\infty(0) = g_\infty$. Here, we used the pointed C^∞ Cheeger–Gromov compactness theorem for Riemannian manifolds and its extension to sequences of *asymptotically Ricci flows*, which hold because of the curvature estimates and the consequent higher derivative estimates (Hamilton's proof [178] easily carries over to this case). Thus, the strong maximum principle contradicts that $R_\infty(p_\infty) = 0$ since there is no flat metric on $\mathbb{S}^2$. This completes the proof of Claim 7.15 and hence also completes the proof of Proposition 7.14. $\qquad \square$

Proof of Theorem 7.8. By the proof of Proposition 7.14, a subsequence of the families of metrics on $\mathbb{S}^2$ defined by (7.50) converges to a Type I ancient solution $g_\infty(t)$ (see Definition 3.18) to the Ricci flow on $\mathbb{S}^2$ with diameter bounded by $C\sqrt{-t}$. This implies that the ancient solution on $\mathbb{S}^2$ is a round shrinking sphere (by a result of Hamilton; see e.g. Theorem 9.14 in [**111**]. Another way to see this is to note that our ancient solution to the Ricci flow on $\mathbb{S}^2$ has a uniformly bounded spatial maximum over minimum scalar curvature.) One can easily deduce from this that the steady GRS is asymptotically cylindrical. It now follows from Brendle's theorem (see Theorem 8.1 in the next chapter), which says that any 3-dimensional κ-noncollapsed nonflat steady GRS must be the Bryant soliton and so consequently that our steady soliton is indeed the Bryant soliton. This completes the proof of Theorem 7.8 in the case where g has positive sectional curvature.

Finally, we need to classify the quotients of $\mathbb{R} \times (\mathbb{R}^2, g_\Sigma)$. Suppose that ψ is an isometry of $\mathbb{R} \times (\mathbb{R}^2, g_\Sigma)$. Then ψ maps any slice $\{x\} \times (\mathbb{R}^2, g_\Sigma)$ isometrically to some slice $\{x + \alpha\} \times (\mathbb{R}^2, g_\Sigma)$, where $\alpha \in \mathbb{R}$ is independent of $x \in \mathbb{R}$. We are assuming that $(\mathcal{M}^3, g)$ is isometric to $\left(\mathbb{R} \times (\mathbb{R}^2, g_\Sigma)\right)/\Gamma$, where Γ is a discrete group of isometries acting properly discontinuously on $\left(\mathbb{R} \times (\mathbb{R}^2, g_\Sigma)\right)/\Gamma$. Since the quotient is a manifold, there exists an isometry $\psi_1 \in \Gamma \setminus \{\mathrm{id}\}$ with $\alpha_1 := \alpha$ such that $|\alpha_1|$ is minimal. It is easy to see that $|\alpha_1| > 0$ since otherwise ψ_1 would have a fixed point. Without loss of generality, we may assume that $\alpha_1 > 0$. Then ψ_1 is unique and generates Γ. This completes the proof of Theorem 7.8. $\qquad\qquad\square$

7.3. Lai's flying wing steady solitons

In this section we discuss one of Lai's results regarding the existence of flying wing steady GRS in dimension 3.

7.3.1. Solitons with symmetries.

In general dimensions, one may say that the "most" symmetric steady and expanding Ricci solitons are the Gaussian steady and expanding GRS $(\mathbb{R}^n, g_{\mathrm{Euc}}, f_{\mathrm{Gau}})$ on Euclidean space given by (2.9), where $f_{\mathrm{Gau}}(x) = \frac{\lambda}{4}|x|^2$ and where λ equals 0 and -1, respectively. Here, the isometry group in each case is the group of Euclidean motions, which has the maximal dimension $\frac{n(n+1)}{2}$. Perhaps the next most symmetric steady and expanding GRS with nontrivial potential functions are the rotationally symmetric Bryant steady and expanding solitons (see Theorems 6.1 and 7.2, respectively), which each have isometry group $\mathrm{O}(n)$.

One can imagine the possible existence of steady GRS with a smaller isometry group, namely $\mathrm{O}(n-1)$, which corresponds to rotational symmetry in one less dimension. It is natural to add to this a reflectional symmetry in the complementary dimension and hence consider steady GRS with the isometry group $O(n-1) \times \mathbb{Z}_2$.[1]

Recall from Theorem 7.2 that the 1-parameter family of Bryant expanding GRS limits at one end of the family to Euclidean space, as the cone angle (equivalently, AVR) increases to that of Euclidean space. Recall also from Theorem 7.5 that Deruelle's expanding GRS can have their asymptotic cones over arbitrary closed $(n-1)$-manifolds with $\mathrm{Rm} > \mathrm{id}$ and that these expanders can be connected by 1-parameter families of expanders to Bryant's rotationally symmetric expanders (Theorem 7.6).

One may ask if one can prove the existence of new nonrotationally symmetric shrinking GRS as the limits of expanding GRS. This is possible, and we discuss in this chapter a construction of Yi Lai [**214**] of the flying wing steady GRS conjectured by Hamilton.

7.3.2. Statements of the results on flying wings.

In this subsection we present some of the statements of Lai's results. In general dimensions, we have the following; see [**214**].

Theorem 7.16 (Nonrotationally symmetric steady GRS). *For every real number $\beta > 1$ there exists an $O(n-1) \times \mathbb{Z}_2$-invariant steady GRS $(\mathbb{R}^n, g, f)$ with $\mathrm{Rm} > 0$ such that $n-1$ of the eigenvalues of Ric at the origin 0^n are equal to β times the nth eigenvalue of Ric.*

We will prove this result for the special case of $n = 3$. Note that this existence result is formulated in terms of the ratio of the two distinct eigenvalues of the Ricci tensor at the origin. One also wishes to formulate results in terms of the asymptotic behavior of the metric. To this end, we say that an $O(n-1) \times \mathbb{Z}_2$-invariant steady GRS is a **flying wing** if its asymptotic cone is the metric cone over a closed interval $[-\alpha, \alpha]$, where $0 < \alpha < \pi/2$ (a.k.a. a sector). Note that the Bryant steady n-soliton has asymptotic cone equal to a ray, i.e., the metric cone over a point (interval with $\alpha = 0$). At the other extreme, the product of $\mathbb{R}$ with a Bryant steady $(n-1)$-soliton has asymptotic cone equal to the metric cone over $[-\pi/2, \pi/2]$, corresponding to $\alpha = \pi/2$.

Lai's existence theorem for flying wings is the following; see [**212**] and [**214**] for the proof.

Theorem 7.17 (Existence of 3-dimensional flying wings). *For any $\alpha \in (0, \pi/2)$ there exists a 3-dimensional flying wing $(\mathbb{R}^3, g, f)$ whose asymptotic cone is the metric cone over $[-\alpha, \alpha]$.*

---

[1]One can make an analogy with adding a reflectional symmetry hypothesis to the class of rotationally symmetric solutions to the Ricci flow.

The angle of the asymptotic cone sector yields the following information on the asymptotic geometry of the flying wing; see [**214**] for the proof.

Theorem 7.18 (Cigar limits of 3-dimensional flying wings). *Suppose that* $(\mathbb{R}^3, g, f)$ *is a 3-dimensional flying wing steady GRS whose asymptotic cone is the metric cone over* $[-\alpha, \alpha]$, $\alpha \in (0, \pi/2)$. *Let* $\gamma : \mathbb{R} \to \mathbb{R}^3$ *be a unit-speed geodesic parametrizing the axis* $\mathbb{R}\mathbf{e}_3 = \{t\mathbf{e}_3 : t \in \mathbb{R}\}$ *and with* $\gamma(0) = 0^3$, *the origin in* $\mathbb{R}^3$. *Then, for any sequence of real numbers* a_i *such that* $|a_i| \to \infty$, *we have that the pointed sequence of Riemannian 3-manifolds* $(\mathbb{R}^3, g, a_i\mathbf{e}_3)$ *converges in the pointed* C^∞ *Cheeger–Gromov sense to the product of* $\mathbb{R}$ *with a cigar steady 2-soliton with scalar curvature at its tip equal to* $R(0^3)\sin^2 \alpha$.

Lai has further proven that all 3-dimensional complete steady GRS must be O(2)-invariant; see Lai [**213**].

7.3.3. $\mathrm{O}(n-1) \times \mathbb{Z}_2$-invariant Riemannian metrics.

We can realize $\mathrm{O}(n-1) \times \mathbb{Z}_2$ as a subgroup of $\mathrm{O}(n)$ explicitly by considering $\mathrm{O}(n-1)$ as $\mathrm{O}(n-1) \times \{\mathrm{id}_\mathbb{R}\}$, which preserves each hyperplane $\mathbb{R}^{n-1} \times \{x^n\}$, and considering $\mathbb{Z}_2$ as generated by the reflection in $\mathbb{R}^n$ about the coordinate hyperplane $\mathbb{R}_0^{n-1} := \mathbb{R}^{n-1} \times \{0\}$; i.e., $\mathbb{Z}_2$ is generated by the isometric involution

$$(7.51) \qquad \iota : (x^1, \ldots, x^{n-1}, x^n) \mapsto (x^1, \ldots, x^{n-1}, -x^n).$$

Note that $\mathbb{R}_0^{n-1}$ is the fixed point set of the $\mathbb{Z}_2$-action and that the x^n-axis

$$\mathbb{R}\mathbf{e}_n := \{(0, \ldots, 0, t) : t \in \mathbb{R}\}$$

is the fixed point set of the $\mathrm{O}(n-1)$-action, where $\{\mathbf{e}_i\}_{i=1}^n$ is the standard Euclidean basis. Define the half-planes

$$(7.52) \quad H_\pm^2 := \{0^{n-2}\} \times \mathbb{R}_\pm^2, \quad \text{where} \quad \mathbb{R}_\pm^2 = \{(x^{n-1}, x^n) : \pm x^{n-1} > 0\},$$

where 0^{n-2} denotes the origin in $\mathbb{R}^{n-2}$. Consider $\mathrm{O}(n-2)$ as the subgroup $\mathrm{O}(n-2) \times \{\mathrm{id}_{\mathbb{R}^2}\}$. Then the fixed point set of $\mathrm{O}(n-2)$ acting on $\mathbb{R}^n$ is the plane $\{0^{n-2}\} \times \mathbb{R}^2 = H_+^2 \cup H_-^2 \cup \mathbb{R}\mathbf{e}_{n-1}$ (expressed as a disjoint union of half-planes and their common boundary line).

Let $(\mathbb{R}^n, g)$ be an $\mathrm{O}(n-1) \times \mathbb{Z}_2$-invariant Riemannian manifold. Then:

(1) $\mathbb{R}\mathbf{e}_n$ is the image of an infinite geodesic in $(\mathbb{R}^n, g)$. To see this, let γ be a unit-speed path parametrizing $\mathbb{R}\mathbf{e}_n$,[2] and let T denote its tangent vector γ'. Let $\alpha \in \mathrm{O}(n-1)$. Then α fixes $\mathbb{R}\mathbf{e}_n$, so that

$$\alpha_*(\nabla_T T) = \nabla_{\alpha_* T}(\alpha_* T) = \nabla_T T.$$

This implies that $\nabla_T T = \phi T$ for some function ϕ along $\mathbb{R}\mathbf{e}_n$. Since T has unit length, we conclude that $\phi \equiv 0$. Hence γ is a geodesic.

---

[2]This is in general different from the Euclidean parametrization $t \mapsto t\mathbf{e}_n$.

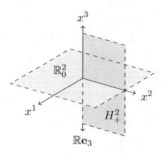

Figure 7.1. The subsets $\mathbb{R}_0^2$, $\mathbb{R}e_3$, and H_+^2 of $\mathbb{R}^3$.

(2) The hypersurface $\mathbb{R}_0^{n-1}$ is rotationally symmetric and totally geodesic in $(\mathbb{R}^n, g)$. These facts are because $\mathrm{O}(n-1)$ preserves $\mathbb{R}_0^{n-1}$ and because $\mathbb{R}_0^{n-1}$ is fixed under the isometric involution $\iota : \mathbb{R}^n \to \mathbb{R}^n$, respectively. Indeed, let N be a smooth choice of unit normal to the hypersurface $\mathbb{R}_0^{n-1}$. Then, for any $X \in T\mathbb{R}_0^{n-1}$,

$$-\mathrm{II}(X) = \nabla_X(-N) = \nabla_{\iota(X)}(\iota(N)) - \iota(\nabla_X N) = \nabla_X N = \mathrm{II}(X),$$

which implies that the second fundamental form vanishes; i.e., $\mathrm{II} = 0$.

(3) Each of the surfaces $H_\pm^2$ is totally geodesic. We leave this as an exercise.

(4) $\mathbb{R}_0^{n-1}$ and $\mathbb{R}e_n$ intersect orthogonally at 0^n, which is the unique fixed point of the $\mathrm{O}(n-1) \times \mathbb{Z}_2$-action on $\mathbb{R}^n$. This is also easy to see.

Define the projection map $\pi : \mathbb{R}^n \setminus \mathbb{R}e_n \to H_+^2$ by

$$(7.53) \qquad \pi(\mathbf{x}) = \mathrm{O}(n-1)(\mathbf{x}) \cap H_+^2,$$

which is a unique point. Note that each $\mathrm{O}(n-1)(\mathbf{x})$ is an $(n-2)$-sphere in $\mathbb{R}^n$, so we may naturally identify $\mathbb{R}^n \setminus \mathbb{R}e_n$ with $\mathbb{S}^{n-2} \times H_+^2$. Let $g_H := g|_{H_+^2}$ be the induced metric on H_+^2. By the $\mathrm{O}(n-1) \times \mathbb{Z}_2$-invariance of g, we have that the map $\pi : (\mathbb{R}^n \setminus \mathbb{R}e_n, g) \to (H_+^2, g_H)$ is a Riemannian submersion. This implies that there exists a function $\varphi : H_+^2 \to \mathbb{R}_+$ such that g is the warped product metric

$$(7.54) \qquad g = \varphi^2 g_{\mathbb{S}^{n-2}} + g_H$$

on $\mathbb{R}^n \setminus \mathbb{R}e_n \cong \mathbb{S}^{n-2} \times H_+^2$, where $g_{\mathbb{S}^{n-2}}$ is the standard metric on $\mathbb{S}^{n-2}$. Since g is $\mathbb{Z}_2$-invariant, we have that φ is $\mathbb{Z}_2$-invariant; that is,

$$(7.55) \qquad\qquad \varphi(x^{n-1}, -x^n) = \varphi(x^{n-1}, x^n).$$

Lemma 7.19. *If g is an $\mathrm{O}(n-1) \times \mathbb{Z}_2$-invariant metric with nonnegative sectional curvature, then the function $\varphi : H_+^2 \to \mathbb{R}_+$ is weakly concave with respect to g_H. Moreover, $|\nabla \varphi| \leq 1$.*

Proof. On $\mathbb{S}^{n-2} \times H_+^2$ let the indices i, j, k, ℓ correspond to local coordinates on H_+^2 and let the indices $\alpha, \beta, \gamma, \delta$ correspond to local coordinates on $\mathbb{S}^{n-2}$. By a standard calculation using the O'Neill formula for the curvatures of a warped product metric (see e.g. [**101**, Lemma 8.40]), we have

$$(7.56a) \qquad\qquad \mathrm{Rm}^g_{ijk\ell} = \mathrm{Rm}^H_{ijk\ell},$$

$$(7.56b) \qquad\qquad \mathrm{Rm}^g_{i\beta\gamma\ell} = -g_{\beta\gamma}\varphi\nabla_i\nabla_\ell\varphi,$$

$$(7.56c) \qquad\qquad \mathrm{Rm}^g_{\alpha\beta\gamma\delta} = (1 - |\nabla\varphi|^2)\varphi^2(g_{\alpha\delta}g_{\beta\gamma} - g_{\alpha\gamma}g_{\beta\delta}),$$

where the covariant derivatives are with respect to g_H. Since g has nonnegative sectional curvature, by (7.56b) we have

$$0 \leq \mathrm{Rm}^g_{i\beta\beta i} = -g_{\beta\beta}\varphi\nabla_i\nabla_i\varphi.$$

Since the local coordinates are arbitrary, on their respective factors, we conclude that $\nabla^2\varphi \leq 0$. Moreover, by (7.56c) we have $|\nabla\varphi| \leq 1$.

Now, by (7.55), we have that $\mathbf{e_n}(\varphi)(x^{n-1}, 0) = 0$ for all $x^{n-1} > 0$. Let $\beta : [0, \infty) \to \mathbb{R}_0^{n-1} \cap \overline{H_+^2} = \{te_{n-1} : t \geq 0\} \subset \mathbb{R}^n$ be a unit-speed geodesic emanating from 0^n. Since $\mathbb{R}_0^{n-1}$ and $\overline{H_+^2}$ are both totally geodesic, they intersect in a geodesic, which is parametrized by β above. Suppose that for some $t_0 > 0$ we have $(\varphi \circ \beta)'(t_0) = 0$. Then, since $(\varphi \circ \beta)''(t) \leq 0$ and $\varphi(t) > 0$ for all $t \in (0, \infty)$, we must have that $\varphi(t) \equiv \varphi(t_0)$ for all $t \geq t_0$. It is easy to obtain a contradiction from this. Hence $(\varphi \circ \beta)'(t) \neq 0$ for all $t > 0$.

On the other hand, we observe that the function φ extends smoothly to the closed half-plane $\overline{H_+^2}$. We have $\varphi(0^n) = 0$ and $\beta'(0)(\varphi) = (\varphi \circ \beta)'(0) = 1 > 0$. Therefore,

$$(7.57) \qquad\qquad 0 < (\varphi \circ \beta)'(t) \leq 1$$

for all $t \geq 0$. See Figure 7.2. $\square$

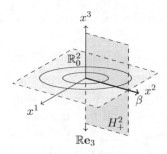

Figure 7.2. The geodesic β. The circles are examples of O(2)-orbits. The metric g and potential function f are invariant under rotations about the x^3-axis and reflections about the plane $\mathbb{R}_0^2$.

7.3.4. Properties of $O(n-1) \times \mathbb{Z}_2$-invariant GRS.

Now let $(\mathcal{M}^n, g, f, 0^n)$ be an $O(n-1) \times \mathbb{Z}_2$-invariant expanding or steady GRS. We claim that the potential function f is $O(n-1) \times \mathbb{Z}_2$-invariant. Indeed, the $O(n-1) \times \mathbb{Z}_2$-invariance of the metric g implies the same for Ric (and R), and hence by the GRS equation the same for $\nabla^2 f$, and hence in turn for f. In more detail regarding the last point, since Ric > 0, we have that $\nabla f = \frac{1}{2}\mathrm{Ric}^{-1}(\nabla R)$ is $O(n-1) \times \mathbb{Z}_2$-invariant. From this we have that integral curves of ∇f map to integral curves of ∇f under the $O(n-1) \times \mathbb{Z}_2$ action, and we obtain the invariance of f by considering such curves emanating from points on $\mathbb{R}\mathbf{e}_n$. This implies that $f(x^1, \ldots, x^{n-1}, x^n)$ is a function of the norms $|(x^1, \ldots, x^{n-1})|$ and $|x^n|$.

Since f is $O(n-1) \times \mathbb{Z}_2$-invariant, 0^n must be a critical point of f. Now assume that for our GRS, either Ric ≥ 0 if it is an expander, or Ric > 0 if it is a steady. Then f is strictly concave, and hence 0^n is the unique critical point of f. Similarly, since R is $O(n-1) \times \mathbb{Z}_2$-invariant, 0^n must be a critical point of R. Assuming that Ric > 0, this is the unique critical point of R since $\nabla R = 2\mathrm{Ric}(\nabla f)$. In particular, on an expander, if $R \to 0$ at infinity (e.g., if it is asymptotically conical), then R attains its unique maximum at 0^n. On the other hand, on a steady, we have $R + |\nabla f|^2 = R(0^n)$ since $\nabla f(0^n) = 0$, which implies that R also attains its unique maximum at 0^n. In the following, we will always assume that we are in such situations.

Proposition 7.20. *There exists a positive constant c depending only on n with the following property. If $(\mathbb{R}^n, g, f, 0^n)$ is an $O(n-1) \times \mathbb{Z}_2$-invariant steady GRS or asymptotically conical expanding GRS with Rm > 0, then*

$$(7.58) \quad \mathrm{inj}(0^n) \geq cR(0^n)^{-1/2} \quad and \quad \mathrm{Vol}\, B_{R(0^n)^{-1/2}}(0^n) \geq cR(0^n)^{-n/2}.$$

Proof. The result follows from Theorem 1.18 since $R \leq R(0^n)$ by the results of the paragraph preceding this proposition. $\square$

7.3.5. Asymptotically conical expanding GRS satisfying the symmetry assumption.

The flying wing *steady* soliton will be constructed as the limit of a suitable sequence of *expanding* solitons.

Given a sequence of rotationally symmetric asymptotically conical expanding GRS with the AVR tending to zero, one can expect that after rescaling by the curvatures at the origins, the GRS may converge to the rotationally symmetric Bryant steady GRS. The reason for this is that in the expanding soliton equation, the left-hand side $\mathrm{Ric} + \nabla^2 f$ is scale-invariant, whereas the right-hand side $-\frac{1}{2}g$ scales like the metric. So, if the curvatures at the origin blow up, then the rescaled metrics, if they converge, should converge to a steady GRS. Similarly, if we have a sequence of asymptotically conical expanding GRS whose asymptotic slices degenerate suitably, one may also expect convergence to a nonrotationally symmetric steady GRS. The question is: What is a suitable degeneration?

In view of Theorems 7.5 and 7.6, we consider sequences of 1-parameter families of Riemannian metrics on $\mathbb{S}^{n-1}$ with positive curvature operators. We now consider just the $n = 3$ case, and we consider a special sequence of (shrinking) round metrics on $\mathbb{S}^2$ being joined by a sequence of families to a sequence of degenerating metrics on $\mathbb{S}^2$.

Lemma 7.21 (Metrics on $\mathbb{S}^2$ for asymptotic slices). *On $\mathbb{S}^2$ there exists a sequence of 1-parameter families of $O(2) \times \mathbb{Z}_2$-invariant Riemannian metrics $h_i(t)$, $i \geq 1$, $t \in [0,1]$, with the following properties:*

(1) *The Gauss curvatures satisfy $K_{h_i(t)} > 1$ for all i, t.*

(2) *$h_i(1)$ has constant positive curvature for each i.*

(3) *$\mathrm{diam}(h_i(0)) \to \pi$ as $i \to \infty$.*

(4) *$\lim_{i \to \infty} \sup_{t \in [0,1]} \mathrm{Area}\big(h_i(t)\big) = 0$.*

See Figure 7.3.

Remark 7.22. (a) Alternatively, for $h_i(0)$ we may consider the $O(2) \times \mathbb{Z}_2$-invariant constant curvature surfaces in $\mathbb{R}^3$ of constant curvature 1 and scale them down a little (i.e., multiply by constants slightly less than 1) and round off their tips; see §3 of Chapter 3 of Volume III of Spivak's comprehensive book [**265**] regarding the aforementioned constant curvature-1 surfaces.

(b) The lemma holds in all dimensions $n \geq 3$, where in (1) the Gauss curvatures > 1 property is replaced by the $\mathrm{Rm} > \mathrm{id}$ property, and in (4) the area is replaced by the $(n-1)$-dimensional volume.

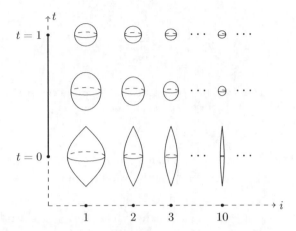

Figure 7.3. The sequence of families of metrics $h_i(t)$, $t \in [0, 1]$, $i \in \mathbb{Z}_+$. In the picture, the metrics $h_i(0)$ should be viewed as rounded off at the tips even though they look sharp.

Proof of Lemma 7.21. We start by considering a surface of revolution in $\mathbb{R}^3$ parametrized by

$$(7.59) \qquad X(z, \theta) = (f(z)\cos(\theta), f(z)\sin(\theta), z).$$

The Gauss curvature is

$$(7.60) \qquad K = -\frac{f''}{f(1 + (f')^2)^2}.$$

In particular, taking $f(z) = i^{-1}\sin(z)$, $z \in [0, \pi]$, $i \in \mathbb{N}$, we have

$$\frac{1}{(1 + i^{-2})^2} \leq K(z, \theta) = \frac{1}{(1 + i^{-2}\cos^2(z))^2} \leq 1.$$

For each $i \geq 1$, this describes a surface homeomorphic to $\mathbb{S}^2$ with an $O(2) \times \mathbb{Z}_2$-invariant metric that is smooth except at the poles $(0, 0, 0)$ and $(0, 0, \pi)$. The cone angles at the two singular points are equal to $\frac{2\pi}{i}\frac{1}{\sqrt{1 + i^{-2}}}$, which tends to 0 as $i \to \infty$.

Provided $|f|$ is small enough, e.g., $|f| \leq \frac{1}{10}$, we have that

$$\text{diam} = \int_0^\pi \sqrt{1 + f'(z)^2}dz,$$

so it is clear that diam $\to \pi$ as $i \to \infty$. We also have that

$$\text{Area} = 2\pi \int_0^\pi f(z)\sqrt{1 + f'(z)^2}dz,$$

so it is also clear that Area $\to 0$ as $i \to \infty$.

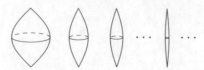

Figure 7.4. The sequence of metrics $h_i(0)$ are obtained by slightly shrinking these cone surfaces and smoothing out the pairs of antipodal cone points.

We define the Riemannian metrics $\tilde{h}_i(0)$ by rescaling the metrics obtained above by first multiplying them by the constants $\frac{1}{(1+i^{-2})^3}$ (this ensures that $K_{h_i(0)} \geq 1 + i^{-2} > 1$), and then we smooth the two cone points, which only increases the curvatures, decreases the areas, and decreases the diameters only by a little bit (an amount that goes to zero as $i \to \infty$). It is clear that we can do all of this in an $O(2) \times \mathbb{Z}_2$-invariant way.

Next we run the Ricci flow starting at the metrics $\tilde{h}_i(0)$ to obtain a sequence of solutions $\tilde{h}_i(t)$ defined on maximal finite time intervals $[0, T_i)$. As $t \to T_i$, the $\tilde{h}_i(t)$ shrink to round points. We proceed to linearly rescale the time parameter, so that now $t \in [0, 1)$.

Let $K_i(t) := K_{\min}\big(\tilde{h}_i(t)\big)$. Let $A_i(t) := \mathrm{Area}\big(\tilde{h}_i(t)\big) = (1-t)\,\mathrm{Area}(\tilde{h}_i(0))$, where the equality on the right is because the area of the Ricci flow on surfaces is a linear function of time.

We rescale the metrics by defining $h_i(t) := c_i(t)\tilde{h}_i(t)$. To prove the lemma we need to show the existence of a smooth function $c_i(t) > 0$, $i \geq 1$, $t \in [0, 1)$, satisfying:

(1) $c_i(0) = 1$.

(2) $c_i(t)^{-1} K_i(t) > 1$.

(3) $\lim_{t \to 1} c_i(t)^{-1} K_i(t) > 1$.

(4) For each i, $h_i(t)$ converges to a constant curvature metric as $t \to 1$.

(5) $\lim_{i \to \infty} \sup_{s \in [0,1]} c_i(t) A_i(t) = 0$.

To this end we define the $c_i(t)$ so that:

(i) $c_i(t) \leq \min\left\{ \frac{K_i(t)}{K_i(0)}, \frac{A_i(0)}{A_i(t)} \right\} =: b_i(t)$ for all $t \in [0, 1)$.

(ii) $c_i(t) = \frac{A_i(0)}{A_i(t)}$ for t sufficiently close to 1.

Note that $b_i(t)$ is a positive locally Lipschitz function with $b_i(0) = 1$. Furthermore, for each i sufficiently large, we have that $A_i(0)K_i(0) < A_i(t)K_i(t)$ for t sufficiently close to 1 (depending on i) since $\lim_{i \to \infty} A_i(0)K_i(0) = 0$ and since $A_i(t)K_i(t) \to 2\pi\chi(\mathbb{S}^2) = 4\pi$ as $t \to 1$ for each i. Thus, for each i we have that $\frac{A_i(0)}{A_i(t)} < \frac{K_i(t)}{K_i(0)}$ for t sufficiently close to 1. From this it is easy to see that we can find C^∞ functions $c_i(t)$ satisfying (i), (ii), and (1): $c_i(0) = 1$.

By (i), we have $c_i(t)^{-1}K_i(t) \geq K_i(0) = K_{\min}(h_i(0)) \geq 1 + i^{-2} > 1$, which is (2). Moreover, by (ii), $\lim_{t\to 1} c_i(t)^{-1}K_i(t) = \lim_{t\to 1} \frac{A_i(t)K_i(t)}{A_i(0)} = \frac{4\pi}{A_i(0)} > 1$, which is (3). This and the fact that the Ricci flow shrinks metrics on $\mathbb{S}^2$ to round points imply (4). Finally, (5) is true since by (i) we have that

$$\lim_{i\to\infty} \sup_{t\in[0,1]} c_i(t)A_i(t) \leq \lim_{i\to\infty} A_i(0) = 0.$$

The proof of Lemma 7.21 is complete. $\qquad\square$

7.3.6. Proof of the $n = 3$ case of Theorem 7.16 on the existence of steady GRS with $O(2) \times \mathbb{Z}_2$ symmetry.

By Theorems 7.5 and 7.6, there exists a sequence of 1-parameter families of asymptotically conical expanding GRS

$$(7.61) \qquad\qquad (\mathbb{R}^3, g_i(s), f_i(s)), \quad s \in [0,1],$$

satisfying:

  (1) The asymptotic cone of $g_i(s)$ is a Riemannian cone over $(\mathbb{S}^2, h_i(s))$, where $h_i(s)$ is the sequence of 1-parameter families of metrics given by Lemma 7.21.

  (2) $\mathrm{Rm}_{g_i(s)} > 0$.

In particular, by Theorem 7.7, $(\mathbb{R}^3, g_i(1), f_i(1))$ is a sequence of rotationally symmetric Bryant expanding GRS.

Claim 7.23. *Each of the expanding GRS $(\mathbb{R}^3, g_i(s), f_i(s))$ is $O(2) \times \mathbb{Z}_2$-invariant.*

Proof of Claim 7.23. Since the metric $h_i(s)$ on $\mathbb{S}^2$ is $O(2) \times \mathbb{Z}_2$-invariant, for each i, we may identify $\mathbb{R}^3 \setminus \{0^3\}$ with the punctured cone $C_0(\mathbb{S}^2, h_i(s)) \setminus \{c_0\}$, where c_0 is the vertex of the cone, in an $O(2) \times \mathbb{Z}_2$-invariant way. Let

$$(7.62) \qquad\qquad t \mapsto g_{i,s}(t) := t\varphi_t^* g_i(s), \quad t > 0,$$

be the canonical form of the expanding GRS $g_i(s)$. Then $g_{i,s}(t)$ converges in C^∞ on compact subsets of $\mathbb{R}^3 \setminus \{0\}$ to $dr^2 + r^2 h_i(s)$ as $t \to 0$. By the uniqueness part of Theorem 7.5, it follows that the metrics $g_{i,s}(t)$ are $O(2) \times \mathbb{Z}_2$-invariant. In particular, the metrics $g_i(s)$ are $O(2) \times \mathbb{Z}_2$-invariant, and the claim is proved.

Since $\mathrm{Rm}_{g_i(s)} > 0$ and $g_i(s)$ is asymptotically conical, $f_i(s)$ has a unique critical point, which we denote by $o_{i,s}$. By (7.7), we have

$$\mathrm{AVR}(g_i(s)) = \frac{1}{3}\,\mathrm{Area}(h_i(s)).$$

Hence, by Lemma 7.21(4), we have

$$\text{(7.63)} \qquad \lim_{i \to \infty} \sup_{s \in [0,1]} \text{AVR}\left(g_i(s)\right) = 0.$$

We have the following compactness theorem in the form of the convergence of a sequence of expanding GRS with the AVR $\to 0$ to a steady GRS.

Proposition 7.24 (Convergence of a sequence of expanders to a steady). *Let $(\mathcal{M}_i^n, g_i, f_i)$ be a sequence of asymptotically conical expanding GRS with $\text{Rm}_{g_i} > 0$ and $\text{AVR}(g_i) \to 0$ as $i \to \infty$. Suppose that for each i, f_i has a critical point o_i, which then must be unique since $\text{Ric}_{g_i} \geq 0$. Then there exists a subsequence of the rescaled manifolds $(\mathcal{M}_i^n, R_{g_i}(o_i)^{-1} g_i, f_i, o_i)$ converging in the pointed C^∞ Cheeger–Gromov sense to a complete steady gradient Ricci soliton*

$$(\mathcal{M}_\infty^n, g_\infty, f_\infty, o_\infty)$$

with nonnegative curvature operator and $R_{g_\infty}(o_\infty) = 1$. If each $(\mathcal{M}_i^n, g_i, f_i)$ is $\text{O}(n-1) \times \mathbb{Z}_2$-invariant, then there exists a further subsequence such that $(\mathcal{M}_\infty^n, g_\infty, f_\infty)$ is $\text{O}(n-1) \times \mathbb{Z}_2$-invariant.

Proof. We first show the following:

Claim 7.25. *Let $R_i := R_{g_i}(o_i)$. Then $R_i \to \infty$ as $i \to \infty$.*

Proof of Claim 7.25. Suppose that the claim is false. Then we may pass to a subsequence for which $R_i \leq C$ for all i and some positive constant C. By our previous discussion, we have that $R_{g_i}(x) \leq R_i$ for all $x \in \mathcal{M}_i^n$. Thus, $0 < \text{Rm}_{g_i} \leq R_{g_i} \text{id} \leq C \, \text{id}$ on $\mathcal{M}_i^n$. By the injectivity radius estimate in Theorem 1.18, we obtain a uniform lower bound for the injectivity radii at o_i (or at any point for that matter). By Shi's local derivative estimates for the canonical forms of the expanding GRS in the sequence, for each $k \geq 0$ we obtain uniform bounds for $|\nabla_{g_i}^k \text{Rm}_{g_i}|$. Thus, from $\text{Ric}_{g_i} + \nabla_{g_i}^2 f_i = -\frac{1}{2} g_i$, we have for each $\ell \geq 2$, uniform bounds for $|\nabla_{g_i}^\ell f_i|$. By adding constants to the f_i, we may also assume that $f_i(o_i) = 0$. This implies that $f_i \leq 0$ on $\mathcal{M}_i^n$. Since $R_{g_i} \geq 0$ on $\mathcal{M}_i^n$, $R_i \leq C$, and $|\nabla_{g_i} f_i|(o_i) = 0$, we have

$$\text{(7.64)} \qquad |\nabla_{g_i}(-f_i + C)|^2 \leq R_{g_i} + |\nabla_{g_i} f_i|^2 = -f_i + R_i \leq -f_i + C.$$

Integrating this inequality over minimal geodesics emanating from o_i yields

$$\text{(7.65)} \qquad -f_i(x) + C \leq \left(\frac{1}{2} d(x, o_i) + C\right)^2.$$

This gives us uniform local bounds on both f_i and $|\nabla_{g_i} f_i|$. Hence, by the pointed C^∞ Cheeger–Gromov compactness theorem (see Theorem 1.2) and passing to a subsequence, we may assume that $(\mathcal{M}_i^n, g_i, f_i, o_i)$ converges to an expanding GRS $(\mathcal{M}^n, g, f, o)$ satisfying $\mathrm{Ric} + \nabla^2 f = -\frac{1}{2}g$ and $\mathrm{Rm} \geq 0$. By Theorem 7.1, we must have

$$\mathrm{AVR}(g) > 0.$$

By definition, the convergence occurs after the pullback by basepoint-preserving diffeomorphisms and in each C^k on compact sets. We now show that we can derive a contradiction.

For each i, let $(\mathcal{M}_i^n, g_i(t), f_i(t))$, $t \in \mathbb{R}_+$, be the canonical form associated to $(\mathcal{M}_i^n, g_i, f_i)$, which is an immortal solution to the Ricci flow (since it is defined for all $t \in \mathbb{R}_+$). Let $(\mathcal{M}^n, g(t), f(t))$, $t \in \mathbb{R}_+$, be the canonical form associated to $(\mathcal{M}^n, g, f)$. We may assume that $(\mathcal{M}_i^n, g_i(t), f_i(t), o_i)$ converges to $(\mathcal{M}^n, g(t), f(t), o)$ as pointed Ricci flow solutions.

Subclaim.

(1) As $t \to 0$, for each i the family $(\mathcal{M}_i^n, g_i(t), o_i)$ converges in the pointed Gromov–Hausdorff sense to a metric cone $(C_0(X_i), z_i^*)$, where X_i is a closed Riemannian $(n-1)$-manifold with $\mathrm{Rm}_{X_i} \geq \mathrm{id}$. In particular, $C_0(X_i)$ is an Alexandrov space with nonnegative sectional curvature.

(2) As $t \to 0$, $(\mathcal{M}^n, g(t), o)$ converges in the pointed Gromov–Hausdorff sense to a metric cone $(C_0(X), z^*)$, where X is a compact length metric space. In fact, both X and $C_0(X)$ are Alexandrov spaces with nonnegative sectional curvature.

(3) $(C_0(X_i), z_i^*)$ subconverges in the pointed Gromov–Hausdorff sense to $(C_0(X), z^*)$.

Proof of the subclaim. (1) Let $(\mathcal{M}_i^n, g_i(t), f_i(t), o_i)$, $t \in \mathbb{R}_+$, be the canonical form of the asymptotically conical expanding GRS $(\mathcal{M}_i^n, g_i, f_i, o_i)$, where $g_i(1) = g_i$ and $f_i(1) = f_i$. Then $g_i(t)$ is a solution to the Ricci flow. Define diffeomorphisms $\varphi_{i,t} : \mathcal{M}_i^n \to \mathcal{M}_i^n$, $t \in \mathbb{R}_+$, by

$$(7.66) \qquad \partial_t \varphi_{i,t}(x) = \frac{1}{t}(\nabla_{g_i} f_i)(\varphi_{i,t}(x)), \qquad \varphi_{i,1} = \mathrm{id}_{\mathcal{M}_i},$$

where $\nabla_{g_i} = \nabla_{y_i(1)}$. We have $g_i(t) = t\varphi_{i,t}^* g_i$, $f_i(x, t) = f_i(\varphi_{i,t}(x))$,

$$\mathrm{Ric}_{g_i(t)} + \nabla^2_{g_i(t)} f_i(t) = -\frac{1}{2t} g_i(t),$$

and

$$\partial_t f_i(x, t) = \frac{1}{t}|\nabla_{g_i} f_i|^2(\varphi_{i,t}(x)) = |\nabla_{g_i(t)} f_i(t)|^2(x).$$

Since g_i is asymptotically conical, we have that $(\mathcal{M}_i^n, tg_i, o_i)$ converges in the pointed Gromov–Hausdorff sense to $(C_0(X_i), z_i^*)$ as $t \to 0$. Note that since $\nabla f_i(o) = 0$, by (7.66), we have that $\varphi_{i,t}(o_i) = o_i$ for all $t > 0$. Thus, we also have that $(\mathcal{M}_i^n, g_i(t), o_i)$ converges in the pointed Gromov–Hausdorff sense to $(C_0(X_i), z_i^*)$ as $t \to 0$.

(3) Note that $|\mathrm{Rm}_{g_i(t)}| \le \frac{C}{t}$ for all t, and hence $|\mathrm{Rm}_{g(t)}| \le \frac{C}{t}$ for all t. Since $\mathrm{Rm}_{g_i} > 0$, we have $\mathrm{Rm}_{X_i} \ge \mathrm{id}$. In particular, $\mathrm{diam}(X_i) \le \pi$ for all i. Thus, by Gromov's precompactness theorem [**162**], after passing to a subsequence, we have that X_i converges in the Gromov–Hausdorff sense to some compact length metric space X with $\mathrm{diam}(X) \le \pi$. This in turn implies that $(C_0(X_i), z_i^*)$ subconverges in the pointed Gromov–Hausdorff sense to $(C_0(X), z^*)$.

(2) Since $\mathrm{Ric}_{g_i(t)} \ge 0$, we have $d_{g_i(t_2)}(x,y) \le d_{g_i(t_1)}(x,y)$ for all $x, y \in \mathcal{M}_i^n$, $t_1 \le t_2$, and i. By Hamilton's changing distances estimate (see e.g. Lemma 8.33 in [**111**]), since $\mathrm{Ric}_{g_i(t)} \le (n-1)\frac{C}{t}$, independent of i, we have

$$(7.67) \qquad d_{g_i(t_2)}(x,y) \ge d_{g_i(t_1)}(x,y) - 4(n-1)\int_{t_1}^{t_2} \sqrt{C/t}\, dt$$

$$= d_{g_i(t_1)}(x,y) - 8(n-1)\sqrt{C}\left(\sqrt{t_2} - \sqrt{t_1}\right)$$

for all $x, y \in \mathcal{M}_i^n$, $t_1 \le t_2$, and i. In particular, for $0 < t \le 1$, we have

$$d_{g_i(1)}(x,y) \le d_{g_i(t)}(x,y) \le d_{g_i(1)}(x,y) + 8(n-1)\sqrt{C}\left(1 - \sqrt{t}\right),$$

where C is independent of i. We also see from (7.67) that the Gromov–Hausdorff convergence as $t \to 0$ in (1) is uniform in i. From this we conclude that, as $t \to 0$, $(\mathcal{M}^n, g(t), o)$ converges in the pointed Gromov–Hausdorff sense to the metric cone $(C_0(X), z^*)$, where X is as in (2). This completes the proof of the subclaim.   □

Given a metric space (M, d), let $\mathcal{H}_M^n$ denote its n-dimensional Hausdorff measure. Since $(C_0(X_i), z_i^*)$ converges in the pointed Gromov–Hausdorff sense to $(C_0(X), z^*)$, we have that the $(\bar{B}_1(z_i^*), z_i^*)$ are compact Alexandrov spaces with nonnegative curvature converging in the pointed Gromov–Hausdorff sense to $(\bar{B}_1(z^*), z^*)$. By [**56**, Theorem 10.10.10] we have that the $\mathcal{H}_{\bar{B}_1(z_i^*)}^n$ **weakly converge** to $\mathcal{H}_{\bar{B}_1(z^*)}^n$. By definition, this means the following. Let $\phi_i : \bar{B}_1(z^*) \to \bar{B}_1(z_i^*)$ satisfy

$$(7.68) \qquad \mathrm{dis}(\phi_i) := \sup_{x,y \in \bar{B}_1(z^*)} \left| d_{\bar{B}_1(z_i^*)}(\phi_i(x), \phi_i(y)) - d_{\bar{B}_1(z^*)}(x,y) \right| < \varepsilon_i,$$

where dis denotes the *distortion* and $\varepsilon_i \to 0$. Then, for any closed subset E of $\bar{B}_1(z^*)$ with $\mathcal{H}_{\bar{B}_1(z^*)}^n(\partial E) = 0$, we have

$$\mathcal{H}_{\bar{B}_1(z_i^*)}^n\left(B_{\varepsilon_i}(\phi_i^{-1}(E))\right) \to \mathcal{H}_{\bar{B}_1(z^*)}^n(E).$$

Hence

$$\mathcal{H}^n_{C_0(X)}\big(B_1(z^*)\big) = \lim_{i\to\infty} \mathcal{H}^n_{C_0(X_i)}\big(B_1(z_i^*)\big)$$
$$= \lim_{i\to\infty} \mathrm{AVR}\,\big(C_0(X_i)\big)$$
$$= \lim_{i\to\infty} \mathrm{AVR}(g_i)$$
$$= 0.$$

On the other hand,

$$\mathcal{H}^n_{C_0(X)}\big(B_1(z^*)\big) = \lim_{t\to 0} \mathcal{H}^n_{(\mathcal{M},g(t))}\big(B_1^{g(t)}(o)\big) \geq \mathrm{AVR}\,\big(g_\infty(t)\big) > 0.$$

We have derived a contradiction, and hence the proof of the claim that $R_i \to \infty$ is complete. Q.E.D. Claim 7.25. $\qquad\square$

Now we can finish the proof of Proposition 7.24. Let $\tilde{g}_i := R_i g_i$ and $\tilde{f}_i := f_i$, so that

$$(7.69) \qquad\qquad \mathrm{Ric}_{\tilde{g}_i} + \nabla^2_{\tilde{g}_i} \tilde{f}_i = -\frac{R_i^{-1}}{2}\tilde{g}_i.$$

Note that $R_{\tilde{g}_i}(o_i) = 1$, and hence $0 < \mathrm{Rm}_{\tilde{g}_i} \leq \hat{C}\,\mathrm{id}$ on $\mathcal{M}_i^n$. Using that $R_i^{-1} \to 0$ and the uniform curvatures and their derivatives bounds, we have that by passing to a further subsequence, $(\mathcal{M}_i^n, \tilde{g}_i, \tilde{f}_i, o_i)$ converges in the pointed C^∞ Cheeger–Gromov sense to a complete steady gradient Ricci soliton $(\mathcal{M}_\infty^n, \tilde{g}_\infty, \tilde{f}_\infty, o_\infty)$ with $\mathrm{Rm}_{\tilde{g}_\infty} \geq 0$ and $R_{\tilde{g}_\infty}(o_\infty) = 1$. This completes the proof of the proposition. $\qquad\sqcup$

Recall that $\lim_{i\to\infty} \mathrm{AVR}(g_i(s_i)) = 0$ by (7.63). Thus, given *any* sequence $\{s_i\}_{i=1}^\infty$ in $[0,1]$, we may apply Proposition 7.24 to the sequence of asymptotically conical expanding gradient Ricci solitons $(\mathbb{R}^3, g_i(s_i), f_i(s_i))$ given by (7.61) to obtain a subsequential pointed C^∞ Cheeger–Gromov limit of $(\mathbb{R}^3, R_{g_i(s_i)}^{-1} g_i(s_i), f_i(s_i), 0^3)$ which is a complete steady gradient Ricci soliton $(\mathcal{M}_\infty^3, g_\infty, f_\infty, o_\infty)$. For example:

(1) If $s_i = 1$ for all i, then we obtain a sequence of rotationally symmetric Bryant expanding GRS with the $\mathrm{AVR} \to 0$ and which after rescaling converges to the Bryant steady GRS.

(2) If $s_i = 0$ for all i, then $(\mathbb{S}^2, h_i(0))$ converges in the Gromov–Hausdorff sense to an interval of length π. Thus, the asymptotic cone of the steady GRS is a closed half-plane $\mathbb{R}^2_+$. We claim that in this case the limit steady GRS is the product of $\mathbb{R}$ with a cigar steady soliton.

To construct 3-dimensional flying wing steady gradient Ricci solitons, fix i and consider the 1-parameter family of $O(2)\times\mathbb{Z}_2$-invariant expanding GRS $(\mathbb{R}^3, g_i(s), f_i(s))$, $s \in [0,1]$. For each metric $g_i(s)$, let $\lambda_1(i,s), \lambda_2(i,s), \lambda_3(i,s)$

denote the eigenvalues of the Ricci tensor at 0^3. From the $O(2) \times \mathbb{Z}_2$-invariance, we have that two of them are equal, say $\lambda_1(i, s) = \lambda_2(i, s)$. Consider the quantity $\frac{\lambda_1(i,s)}{\lambda_3(i,s)}$. By (1) and (2) above, we have that

$$\frac{\lambda_1(i, 1)}{\lambda_3(i, 1)} \equiv 1 \quad \text{and} \quad \frac{\lambda_1(i, 0)}{\lambda_3(i, 0)} \to \infty \text{ as } i \to \infty.$$

Hence, for any $\beta > 1$, we have that $\frac{\lambda_1(i,0)}{\lambda_3(i,0)} > \beta$ for i sufficiently large. Thus, for each i sufficiently large, by the continuity of $s \mapsto \frac{\lambda_1(i,s)}{\lambda_3(i,s)}$, there exists $s_i \in (0, 1)$ such that the eigenvalues of $\mathrm{Ric}_{g_i(s_i)}$ at 0^3 satisfy

$$(7.70) \qquad\qquad \frac{\lambda_1(i, s_i)}{\lambda_3(i, s_i)} = \beta.$$

As we remarked above, by applying Proposition 7.24 to the pointed sequence $(\mathbb{R}^3, R_{g_i(s_i)}^{-1} g_i(s_i), f_i(s_i), 0^3)$, we obtain a limit $O(2) \times \mathbb{Z}_2$-invariant complete steady GRS $(\mathcal{M}_\infty^3, g_\infty, f_\infty, o_\infty)$ with $R_{g_\infty}(o_\infty) = 1$. By (7.70), the eigenvalues $\lambda_1, \lambda_2, \lambda_3$ of the Ricci tensor of g_∞ at o_∞ satisfy $\lambda_1 = \lambda_2 = \beta\lambda_3 > 0$. Note that since g_∞ has nonnegative sectional curvature, by the strong maximum principle we have that the sectional curvatures of g_∞ are positive. This completes the proof of Theorem 7.16 in dimension 3.

7.4. Notes and commentary

Schulze and Simon [263] prove the existence of expanding solitons with nonnegative curvature operator coming out of their asymptotic cones. We remark that for expanding solitons, quadratic curvature decay (with derivatives) is equivalent to asymptotically conical in a strong and precise sense; see Conlon, Deruelle, and Sun [122, Theorem 3.8].

7.5. Exercises

7.5.1. Volume growth of steadies.

Exercise 7.1 (Yau's Ric ≥ 0 linear volume growth estimate). Let $(\mathcal{M}^n, g, p)$ be a complete, noncompact, pointed Riemannian manifold with Ric ≥ 0. In this case, the **Bishop–Gromov relative volume comparison theorem** (see [85] and the references therein) says that if $0 \leq r \leq R \leq S$ and $r \leq s \leq S$ are nonnegative real numbers, then

$$(7.71) \qquad \frac{\mathrm{Vol}(B_S(p)) - \mathrm{Vol}(B_s(p))}{\mathrm{Vol}(B_R(p)) - \mathrm{Vol}(B_r(p))} \leq \frac{S^n - s^n}{R^n - r^n}.$$

Using this fact, prove that there exists a constant $C(n)$ such that for any $x \in \mathcal{M}$ and $r \geq 1$,

$$\frac{\text{Vol}\left(B_{r+1}\left(x\right)\right) - \text{Vol}\left(B_{r-1}\left(x\right)\right)}{\text{Vol}\left(B_{r-1}\left(x\right)\right)} \leq \frac{C(n)}{r},$$

and hence there exists $c(n) > 0$ such that

(7.72) $$\text{Vol}\left(B_r(p)\right) \geq c(n)r.$$

Exercise 7.2 (Munteanu and Wang's steady soliton linear volume growth estimate). Let $(\mathcal{M}^n, g, f)$ be a steady GRS. Prove that for any $p \in \mathcal{M}^n$ and $r > 0$,

(7.73) $$\int_{B_r(p)} R d\mu = -\int_{\partial B_r(p)} \langle \nabla f, \nu \rangle d\sigma.$$

Use this to prove that if g is not Ricci-flat, then there exists $c > 0$ such that for any $r \geq 1$,

(7.74) $$\text{Area}(\partial B_r(p)) \geq c.$$

Prove that any steady GRS has at least linear volume growth.

Exercise 7.3. Let $(\mathcal{M}^n, g, f)$ be a complete steady GRS that is also the singularity model of some finite-time singular solution to the Ricci flow on a closed manifold. Prove that there exists $\kappa > 0$ such that

(7.75) $$\text{Vol}\, B_r\left(x\right) \geq \kappa r^n \quad \text{for all } 0 < r \leq 1 \text{ and } x \in \mathcal{M}^n.$$

In particular,

(7.76) $$\text{Vol}\, B_1\left(x\right) \geq \kappa \quad \text{for all } x \in \mathcal{M}^n.$$

Exercise 7.4. In addition to the assumptions of the previous exercise, assume that $R(x) \leq C/d(x,p)$ for some positive constant C. Prove that there exists $c > 0$ such that

(7.77) $$\text{Vol}\, B_s(p) \geq c s^{(n+1)/2} \quad \text{for } s \geq 1.$$

Exercise 7.5. Let $(\mathcal{M}^n, g)$ be a complete noncompact Ricci-flat (a.k.a. trivial steady soliton) singularity model. Prove that the $\text{AVR}(g) > 0$, where the AVR is the asymptotic volume ratio.

7.5.2. Steadies with nonnegative Ricci curvature.

Exercise 7.6 (Carillo and Ni [**73**]). Let $(\mathcal{M}^n, g, f)$ be a steady GRS with $\text{Ric} \geq 0$ and suppose that $\sup_{\mathcal{M}} f$ is attained at a point o. Prove that for any $c_1 < f(o)$ there exists $\varepsilon > 0$ such that

(7.78) $$|\nabla f| \geq \varepsilon \quad \text{on } \{f \leq c_1\}.$$

7.5.3. Scalar curvature lower bound for steadies.

Exercise 7.7 (Lu, Yang, and the author [112]). Let $(\mathcal{M}^n, g, f)$ be a steady GRS, and let $a \in \mathbb{R}$. Prove that

$$(7.79) \qquad \Delta_f \left(R - ae^f \right) \leq -\frac{2}{n}R^2 + aRe^f.$$

Define $G := R - ae^f - \frac{1-2a}{4}e^{2f}$. Suppose that $\lim_{x \to \infty} f(x) = -\infty$. By applying the elliptic maximum principle to the equation for $\Delta_f G$, prove that there exists a constant $c > 0$ such that $R \geq ce^f$.

   Remark: Munteanu, Sung, and Wang [233] prove the same conclusion under the much weaker assumption that f is bounded from above.

Chapter 8

Brendle's Theorem on the Uniqueness of 3-Dimensional Steadies

In this chapter we discuss the complete classification of 3-dimensional non-collapsed steady gradient Ricci solitons.

8.1. Statement and outline of the proof of the main theorem

8.1.1. Statement of the main theorem.

In 2013 Brendle [**42**] gave a complete and remarkable proof of the following classification of 3-dimensional κ-noncollapsed steady GRS.

Theorem 8.1. *Any 3-dimensional κ-noncollapsed nonflat steady GRS must be the Bryant soliton.*

This result was originally asserted by Perelman [**250**, §11.9], where he wrote:

"... I can prove uniqueness in the class of gradient steady solitons."

However, he did not indicate a proof of this evidently deep result in his paper.

By Corollary 5.4, any 3-dimensional steady GRS has nonnegative sectional curvature. Hence we may apply Corollary 5.6 (the strong maximum principle for systems) and the proof of Theorem 5.7 (classification of 3-dimensional shrinkers) to conclude that its canonical form $\left(\mathcal{M}^3, g\left(t\right), f(t)\right)$ given by Proposition 4.19 is either flat, has positive sectional curvature (which implies that $\mathcal{M}^3$ is noncompact by Theorem 2.29), or its universal cover splits as the product of a line with a surface solution, which is a nonflat steady GRS and hence is the cigar soliton by Theorem 3.11. Since the cigar soliton is not κ-noncollapsed, Theorem 8.1 reduces to proving that a 3-dimensional noncompact κ-noncollapsed steady GRS with positive sectional curvature, *an assumption we make for the rest of this chapter*, must be the Bryant soliton.

8.1.2. Preliminaries on Killing vector fields on Riemannian manifolds.

Motivated by the fact that the Bryant soliton is rotationally symmetric, we consider Killing vector fields. Let $(\mathcal{M}^n, g)$ be a Riemannian manifold. The Lie bracket of two Killing vector fields is a Killing vector field. Moreover, if $(\mathcal{M}^n, g)$ is complete, then any Killing vector field is complete (see Theorem 1.83 in [**34**]). Thus, the Lie algebra of Killing vector fields is naturally isomorphic to the Lie algebra of the Lie group of isometries of $(\mathcal{M}^n, g)$.

Let V be a Killing vector field; that is,

$$(8.1) \qquad \nabla_j V_k + \nabla_k V_j = 0.$$

Note that we also let V denote the 1-form dual to V. If $\gamma(s)$ is a unit-speed geodesic with tangent vector S, then $V(\gamma(s))$ satisfies

$$\frac{d}{ds} \langle V, S \rangle = \langle \nabla_S V, S \rangle + \langle V, \nabla_S S \rangle = 0.$$

Hence $\langle V, S \rangle$ is constant along $\gamma(s)$.

Define

$$(8.2) \qquad \begin{aligned} A_{ijk} &:= \nabla_i \nabla_j V_k + \nabla_j \nabla_k V_i = \nabla_i \nabla_j V_k - \nabla_j \nabla_i V_k = -R_{ijk\ell} V^\ell \\ &:= -R_{ijk\ell} g^{\ell m} V_m. \end{aligned}$$

This defines a 3-tensor which is antisymmetric in its first two components. The consideration of $A_{ijk} + A_{jki} - A_{kij}$ leads to (see Exercise 8.1)

$$(8.3) \qquad \nabla_j \nabla_k V_i = R_{kij\ell} V^\ell.$$

Notice that tracing this yields the well-known Bochner formula for Killing vector fields[1]

$$(8.4) \qquad (\Delta V)_i + R_{i\ell} V_\ell = 0.$$

Let

$$\nabla_i V_j =: L_{ij},$$

which is antisymmetric in i and j. Then (8.3) says that

$$\nabla_j L_{ki} = R_{kij\ell} V_\ell.$$

---

[1]For a simpler proof of (8.4), see the solution to Exercise 1.36.

In particular, if γ is any path with tangent vector T, then along γ we have the following first-order system of ODEs for the 1-form and 2-tensor pair (V_j, L_{ki}):

$$T^i \partial_i V_j = T^i (L_{ij} + \Gamma_{ij}^k V_k),$$
$$T^j \partial_j L_{ki} = T^j (R_{kij\ell} V_\ell + \Gamma_{jk}^\ell L_{\ell i} + \Gamma_{ji}^\ell L_{k\ell}).$$

For any $p \in \mathcal{M}^n$, $V(p)$, $L(p) = \nabla V(p)$, and any path γ starting at p, there is a unique solution to this ODE. Hence $V(p)$ and $\nabla V(p)$ determine V on all of $\mathcal{M}^n$. Since $L_{ij} = \nabla_i V_j$ is antisymmetric in i and j, the Lie algebra of Killing vector fields has at most dimension $n + \frac{n(n-1)}{2} = \frac{n(n+1)}{2}$. Hence the isometry group of $(\mathcal{M}^n, g)$ has at most dimension $\frac{n(n+1)}{2}$. This is sharp since for the unit n-sphere $\mathbb{S}^n$, $\mathrm{Isom}(\mathbb{S}^n) = O(n+1)$, which has dimension $\frac{n(n+1)}{2}$.

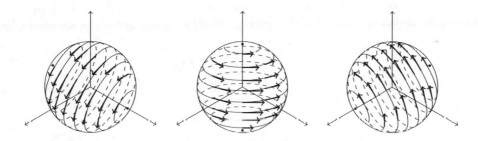

Figure 8.1. The flows of three linearly independent Killing vector fields on $\mathbb{S}^2$. Each vector field generates the circle of rotations about one of the three axes.

We now specialize to GRS.

Lemma 8.2. *If V is a Killing vector field on a GRS $(\mathcal{M}^n, g, f, \lambda)$, then $\nabla \langle \nabla f, V \rangle$ is a parallel gradient vector field.*

Proof. Since $\mathcal{L}_V g = 0$ and by the diffeomorphism invariance of the Ricci tensor, we have

$$(8.5) \qquad\qquad 0 = \mathcal{L}_V \mathrm{Ric} = \mathcal{L}_V (\nabla^2 f).$$

Now

$$\mathcal{L}_V (\nabla^2 f) = \frac{1}{2} \mathcal{L}_V (\mathcal{L}_{\nabla f} g) = \frac{1}{2} \mathcal{L}_{[V, \nabla f]} g$$

since $\mathcal{L}_{\nabla f}(\mathcal{L}_V g) = 0$. Using $\mathcal{L}_V g = 0$ again and the fact that $\nabla^2 f$ is symmetric, we have

$$(8.6) \qquad\qquad [V, \nabla f] = \nabla_V \nabla f - \nabla_{\nabla f} V = \nabla \langle \nabla f, V \rangle.$$

We conclude that[2]

$$(8.7) \qquad 0 = \mathcal{L}_V(\nabla^2 f) = \frac{1}{2}\mathcal{L}_{\nabla\langle\nabla f, V\rangle} g = \nabla^2\langle\nabla f, V\rangle. \qquad \square$$

Corollary 8.3. *If a GRS $(\mathcal{M}^n, g, f, \lambda)$ does not split off a line, then*

$$(8.8) \qquad\qquad \nabla\langle\nabla f, V\rangle = 0$$

on $\mathcal{M}^n$ for all Killing vector fields V.

Proof. Suppose that $\langle\nabla f, V\rangle$ is not a constant function. Then $\nabla\langle\nabla f, V\rangle$ is a nonzero parallel gradient vector field. Hence $(\mathcal{M}^n, g)$ splits as the Riemannian product of the Euclidean line $\mathbb{R}$ with a Riemannian manifold $(\bar{\mathcal{M}}^{n-1}, \bar{g})$ (see the proof of Proposition 2.6). Note that by Lemma 2.5 there exists $\bar{f} : \bar{\mathcal{M}}^{n-1} \to \mathbb{R}$ such that $(\bar{\mathcal{M}}^{n-1}, \bar{g}, \bar{f}, \lambda)$ is a GRS and $f(x, y) = \frac{\lambda}{4}x^2 + \bar{f}(y)$. $\qquad\square$

Now let $(\mathcal{M}^n, g, f)$ be a steady GRS, so that

$$(8.9) \qquad\qquad \mathrm{Ric} + \nabla^2 f = 0.$$

Let $\Delta_f = \Delta - \nabla_{\nabla f}$ be the f-Laplacian acting on vector fields. Given a Killing vector field V, the Bochner formula (8.4) may be rewritten as (see Exercise 8.2)

$$(8.10) \qquad\qquad \Delta_f V - \nabla\langle\nabla f, V\rangle = 0.$$

By (8.8), if $(\mathcal{M}^n, g)$ does not split off a line, then $\nabla\langle\nabla f, V\rangle = 0$. Hence, in this case (8.10) is equivalent to

$$(8.11) \qquad\qquad \Delta_f V = 0.$$

Note that if we further assume that f has a critical point (e.g., as for the Bryant soliton), then $\langle\nabla f, V\rangle = 0$ on $\mathcal{M}^n$; i.e., Killing vector fields preserve the level sets of f.

Conversely, we may ask: Given a vector field satisfying $\Delta_f V = 0$, when is V a Killing vector field? Remarkably, in the case we discuss, we can ask and answer this in a pertinent way.

---

[2]Alternatively, the calculation can be justified by observing that if V is a Killing vector field, then $\mathcal{L}_V$ commutes with covariant differentiation, so that $\mathcal{L}_V(\nabla^2 f) = \nabla^2(\mathcal{L}_V f)$.

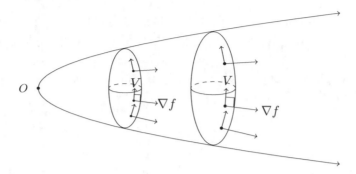

Figure 8.2. At each point the gradient of the potential function ∇f and any Killing vector field V are orthogonal on the Bryant steady GRS.

8.1.3. Outline of the proof of the main theorem.

The goal is to prove rotational symmetry of the 3-dimensional steady GRS, which corresponds to the isometry group of $(\mathcal{M}^3, g)$ containing a Lie subgroup isomorphic to $\mathrm{SO}(3)$. We shall prove the existence on $(\mathcal{M}^3, g)$ of Killing vector fields spanning a Lie algebra isomorphic to $\mathfrak{so}(3)$. This is based on studying Δ_f acting on vector fields and $\Delta_{L,f}$ acting on symmetric 2-tensors.

Let

(8.12) $$\Delta_{L,f} := \Delta_L - \mathcal{L}_{\nabla f}$$

be the f-**Lichnerowicz Laplacian** acting on symmetric 2-tensors, where Δ_L is defined by (4.79). Let r denote the distance to a fixed point. The proof of Theorem 8.1 is predicated on the following uniqueness result (restated as Proposition 8.40 below):

If $\Delta_{L,f} h = 0$ and $|h| \le \mathrm{O}(r^{-\varepsilon})$ for some $\varepsilon > 0$, then $h = b\,\mathrm{Ric}$ for some $b \in \mathbb{R}$.

A key link to this is the fact that if $\Delta_f W = 0$, then $\Delta_{L,f}(\mathcal{L}_W g) = 0$. Hence, if W satisfies $\Delta_f W = 0$ and is an almost Killing vector field in that it also satisfies $|\mathcal{L}_W g| \le \mathrm{O}(r^{-\varepsilon})$, then the uniqueness result yields

$$\mathcal{L}_W g = b\,\mathrm{Ric} = -\frac{b}{2}\mathcal{L}_{\nabla f} g$$

for some $b \in \mathbb{R}$; that is, $W + \frac{b}{2}\nabla f$ is a Killing vector field. Note that $\nabla f = \mathrm{O}(1)$.

Now the idea is to use the asymptotic rotational symmetry of g and f to construct three independent vector fields U_a satisfying

$$|U_a| = \mathrm{O}(r^{1/2}), \quad |\mathcal{L}_{U_a} g| \le \mathrm{O}(r^{-2\varepsilon}), \quad |\Delta_f U_a| \le \mathrm{O}(r^{-\frac{1}{2}-2\varepsilon}),$$

for $a = 1, 2, 3$ and some $\varepsilon > 0$. For each U_a, we show that there exists a vector field V_a with $\Delta_f V_a = \Delta_f U_a$ and $|V_a| \leq \mathrm{O}(r^{\frac{1}{2}-\varepsilon})$. Thus $W_a := U_a - V_a$ satisfies $\Delta_f W_a = 0$, where W_a is close to U_a (after scaling). Moreover, we can show that $|\mathcal{L}_{W_a} g| \leq \mathrm{O}(r^{-\varepsilon})$. From the aforementioned uniqueness result, we conclude that each

$$(8.13) \qquad \widehat{U}_a := W_a + \frac{b_a}{2} \nabla f$$

is a Killing vector field close to U_a. We can also show that $\langle \widehat{U}_a, \nabla f \rangle = 0$, $[\widehat{U}_a, \nabla f] = 0$, and as symmetric 2-tensors on the level set $\Sigma_s := f^{-1}(s)$,

$$(8.14) \qquad \sum_{a=1}^{3} \widehat{U}_a \otimes \widehat{U}_a = |s|(g_{\Sigma_s} + \mathrm{O}(|s|^{-\varepsilon}))$$

for some $\varepsilon > 0$, where g_{Σ_s} denotes the induced metric on Σ_s. Moreover, the Lie algebra generated by $\widehat{U}_1, \widehat{U}_2, \widehat{U}_3$ is 3-dimensional and isomorphic to $\mathfrak{so}(3)$. Corresponding to this Lie subalgebra of the Lie algebra of all Killing vector fields is a Lie subgroup of the isometry group of $(\mathcal{M}^3, g)$ which is isomorphic to $\mathrm{SO}(3)$. Moreover, since $\langle \widehat{U}_a, \nabla f \rangle = 0$, we have that $\widehat{U}_a$ is tangent to the level set Σ_s. Hence, each $\widehat{U}_a$ is a Killing vector field for each (Σ_s, g_{Σ_s}). We conclude that each (Σ_s, g_{Σ_s}) is a round 2-sphere and that g is rotationally symmetric.

We summarize the remaining sections in this chapter, which detail the proof of the main theorem. In §8.2 and §8.3 we show that the steady GRS must be asymptotically cylindrical in a strong sense. Using this, in §8.4 we construct the almost Killing vector fields U_a. In §8.5 we show that $\Delta_f W = 0$ implies $\Delta_{L,f}(\mathcal{L}_W g) = 0$. In §8.6 we solve $\Delta_f V = Q$ for $|Q| \leq \mathrm{O}(r^{-\frac{1}{2}-2\varepsilon})$ with $|V| \leq \mathrm{O}(r^{-\frac{1}{2}-\varepsilon})$ and we apply this to $Q = \Delta_f U_a$ to obtain the solution V_a and hence determine that $W_a := U_a - V_a$ satisfies $\Delta_f W_a = 0$. In §8.7 we prove the uniqueness theorem for solutions to $\Delta_{L,f} h = 0$. Applying this to $\mathcal{L}_{W_a} g$, we conclude that $\widehat{U}_a = W_a + \frac{b_a}{2} \nabla f$ is a Killing vector field. In §8.8 we finish the proof of the main theorem.

8.1.4. Notation.

Let $(\mathcal{M}^n, g)$ be a Riemannian manifold and let $\pi : \mathcal{E} \to \mathcal{M}^n$ be a C^∞ real vector bundle with a fiber metric and a compatible connection, such as a tensor bundle. For $\mathcal{E}_\ell := \mathcal{E} \otimes \bigotimes^\ell T^* \mathcal{M}$ we have the covariant derivative $\nabla : \mathcal{E}_\ell \to \mathcal{E}_{\ell+1}$. Given a nonnegative integer k, let $C^k_{\mathrm{loc}}(\mathcal{E})$ be the set of sections ϕ of $\mathcal{E}$ such that for each $0 \leq j \leq k$ the jth covariant derivative $\nabla^j \phi$ exists and is continuous. Define the seminorms $[\phi]_j := \sup_{x \in \mathcal{M}} |\nabla^j \phi(x)|$ for $j \geq 0$. The space $C^k(\mathcal{E})$ is the Banach space of all sections $\phi \in C^k_{\mathrm{loc}}(\mathcal{E})$

for which the norm

$$\|\phi\|_{C^k(\mathcal{E})} := \sum_{j=0}^{k} [\phi]_j$$

is finite.

8.2. Measures of the asymptotic difference from being cylindrical

Let $(\mathcal{M}^3, g, f)$ be a 3-dimensional noncompact κ-noncollapsed steady GRS with positive sectional curvature, so that $\mathrm{Ric} + \nabla^2 f = 0$. Assume the normalization

$$R + |\nabla f|^2 = 1$$

as in (2.55). By Theorem 1.18, $\mathcal{M}^3$ is diffeomorphic to $\mathbb{R}^3$. Since $\mathrm{Rm} > 0$, the quantities R, $|\mathrm{Ric}|$, and $|\mathrm{Rm}|$ are all uniformly equivalent. In particular, $R \leq 1$ implies that g has bounded sectional curvature. Moreover,

$$\nabla^2 f = -\mathrm{Ric} < 0$$

yields that f is strictly concave.

By Proposition 4.19, associated to our steady GRS is its canonical form $(\mathcal{M}^3, g(t), f(t), \varphi_t)$, $t \in (-\infty, \infty)$, satisfying $\frac{\partial}{\partial t} g(t) = -2 \mathrm{Ric}_{g(t)}$, $\frac{\partial}{\partial t} \varphi_t(x) = \nabla f_{\varphi_t(x)}$, $\varphi_0 = \mathrm{id}$, $g(t) = \varphi_t^* g$, $f(t) = f \circ \varphi_t$, $\mathrm{Ric}_{g(t)} + \nabla^2_{g(t)} f(t) = 0$, and $\frac{\partial f}{\partial t}(t) = |\nabla f(t)|^2_{g(t)}$. (In (8.27) below we shall show that f actually attains its supremum.) Moreover, by (2.60), we have

$$(8.15) \qquad \Delta_f R + 2|\mathrm{Ric}|^2 = 0.$$

In this section we will show the following estimates that measure the asymptotic difference from being cylindrical:

$$(8.16) \qquad |f| = \mathrm{O}(r),$$

where r is the distance to a fixed point,

$$(8.17) \qquad |f| R(x) \to 1 \quad \text{as} \quad x \to \infty,$$

and

$$(8.18) \qquad |\nabla^k \mathrm{Rm}| \leq \mathrm{O}(|f|^{-1-\frac{k}{2}}).$$

In the subsequent section, we will improve these estimates.

8.2.1. Perelman's results.

In this subsection we assume that we are on a complete noncompact 3-dimensional ancient solution that is κ-noncollapsed on all scales and has $\mathrm{Rm} \geq 0$. We state the fundamental estimates of Perelman that we will use. Terse proofs are in the original papers [**250, 251**] and more detailed proofs that follow Perelman are in [**201**] and [**227**] for example. By considering the Bryant soliton and the more trivial cylinders $\mathbb{S}^2 \times \mathbb{R}$ as the model cases for the estimates, these statements appear more natural and are more easily digested.

Definition 8.4 (ε-neck). Let $(\mathcal{M}^3, g)$ be a complete Riemannian manifold. Given $\varepsilon > 0$ and $p \in \mathcal{M}^3$ with $R(p) > 0$, let $r := \sqrt{2} R(p)^{-1/2}$. A geodesic ball $B_{\varepsilon^{-1} r}(p) =: \mathfrak{N}$ in $(\mathcal{M}^3, g)$ is called an ε-**neck**, if, after scaling g by r^{-2}, the metric on $B_{\varepsilon^{-1} r}(p)$ is ε-close in the $2C^{\lceil \varepsilon^{-1} \rceil + 1}$-topology to a piece of the cylindrical metric $g_{\mathbb{S}^2} + du^2$ on $\mathbb{S}^2 \times \mathbb{R}$, where $\lceil \varepsilon^{-1} \rceil$ denotes the least integer greater than or equal to ε^{-1}. More precisely, there exists an embedding

$$\varphi : B_{\varepsilon^{-1} r}(p) \to \mathbb{S}^2 \times \mathbb{R}$$

with $\varphi(p) \in \mathbb{S}^2 \times \{0\}$ such that on $B_{\varepsilon^{-1} r}(p)$,

$$\left\| \varphi^* \left(g_{\mathbb{S}^2} + du^2 \right) - r^{-2} g \right\|_{C^{\lceil \varepsilon^{-1} \rceil + 1}(r^{-2} g)} < \varepsilon.$$

We call a point in $\varphi^{-1}\left(\mathbb{S}^2 \times \{0\} \right)$ a **center** of the ε-neck.

For the standard cylinder $\mathbb{S}^2 \times \mathbb{R}$, every point of $\mathbb{S}^2 \times \mathbb{R}$ is the center of an ε-neck for all $\varepsilon > 0$.

It is not difficult to see that for the 3-dimensional Bryant soliton, for any $\varepsilon > 0$, there exists a radius r such that every point in the complement of the ball of radius r centered at 0^3 is the center of an ε-neck.

Henceforth, we assume that $(\mathcal{M}^3, g(t)), t \in (-\infty, 0)$, **is a 3-dimensional complete noncompact ancient solution that is κ-noncollapsed on all scales and has bounded** $\mathrm{Rm} > 0$. We call such a solution a noncompact **ancient κ-solution** with $\mathrm{Rm} > 0$.

Let $\mathcal{M}_\varepsilon(t)$ denote the set of points in $\mathcal{M}^3$ which are *not* the centers of ε-necks for the metric $g(t)$. We have the following result of Perelman [**250**, §11.8] which says in a quantitative way how spatial infinity is necklike.

Proposition 8.5 (Compact control of non-ε-neck points). *For each $\varepsilon > 0$ sufficiently small, there exists $C = C(\varepsilon) < \infty$ which has the following property. Let $(\mathcal{M}^3, g(t))$ be a 3-dimensional ancient κ-solution with $\mathrm{Rm} \geq 0$. Then $\mathcal{M}_\varepsilon(t)$ is compact, nonempty, and for each $t \in (-\infty, 0)$ there exists*

$x_{0,\varepsilon}(t) \in \partial \mathcal{M}_\varepsilon(t)$ *such that*

(8.19a) $$\operatorname{diam} \mathcal{M}_\varepsilon(t) \leq C R^{-1/2}(x_{0,\varepsilon}(t), t),$$

(8.19b) $$C^{-1} \sup_{x \in \mathcal{M}} R(x, t) \leq R(x_{0,\varepsilon}(t), t) \leq C \inf_{x \in \mathcal{M}_\varepsilon(t)} R(x, t).$$

The following is Perelman's κ-**compactness theorem** (see [**250**, §11.7]).

Theorem 8.6. *Let* $\kappa > 0$. *Then for any sequence* $\{(\mathcal{M}_i^3, g_i(t))\}$ *of noncompact 3-dimensional ancient κ-solutions with* $\operatorname{Rm} > 0$ *to the Ricci flow and for any* $(x_i, t_i) \in \mathcal{M}_i^3 \times (-\infty, 0]$, *there exists a subsequence such that* $(\mathcal{M}_i^3, g_i(t), x_i)$, *where*

(8.20) $$g_i(t) := R(x_i, t_i) g(t_i + t R^{-1}(x_i, t_i)), \quad t \in (-\infty, 0],$$

converges in the pointed C^∞ Cheeger–Gromov sense to an ancient κ-solution with $\operatorname{Rm} \geq 0$.

The following are Perelman's consequent κ-**derivative estimates** (see [**251**, §1.5]).

Theorem 8.7. *Given* $\kappa > 0$ *and nonnegative integers* $\ell \geq 0$ *and* $m \geq 0$, *there exist constants* $C_{\ell,m}$ *depending only on ℓ and m such that for any 3-dimensional ancient κ-solution* $(\mathcal{M}^3, g(t))$ *with* $\operatorname{Rm} \geq 0$, *we have*

(8.21) $$\left| \frac{\partial^\ell}{\partial t^\ell} \nabla^m \operatorname{Rm} \right| \lesssim C_{\ell,m} R^{1+\ell+\frac{m}{2}} \quad on \; \mathcal{M}^3 \times (-\infty, 0].$$

8.2.2. Blow-down cylinder limit for 3-dimensional noncompact ancient κ-solutions.

In view of the fact that the canonical form of our steady GRS is a 3-dimensional ancient κ-solution with $\operatorname{Rm} > 0$, we can invoke the results of Perelman stated in the previous subsection.

Let $p_k \in \mathcal{M}^3$ and let $R_k := R(p_k, 0)$. Define

$$\hat{g}^{(R_k)}(t) = R_k g\left(R_k^{-1} t\right).$$

By Perelman's compactness theorem (Theorem 8.6), there exists a subsequence of $(\mathcal{M}^3, \hat{g}^{(R_k)}(t), p_k)$ which converges to a noncompact ancient κ-solution $(\mathcal{M}_\infty^3, g_\infty(t), p_\infty)$ with $\operatorname{Rm} \geq 0$.

By Proposition 8.5, we conclude that if $p_k \to \infty$, then $(\mathcal{M}_\infty^3, g_\infty(t))$ is isometric to the shrinking round cylinder $(\mathbb{S}^2 \times \mathbb{R}, \bar{g}(t))$.

Remark 8.8. Since the exact cylinder satisfies $\Delta R = 0$ and $2|\operatorname{Ric}|^2 = R^2$, we have at spatial infinity of each time slice that $\Delta R = \operatorname{o}(R^2)$ and $2|\operatorname{Ric}|^2 = R^2 + \operatorname{o}(R^2)$, so that

$$\frac{\partial R}{\partial t} = R^2 + \operatorname{o}(R^2).$$

Remark 8.9. We also have Perelman's "asymptotic soliton" (see [**250**, §11.2]. Namely, for any $\tau_i \to \infty$ and $q_i \in \mathcal{M}^3$ such that the **reduced distance**

$$\ell^{\tilde{g}}(q_i, \tau_i) \leq C,$$

there exists a subsequence such that $\left(\mathcal{M}^3, \tau_i^{-1}\tilde{g}(\tau_i\tau), q_i\right)$ converges to the shrinking round cylinder.

8.2.3. Derivative estimates for noncollapsed steadies.

Since the canonical form from Proposition 4.19 of our steady GRS is a 3-dimensional complete noncompact ancient solution that is κ-noncollapsed with bounded Rm > 0, by the conclusion of Theorem 8.7 at $t = 0$, we have that the inequalities in (8.21) hold for $(\mathcal{M}^3, g, f)$.

8.2.4. Asymptotic behavior of the curvature.

Throughout the rest of this chapter as noted above, $(\mathcal{M}^3, g, f)$ shall be a 3-dimensional noncompact κ-noncollapsed steady GRS with positive sectional curvature. The reader would do well to again keep in mind the qualitative behavior of the model case of the Bryant soliton from Chapter 6. We have the following preliminary estimates reflecting that the steady GRS is asymptotically cylindrical.

Lemma 8.10. *Fix a point $p \in \mathcal{M}^3$ and suppose that $x_i \to \infty$ satisfies $R(x_i)d^2(x_i, p) \to \infty$. Then*

$$(8.22) \qquad \lim_{i \to \infty} \frac{|\nabla R|}{R^{3/2}}(x_i) = 0 \quad and \quad \lim_{i \to \infty} \frac{\left|\langle \nabla f, \nabla R\rangle - R^2\right|}{R^2}(x_i) = 0.$$

Proof. Firstly, observe that it suffices to show that for any subsequence of a sequence $x_i \to \infty$ with $R(x_i)d^2(x_i, p) \to \infty$ there exists a further subsequence such that (8.22) holds.

Let $(\mathcal{M}^3, g(t), f(t))$, $t \in (-\infty, \infty)$, be the canonical form associated to $(\mathcal{M}^3, g, f)$. Since $(\mathcal{M}^3, g(t))$ is a 3-dimensional ancient κ-solution, by (8.21) there exists a constant C such that

$$(8.23) \quad |\langle \nabla f, \nabla R\rangle|(x, t) = \left|\frac{\partial R}{\partial t}\right|(x, t) \leq CR^2(x, t) \quad \text{on } \mathcal{M}^3 \times (-\infty, 0],$$

where we used Perelman's derivative estimates (Theorem 8.7) to obtain the last inequality. In particular,

$$(8.24) \qquad\qquad |\langle \nabla f, \nabla R\rangle| \leq CR^2 \quad \text{for } (\mathcal{M}^3, g, f).$$

Pass to any subsequence and define

$$g_i(t) := R(x_i)g(tR^{-1}(x_i)).$$

Since $\{(\mathcal{M}^3, g_i(t))\}$, $i \in \mathbb{N}$, are noncompact 3-dimensional ancient κ-solutions, by Perelman's compactness theorem (Theorem 8.6), there exists a subsequence such that $\{(\mathcal{M}^3, g_i(t), x_i)\}$ converges in the pointed C^∞ Cheeger–Gromov sense to an orientable ancient κ-solution $(\mathcal{M}_\infty^3, g_\infty(t), x_\infty)$.

Since $R(x_i)d^2(x_i, p) \to \infty$, we may apply Theorem 1.19 to obtain that $(\mathcal{M}_\infty^3, g_\infty(t))$ splits as the product of $\mathbb{R}$ and a 2-dimensional ancient κ-solution, which must be the shrinking round 2-sphere by Hamilton's classification theorem (the real projective plane does not embed in $\mathcal{M}^3$, which is diffeomorphic to $\mathbb{R}^3$). From $|\nabla R| = \Delta R = 0$ and $2|\mathrm{Ric}|^2 = R^2$ on the cylinder $(\mathcal{M}_\infty^3, g_\infty(t))$ and from

$$\langle \nabla f, \nabla R \rangle = \Delta R + 2|\mathrm{Ric}|^2$$

on $(\mathcal{M}^3, g, f)$, we conclude the estimates for the scale-invariant quantities in (8.22). $\qquad\square$

Recall Perelman's *bounded curvature at bounded distance* result.

Theorem 8.11 ([**250**, §11.7]; see also [**227**, Lemma 9.65]). *For each $r \in (0, \infty)$ there exists $C(r)$ with the following property. If $(\mathcal{M}^3, g(t))$, $t \in (-\infty, 0]$, is an ancient κ-solution and $x \in \mathcal{M}^3$ satisfies $R(x, 0) = 1$, then $R(p, 0) \le C(r)$ for all $p \in B_r^{g(0)}(x)$.*

Invoking this, one can show that for any sequence $x_i \to \infty$ in the steady GRS we have $R(x_i)d^2(x_i, p) \to \infty$ (see Exercise 8.7).

A contradiction argument yields the following result for our steady GRS.

Lemma 8.12. *The curvature of $(\mathcal{M}^3, g)$ tends to zero at infinity:*

$$(8.25) \qquad \lim_{x \to \infty} R(x) = 0 \quad \text{and hence} \quad \lim_{x \to \infty} |\nabla f|(x) = 1.$$

Proof. Suppose that the statement $\lim_{x \to \infty} R(x) = 0$ were false. Then there would exist a sequence of points $x_i \to \infty$ and $c > 0$ such that $R(x_i) \ge c$ for all i. Applying (8.22) to the sequence $\{x_i\}$, we obtain

$$(8.26) \qquad \lim_{i \to \infty} |\nabla R|(x_i) = 0 \quad \text{and} \quad \lim_{i \to \infty} \left| \langle \nabla f, \nabla R \rangle - R^2 \right|(x_i) = 0.$$

Since $|\nabla f| \le 1$, this implies that $\lim_{i \to \infty} R^2(x_i) = 0$, which is a contradiction. $\qquad\square$

Since $\lim_{x \to \infty} R(x) = 0$ and $R > 0$, we have that R_{sup} is attained at some point which we call o. Since $\mathrm{Ric} > 0$, $\nabla R = 2\,\mathrm{Ric}\,(\nabla f)$ implies that $\nabla f(o) = 0$. Let

$$r(x) := d(x, o).$$

Normalize f so that $f(o) = -1$. Since f is strictly concave, it follows that the maximum of f is attained at o, i.e.,

$$(8.27) \qquad \sup_{x \in \mathcal{M}} f(x) = f(o) = -1,$$

and that there exists $C > 0$ such that
$$(8.28)$$
$$|\nabla f| \geq C^{-1} \quad \text{and} \quad C^{-1} r(x) \leq |f|(x) = -f(x) \leq C r(x) \quad \text{on } \mathcal{M}^3 \setminus B_1(o).$$

In particular, f is proper. Define the level sets

$$\Sigma_s := f^{-1}(s)$$

for $s \leq -1$. Since $\mathcal{M}^3$ is diffeomorphic to $\mathbb{R}^3$ and since f is proper with exactly one critical point o, we have that each level set Σ_s is diffeomorphic to $\mathbb{S}^2$ for $s < -1$.

Because of (8.28), when considering the order of growth of quantities, we may use $r = r(x)$ and $|s| = |f|(x)$ interchangeably.

The curvature of $(\mathcal{M}^3, g)$ decays inverse linearly:

Lemma 8.13. *On our steady GRS,*

$$(8.29) \qquad \lim_{x \to \infty} (|f| R)(x) = 1,$$

and consequently

$$|\operatorname{Rm}|(x) = O(r^{-1}(x)) \quad \text{as } x \to \infty.$$

Proof. We compare R^{-1} to f. By (8.25), there exists a constant r_0 such that $|\nabla f|^2 \geq \frac{1}{2}$ on $\mathcal{M}^3 \setminus B_{r_0}(o)$. Let C be as in (8.24). We compute on $\mathcal{M}^3 \setminus B_{r_0}(o)$ that

$$(8.30) \qquad \langle \nabla f, \nabla (R^{-1} + 2Cf) \rangle = 2C |\nabla f|^2 - \frac{\langle \nabla f, \nabla R \rangle}{R^2}$$
$$\geq C - \frac{|\langle \nabla f, \nabla R \rangle|}{R^2}$$
$$\geq 0.$$

Let $x \in \mathcal{M}^3 \setminus B_{r_0}(o)$ and let $\sigma : (-\infty, \infty) \to \mathcal{M}^3$ be the integral curve of the vector field ∇f with $\sigma(0) = x$. Since $|\nabla f| \geq \frac{1}{\sqrt{2}}$, there exists $u_0 > 0$ such that $\sigma(u_0) \in B_{r_0}(o)$. From this and (8.30), we conclude that

$$\sup_{\mathcal{M}} (R^{-1} + 2Cf) \leq \sup_{B_{r_0}(o)} (R^{-1} + 2Cf) < \infty.$$

Hence $R^{-1} \leq C|f|$ for some constant C, so we have the lower bound:

$$(8.31) \qquad R \geq \frac{1}{C|f|} \quad \text{on } \mathcal{M}^3.$$

Since we now know that $R(x)d^2(x,o) \to \infty$ as $x \to \infty$, we can apply the proof of Lemma 8.10 to improve (8.22) to

$$(8.32) \qquad \lim_{x \to \infty} \frac{\left| \langle \nabla f, \nabla R \rangle - R^2 \right|}{R^2}(x) = 0.$$

Therefore, from

$$(8.33) \quad \left\langle \nabla f, \nabla \left(R^{-1} + f \right) \right\rangle = |\nabla f|^2 - \frac{\langle \nabla f, \nabla R \rangle}{R^2} = -R - \frac{\langle \nabla f, \nabla R \rangle - R^2}{R^2},$$

we obtain the decay estimate

$$\lim_{x \to \infty} \left\langle \nabla f, \nabla \left(R^{-1} + f \right) \right\rangle (x) = 0.$$

Let $x \in \mathcal{M}^3 \setminus B_1(o)$ and let $\beta(t)$ be the integral curve of ∇f with $\beta(0) = x$. Let $t_0 > 0$ be the first time such that $r(\beta(t_0)) = 1$. By (8.28), t_0 exists and $t_0 \le Cr(x)$, where $C < \infty$. Hence integrating along β yields

$$\left(R^{-1} + f \right)(x) = \left(R^{-1} + f \right)(\beta(t_0)) - \int_0^{t_0} \left\langle \nabla f, \nabla \left(R^{-1} + f \right) \right\rangle (\beta(s)) ds$$

$$= o(|f|(x)) \quad \text{as } x \to \infty.$$

The lemma follows. $\square$

Since we now know that $R(x_i)d^2(x_i, o) \to \infty$ and $|f|R(x_i) \to 1$ for any sequence $x_i \to \infty$, by Perelman's argument (see the proof of Lemma 8.10) we have the following.

Corollary 8.14 (Dimension reduction). *On our steady GRS, for any $x_i \to \infty$ there exists a subsequence such that $\{(\mathcal{M}^3, \hat{g}_i(t), x_i)\}$, $t \in (-\infty, 1)$, where*

$$(8.34) \qquad \hat{g}_i(t) := |f(x_i)|^{-1} g(|f(x_i)|t),$$

converges in the pointed C^∞ Cheeger–Gromov sense to a shrinking round cylinder $\left(\mathbb{S}^2 \times \mathbb{R}, \bar{g}(t) \right)$, $t \in (-\infty, 1)$, where

$$(8.35) \qquad \bar{g}(t) := 2(1 - t)g_{\mathbb{S}^2} + dz^2.$$

We observe some further consequences of the GRS being asymptotically cylindrical.

Lemma 8.15. *On any 3-dimensional κ-noncollapsed nonflat steady GRS $(\mathcal{M}^3, g, f)$, we have*

$$(8.36) \qquad \lim_{x \to \infty} \frac{|\mathrm{Ric}|^2}{R^2}(x) = \frac{1}{2} \quad and \quad \lim_{x \to \infty} \frac{|\nabla^k \mathrm{Rm}|}{R^{1+(k/2)}}(x) = 0 \text{ for } k \ge 1.$$

Note that since $R > 0$ is bounded from above, the last inequality implies that $\lim_{x \to \infty} |\nabla^k \mathrm{Rm}|(x) = 0$ for $k \ge 1$.

Proof. Suppose that the lemma were false. Then there would exist a sequence $\{x_i\}$ in $\mathcal{M}^3$ with $x_i \to \infty$ such that $\lim_{i\to\infty} \frac{|\mathrm{Ric}|^2}{R^2}(x_i) \neq \frac{1}{2}$ or $\lim_{i\to\infty} \frac{|\nabla^k \mathrm{Rm}|}{R^{1+(k/2)}}(x_i) \neq 0$ for some $k \geq 1$, if the limit exists. Let

$$g_i(t) = R(x_i)g(tR^{-1}(x_i)),$$

where $(\mathcal{M}^3, g(t), f(t))$ is the canonical form ($g = g(0)$). By Perelman's compactness theorem (Theorem 8.6) and by Corollary 8.14, i.e., their consequence at $t = 0$, there exists a subsequence such that $(\mathcal{M}^3, g_i, x_i)$, where $g_i := R(x_i)g$, converges in the pointed C^∞ Cheeger–Gromov sense to a round cylinder $(\mathbb{S}^2 \times \mathbb{R}, g_{\mathbb{S}^2} + dz^2, (y_\infty, 0))$. On this limit we have scale-invariant quantities satisfying $\frac{|\mathrm{Ric}|^2}{R^2} = \frac{1}{2}$ and $\frac{|\nabla^k \mathrm{Rm}|}{R^{1+(k/2)}} = 0$ for $k \geq 1$. This leads to a contradiction, thereby proving (8.36). $\qquad\square$

By applying Lemma 8.13 to Lemma 8.15, we now obtain both

$$(8.37) \qquad \lim_{x\to\infty} r(x)^{1+(k/2)}|\nabla^k \mathrm{Rm}|(x) = 0 \quad \text{for } k \geq 1$$

and

(8.38)

$$\lim_{x\to\infty} 2|f|^2 \mathrm{Ric}(\nabla f, \nabla f) = \lim_{x\to\infty} |f|^2 \langle \nabla f, \nabla R \rangle = \lim_{x\to\infty} \frac{\Delta R + 2|\mathrm{Ric}|^2}{R^2} = 1.$$

In particular,

$$(8.39) \qquad \mathrm{Ric}(\nabla f, \nabla f) = \frac{1}{2}\langle \nabla f, \nabla R \rangle = \mathrm{O}(r^{-2}).$$

Note that by applying (8.37) to (4.99), we have

$$(8.40) \qquad |R_{ijk\ell}\nabla_\ell f| = |\nabla_i R_{jk} - \nabla_j R_{ik}| = \mathrm{o}(r^{-3/2}).$$

In summary, for each covariant derivative of Rm, the rate of decay improves by the power $-\frac{1}{2}$ of r. For this reason, equations derived from the GRS equation which relate curvatures to its derivatives, such as (4.99), can be used to improve the decay rate.

We conclude this section with a heuristic calculation on the Bryant soliton which exhibits the qualitative behavior of dimension reduction to a shrinking cylinder.

Example 8.16. Take any sequence of points $p_k \in \mathcal{M}^3$ in the Bryant soliton $(\mathcal{M}^3, g, f)$ with $r_k := d(p_k, o) \to \infty$, where o is the origin. Given $\delta > 0$, g is an approximate cylinder on the region $r_k - \delta^{-1}r_k^{1/2} \leq r \leq r_k + \delta^{-1}r_k^{1/2}$ for k sufficiently large. We have $r_k R(p_k) \to 1$, so the radius of this cylinder is approximately $(2r_k)^{1/2}$. Let $(\mathcal{M}^3, g(t), f(t))$ be the associated canonical form. Under the Ricci flow (i.e., flowing by diffeomorphisms generated by ∇f), since $|\nabla f| \sim 1$, after time t a point at large distance r to the origin goes to a point at distance $r - t$ to the origin. Rescale the solution by multiplying the metric by $R(p_k) \sim r_k^{-1}$ and multiplying time by r_k^{-1}. Since a point at large distance r goes to a point with $R \sim (r-t)^{-1}$, after rescaling, a point at distance $\frac{r}{r_k}$, where $r \in [r_k - \delta^{-1} r_k^{1/2}, r_k + \delta^{-1} r_k^{1/2}]$, evolves to a point with

$$R \sim \left(\frac{(1-t)r_k \pm \delta^{-1} r_k^{1/2}}{r_k} \right)^{-1} = (1-t)^{-1} + O(r_k^{-1/2}).$$

This rough asymptotic calculation agrees with the fact that for the shrinking 2-sphere $(\mathbb{S}^2, \bar{g}(t))$ of the cylinder limit, the scalar curvature satisfies $R_{\bar{g}(t)} = (1-t)^{-1}$; compare with Remark 6.15.

8.3. Improved estimates for the curvature and its derivatives

In this section we obtain improved asymptotic estimates for the curvature as well as derive strong estimates for the intrinsic and extrinsic geometries of the level surfaces of the potential function. These estimates are more qualitatively precise than the ones in the previous section and they reflect the asymptotic rotational symmetry of our steady GRS. In particular, we shall show that $|f| R = 1 + O(r^{-1/4})$ and $|\nabla R| = O(r^{-7/4})$ as well as verify estimates for the second fundamental form and intrinsic Gauss curvature of the level sets of f.

8.3.1. Level surfaces of f are asymptotically umbilical.

In the rotationally symmetric case, the level surfaces are totally umbilical. For the steady GRS, we consider the asymptotic behavior of the level sets of f. Let $x_i \to \infty$ and let $K_i := |f(x_i)|^{-1}$. Considering time $t = 0$ in (8.34) together with the associated rescaled potential function, we define

$$(8.41) \qquad \hat{g}_i := K_i g \quad \text{and} \quad f_i := K_i^{1/2}(f - f(x_i)).$$

By Corollary 8.14, the sequence of pointed Riemannian manifolds $(\mathcal{M}^3, \hat{g}_i, x_i)$ subconverges in the C^∞ Cheeger–Gromov sense to $(\mathbb{S}^2 \times \mathbb{R}, \bar{g}, (y_\infty, 0))$, where $\bar{g} := 2g_{\mathbb{S}^2} + dz$.

Lemma 8.17. *For our steady GRS, under the pointed C^∞ Cheeger–Gromov convergence of $(\mathcal{M}^3, \hat{g}_i, x_i)$ to the round cylinder $(\mathbb{S}^2 \times \mathbb{R}, \bar{g}, (y_\infty, 0))$, the vector fields $\nabla_{\hat{g}_i} f_i = |f(x_i)|^{1/2} \nabla_g f$ subconverge to $\nabla_{\bar{g}} f_\infty = \frac{\partial}{\partial z}$, where $f_\infty(y, z) := z$. Hence the level surfaces $\Sigma_{f(x_i)}$ converge to the slice $\mathbb{S}^2 \times \{0\}$.*

Proof. By (8.41), we have

$$|\nabla f_i|^2_{\hat{g}_i} = |\nabla f|^2_g,$$
$$|\nabla^k f_i|^2_{\hat{g}_i} = K_i^{1-k} |\nabla^k f|^2_g = K_i^{1-k} |\nabla^{k-2} \operatorname{Ric}_g|^2_g \quad \text{for } k \geq 2.$$

From this, $f_i(x_i) = 0$, $|\nabla f| \to 1$, and $\left|\nabla^\ell \operatorname{Ric}_g\right|_g = \mathrm{O}(|f|^{-1-\frac{\ell}{2}})$ for $\ell \geq 0$, we see under the above convergence that the f_i converge to a C^∞ function f_∞ on $\mathbb{S}^2 \times \mathbb{R}$ satisfying

$$f_\infty(y_\infty, 0) = 0, \quad |\nabla f_\infty|_{\bar g} \equiv 1, \quad \text{and} \quad |\nabla^k f_\infty|_{\bar g} \equiv 0 \quad \text{for } k \geq 2.$$

We conclude, without loss of generality, that $f_\infty(y, z) = +z$.[3] $\square$

From this lemma it follows that the level sets are asymptotically umbilical. Below, we shall make this more precise by estimating a suitable 2-tensor. The basic ideas we use are that we are in dimension 3 and that a contraction of Rm and ∇f yields ∇ Ric terms, providing better decay by the derivative estimates. Then the GRS equation ties this in with the second fundamental form of the level sets of f.

For a round cylinder $\mathbb{S}^2 \times \mathbb{R}$ we have

$$2 \operatorname{Ric} - Rg + R\, du \otimes du = 0,$$

where u is the coordinate on $\mathbb{R}$. Motivated by this and the fact that $|df| \to 1$, we define the symmetric 2-tensor

(8.42) $$T := 2 \operatorname{Ric} - Rg + R\, df \otimes df.$$

If, in the definition of T, we replace df by $\frac{df}{|df|}$, then we obtain a 2-tensor whose norm measures how far Rm is from being the curvature operator of a cylinder, with $\frac{df}{|df|}$ corresponding to the radial unit 1-form. From expanding the expression for $|T|^2$ and by applying Lemmas 8.13 and 8.15, one can derive that $|T| = \mathrm{o}(r^{-1})$. We now improve this estimate by using equations satisfied by GRS. Let $\nu = -\frac{\nabla f}{|\nabla f|}$ be the outward unit normal to $\Sigma_s = \partial\{f \geq s\}$.

Remark 8.18. On the Bryant soliton we have

$$T(\nu, \nu) = 2 \operatorname{Ric}(\nu, \nu) - R + R|\nabla f|^2 = 2 \operatorname{Ric}(\nu, \nu) - R^2 = \mathrm{O}(r^{-2})$$

and for any unit vector V tangent to Σ_s,

$$T(V, V) = 2 \operatorname{Ric}(V, V) - R = -\operatorname{Ric}(\nu, \nu) = \mathrm{O}(r^{-2}).$$

Hence $|T| = \mathrm{O}(r^{-2})$ decays quadratically on the Bryant soliton.

We now establish the decay estimate for $|T|$ on our steady soliton. Since $n = 3$, from (4.99), from

(8.43) $$R_{ijk\ell} = R_{i\ell}g_{jk} + R_{jk}g_{i\ell} - R_{ik}g_{j\ell} - R_{j\ell}g_{ik} - \frac{R}{2}\left(g_{i\ell}g_{jk} - g_{ik}g_{j\ell}\right),$$

---

[3]Of the two possibilities $\pm$ we choose $+$.

and from (2.40), we have

$$\nabla_i R_{jk} - \nabla_j R_{ik} = R_{ijk\ell} \nabla_\ell f$$
$$= R_{jk} \nabla_i f - R_{ik} \nabla_j f$$
$$+ \frac{1}{2} \left(\nabla_i R - R \nabla_i f \right) g_{jk} - \frac{1}{2} \left(\nabla_j R - R \nabla_j f \right) g_{ik}.$$

By this, (2.40) again, the definition of T, and $1 - |\nabla f|^2 = R$, we obtain

$$(8.44) \qquad 2 \left(\nabla_i R_{jk} - \nabla_j R_{ik} \right) \nabla_i f = T_{jk} |\nabla f|^2 - \nabla_k R \nabla_j f - \nabla_j R \nabla_k f$$
$$+ \langle \nabla f, \nabla R \rangle g_{jk} + R^2 \nabla_j f \nabla_k f.$$

Since $R = \mathrm{O}(r^{-1})$, $|\nabla \operatorname{Ric}| = \mathrm{o}(r^{-3/2})$ by (8.37), $\langle \nabla f, \nabla R \rangle = \mathrm{O}(r^{-2})$ by (8.38), and $|\nabla f| \to 1$, we conclude that

$$(8.45) \qquad |T| = \mathrm{o}(r^{-3/2}).$$

Note that the order of decay is better by $1/2$ compared to that from scaling (and that we have little o instead of big O).

Therefore, by expanding $|T|^2$ while using (2.41) and (2.55), we have

$$\mathrm{o}(r^{-3}) = |T|^2 = 4 |\operatorname{Ric}|^2 - 2R^2 + R^4 + 2R \langle \nabla f, \nabla R \rangle,$$

which implies[4]

$$(8.46) \qquad \left| |\operatorname{Ric}|^2 - \frac{1}{2} R^2 \right| = \mathrm{O}(r^{-3}).$$

One checks that covariantly differentiating (8.44) decreases the order of each term by at least $\mathrm{O}(|f|^{-1/2})$. Hence

$$(8.47) \qquad |\nabla T| = \mathrm{o}(r^{-2}).$$

Now consider the geometry of the level sets of the potential function. Each level set $\Sigma_s := f^{-1}(s)$ has unit outward normal $\nu = e_3 := -\frac{\nabla f}{|\nabla f|}$. By (2.35), its second fundamental form is the symmetric 2-tensor on Σ_s given by

$$(8.48) \qquad \mathrm{II} = \frac{-(\nabla^2 f)|_{\Sigma_s}}{|\nabla f|} = \frac{\operatorname{Ric}|_{\Sigma_s}}{|\nabla f|} = \frac{(T + Rg)|_{\Sigma_s}}{2 |\nabla f|},$$

where the last equality is by the definition of T. This relates the extrinsic geometry of Σ_s to the intrinsic geometry of $\mathcal{M}^3$. The mean curvature is

$$(8.49) \qquad H = \frac{R}{|\nabla f|} - \frac{\operatorname{Ric}(\nabla f, \nabla f)}{|\nabla f|^3}.$$

---

[4]Note that on the Bryant soliton,

$$|\operatorname{Ric}|^2 - \frac{1}{2} R^2 = \frac{1}{2} \operatorname{Ric}(\nu, \nu)^2 - 2 \operatorname{Ric}(V, V) \operatorname{Ric}(\nu, \nu) = \mathrm{O}(r^{-3})$$

for any unit vector V tangent to Σ_s.

Hence, by (8.29) and by $\mathrm{Ric}(e_3, e_3) = \mathrm{O}(|f|^{-2})$ from (8.39), we have

$$(8.50) \qquad\qquad \lim_{x \to \infty} |f| \, H = 1.$$

In the model Bryant soliton case, the Σ_s are totally umbilical and have constant intrinsic curvature and R is constant on each Σ_s. By (8.48) and (8.49), we have that Σ_s tends toward being totally umbilical:

$$(8.51) \qquad \left| \mathrm{II} - \frac{H}{2} g|_{\Sigma_s} \right| = \frac{1}{2 |\nabla f|} \left| T|_{\Sigma_s} + \frac{\mathrm{Ric}(\nabla f, \nabla f)}{|\nabla f|^2} g|_{\Sigma_s} \right| = \mathrm{o}(r^{-3/2}),$$

where we used $|T| = \mathrm{o}(r^{-3/2})$ and (8.39). Moreover, by this and (8.50),

$$(8.52) \qquad\qquad |\mathrm{II}| \le \mathrm{O}(r^{-1}).$$

From the proof of Lemma 8.15 and from $|f| R \to 1$, one can deduce that

$$(8.53) \qquad \mathrm{diam}(\Sigma_s) = \mathrm{O}(\sqrt{|s|}) \quad \text{and} \quad \mathrm{Area}(\Sigma_s) = \mathrm{O}(|s|).$$

Let $\bar{R}(s)$ denote the average of R on Σ_s. Using $|\nabla R| = \mathrm{o}(r^{-3/2})$, one concludes that

$$(8.54) \qquad\qquad \left| R - \bar{R}(s) \right| = \mathrm{o}(|s|^{-1}) \quad \text{on } \Sigma_s.$$

Again, this is an estimate that we shall improve (see Proposition 8.21 below).

8.3.2. Refined asymptotics for specific derivatives of curvature.

In this subsection we begin the estimation of the oscillation and derivatives of R restricted to a level set of f by considering specific quantities which will be useful for the general estimates we prove later.

Given a function ϕ on $\mathcal{M}^3$, let $\nabla_{\Sigma_s} \phi$ denote the gradient of ϕ restricted to a level set Σ_s; that is, $\nabla_{\Sigma_s} \phi = \nabla \phi - \frac{1}{|\nabla f|^2} \langle \nabla \phi, \nabla f \rangle \nabla f$. Let $\Delta_{\Sigma_s} \phi$ denote the Laplacian on (Σ_s, g_{Σ_s}) of ϕ restricted to Σ_s. Since the unit normal to Σ_s is $\nu = -\frac{\nabla f}{|\nabla f|}$, we have

$$(8.55) \qquad \Delta_{\Sigma_s} \phi = \Delta \phi - \frac{1}{|\nabla f|^2} \nabla^2 \phi(\nabla f, \nabla f) + \frac{H}{|\nabla f|} \langle \nabla f, \nabla \phi \rangle.$$

Hence

$$\nabla^2 R(\nabla f, \nabla f) = |\nabla f|^2 (\Delta R - \Delta_{\Sigma_s} R) + |\nabla f| \, H \, \langle \nabla f, \nabla R \rangle.$$

On the other hand, we compute that

$$\nabla^2 R(\nabla f, \nabla f) = \langle \nabla f, \nabla \langle \nabla f, \nabla R \rangle \rangle - \nabla^2 f(\nabla f, \nabla R)$$

$$= \langle \nabla f, \nabla(\Delta R + 2 |\mathrm{Ric}|^2) \rangle + \frac{1}{2} |\nabla R|^2,$$

since $\nabla^2 f(\nabla f) = \frac{1}{2} \nabla |\nabla f|^2 = -\frac{1}{2} \nabla R$. Thus, from (8.37) we have

$$\left| \nabla^2 R(\nabla f, \nabla f) \right| = \mathrm{o}(r^{-5/2}).$$

Since $\frac{H}{|\nabla f|} \langle \nabla f, \nabla R \rangle = O(r^{-3})$, from (8.55) with $\phi = R$ we then obtain

$$|\Delta R - \Delta_{\Sigma_s} R| = o(r^{-5/2}).$$

Combining this with (2.60) and applying (8.46) yields

$$(8.56) \qquad \left| \Delta_{\Sigma_s} R - \langle \nabla f, \nabla R \rangle + R^2 \right| = o(r^{-5/2}).$$

Below, we shall also use the following variant of this:

$$(8.57) \qquad \left| \Delta_{\Sigma_s} R - \frac{\langle \nabla f, \nabla R \rangle}{|\nabla f|^2} + R^2 \right| = o(r^{-5/2}),$$

which follows from $\left| \langle \nabla f, \nabla R \rangle \right| \left| 1 - \frac{1}{|\nabla f|^2} \right| \leq C r^{-3}$.

8.3.3. A Poincaré-type inequality.

To convert the above specific derivative estimates to the oscillation and general derivative estimates we seek, we shall employ a Poincaré-type inequality on the level sets of the potential function.

Let K denote the intrinsic Gauss curvature of Σ_s and let $\{e_1, e_2, e_3\}$ be an orthonormal frame with $e_3 = \nu$. By applying $R_{33} := \text{Ric}(e_3, e_3) - O(r^{-2})$ and $\det(\text{II}) = O(r^{-2})$ (from (8.52)) to the Gauss equation $K = \frac{R}{2} - R_{33} + \det(\text{II})$, we obtain

$$(8.58) \qquad \left| K - \frac{R}{2} \right| - O(r^{-2}).$$

Since $R = |f|^{-1} + o(|f|^{-1})$, we have $K = \frac{1}{2}|f|^{-1} + o(|f|^{-1})$. By this and the Gauss–Bonnet formula, we obtain $\text{Area}(\Sigma_s) = O(|s|)$. Hence

$$\int_{\Sigma_s} \left(K - \frac{R}{2} \right)^2 d\sigma_s = O(|s|^{-3}),$$

where $d\sigma_s$ denotes the area form of Σ_s.

For the 2-sphere $(\mathbb{S}^2, g)$ with constant Gauss curvature $\frac{1}{2|s|}$, we have that the lowest four eigenvalues of the Laplacian satisfy $\lambda_1(\Delta_g) = \lambda_2(\Delta_g) = \lambda_3(\Delta_g) = \frac{1}{|s|}$, $\lambda_4(\Delta_g) = \frac{3}{|s|}$, and

$$\int_{\mathbb{S}^2} (u - \bar{u})^2 dA_g \leq |s| \int_{\mathbb{S}^2} |\nabla u|^2 dA_g$$

for any function u on $\mathbb{S}^2$, where $\bar{u}$ denotes the average of u on $(\mathbb{S}^2, g)$. Since Σ_s is C^k-close to the 2-sphere of constant Gauss curvature $\frac{1}{2|s|}$ for any k, we have

$$(8.59) \qquad \lambda_1(\Delta_{\Sigma_s}) = \frac{1 + o(1)}{|s|}, \quad \lambda_2(\Delta_{\Sigma_s}) = \frac{1 + o(1)}{|s|},$$

$$\lambda_3(\Delta_{\Sigma_s}) = \frac{1 + o(1)}{|s|}, \quad \lambda_4(\Delta_{\Sigma_s}) = \frac{3 + o(1)}{|s|},$$

and the corresponding Poincaré inequality is of the form

$$(8.60) \qquad \int_{\Sigma_s} (u - \bar{u})^2 d\sigma_s \leq (1 + o(1)) |s| \int_{\Sigma_s} |\nabla_{\Sigma_s} u|^2 d\sigma_s$$

for any function u on Σ_s.

We have the following refined estimate when u is replaced by R.

Lemma 8.19. *For our steady GRS,*

$$(8.61) \qquad \int_{\Sigma_s} |\nabla_{\Sigma_s} R|^2 d\sigma_s \geq \frac{(3 - o(1))}{|s|} \int_{\Sigma_s} (R - \bar{R})^2 d\sigma_s - O(|s|^{-4}).$$

Proof. Let $0 < \lambda_1 \leq \lambda_2 \leq \cdots$ denote the nonzero eigenvalues of Δ_{Σ_s}, with associated eigenfunctions $\varphi_1, \varphi_2, \ldots$ normalized by $\int_{\Sigma_s} \varphi_i \varphi_j d\sigma_s = \delta_{ij}$. By the L^2-expansion for any smooth function u on Σ_s we have

$$\int_{\Sigma_s} (u - \bar{u})^2 d\sigma_s = \sum_{i=1}^{\infty} \left(\int_{\Sigma_s} (u - \bar{u}) \varphi_i d\sigma_s \right)^2$$

and

$$\int_{\Sigma_s} |\nabla_{\Sigma_s} u|^2 d\sigma_s = \sum_{i=1}^{\infty} \lambda_i \left(\int_{\Sigma_s} (u - \bar{u}) \varphi_i d\sigma_s \right)^2 .$$

Hence

$$
\begin{aligned}
\int_{\Sigma_s} |\nabla_{\Sigma_s} u|^2 & d\sigma_s - \lambda_4 \int_{\Sigma_s} (u - \bar{u})^2 d\sigma_s \\
&= \sum_{i=1}^{\infty} (\lambda_i - \lambda_4) \left(\int_{\Sigma_s} (u - \bar{u}) \varphi_i d\sigma_s \right)^2 \\
&\geq \sum_{i=1}^{3} (\lambda_i - \lambda_4) \left(\int_{\Sigma_s} (u - \bar{u}) \varphi_i d\sigma_s \right)^2 \\
&\geq -\frac{2.1}{|s|} \sum_{i=1}^{3} \left(\int_{\Sigma_s} (u - \bar{u}) \varphi_i d\sigma_s \right)^2
\end{aligned}
$$

(8.62)

by (8.59), for $|s|$ sufficiently large. In the following, we shall estimate the right-hand side when $u = R$.

For a constant Gauss curvature 2-sphere, the eigenspace corresponding to the first eigenvalue is the 3-dimensional space of linear functions, whose gradients are conformal vector fields. For $|s|$ large, (Σ_s, g_{Σ_s}) is C^k-close to a 2-sphere of constant Gauss curvature $\frac{1}{2|s|}$. In particular, we may write $g_{\Sigma_s} = v \bar{g}$, where the metric $\bar{g}$ on Σ_s has constant Gauss curvature $\frac{1}{2|s|}$ and where v is C^k-close to the constant function 1. By the Kazdan–Warner identity (A.1) and integrating by parts, we see that

$$0 = \int_{\Sigma_s} g_{\Sigma_s}(\nabla_{\Sigma_s} K, \bar\nabla \ell_s) d\sigma_s = \int_{\Sigma_s} (K - \bar K) \operatorname{tr}_{g_{\Sigma_s}}(\nabla_{\Sigma_s} \bar\nabla \ell_s) d\sigma_s$$

for any first eigenfunction ℓ_s of $\Delta_{\bar g}$ on Σ_s (if we isometrically embed $(\Sigma_s, \bar g)$ into $\mathbb{R}^3$, then ℓ_s is a linear function). Hence

$$\int_{\Sigma_s} (K - \bar K)\varphi_i d\sigma_s = -\lambda_i^{-1} \int_{\Sigma_s} (K - \bar K)\Delta_{\Sigma_s}\varphi_i d\sigma_s$$

$$= -\lambda_i^{-1} \int_{\Sigma_s} (K - \bar K) \operatorname{tr}_{g_{\Sigma_s}}(\nabla_{\Sigma_s}(\nabla_{\Sigma_s}\varphi_i - \bar\nabla\ell_s)) d\sigma_s,$$

which implies that

(8.63)

$$\left| \int_{\Sigma_s} (K - \bar K)\varphi_i d\sigma_s \right|$$

$$\leq \lambda_i^{-1} \left(\int_{\Sigma_s} (K - \bar K)^2 d\sigma_s \right)^{1/2} \left(\int_{\Sigma_s} \left(\operatorname{tr}_{g_{\Sigma_s}}(\nabla_{\Sigma_s}(\nabla_{\Sigma_s}\varphi_i - \bar\nabla\ell_s)) \right)^2 d\sigma_s \right)^{1/2}.$$

We claim that the last factor on the right-hand side is small.

Regarding the metric dependence of eigenfunctions of the Laplacian, we recall the resolvent formalism; see Dunford and Schwartz [**142**]. Recall that for a closed Riemannian manifold $(\mathcal{M}^n, g)$, the **spectrum** $\sigma(\Delta_g)$ is the set of $\lambda \in \mathbb{C}$ such that $\Delta_g - \lambda \operatorname{id}$ does not have a bounded inverse. The **resolvent set** is $\rho(\Delta_g) = \mathbb{C} \setminus \sigma(\Delta_g)$. Define

$$\mathbf{R}_g(\lambda) = (\Delta_g - \lambda \operatorname{id})^{-1},$$

which is called the **resolvent** of Δ_g at λ. The eigenvalues of $\Delta_{2|s|\bar g}$ are $\lambda_1 = \lambda_2 = \lambda_3 = \frac{1}{|s|} < \lambda_4 = \frac{3}{|s|}$. Let B be a small open ball in $\mathbb{C}$ centered at $\lambda = \frac{1}{|s|}$. For g on $\mathbb{S}^2$ sufficiently close to $2|s|\bar g$, the eigenvalues $\lambda_1(\Delta_g), \lambda_2(\Delta_g), \lambda_3(\Delta_g)$, which are all close to $\frac{1}{|s|}$, are the only eigenvalues of Δ_g contained in B. The projection $\mathcal{P}_g$ onto the subspace spanned by eigenspaces of $\lambda_1(\Delta_g), \lambda_2(\Delta_g), \lambda_3(\Delta_g)$ is given by the contour integral over ∂B

$$\mathcal{P}_g = \frac{1}{2\pi i} \oint \mathbf{R}_g(\lambda) \, d\lambda.$$

Hence, from the dependence of $\mathbf{R}_g(\lambda)$ on g, if $\frac{1}{2|s|} g_{\Sigma_s}$ is C^k-close to $\bar g$, then for $1 \leq i \leq 3$ the eigenfunction φ_i of g_{Σ_s} is C^2-close to an eigenfunction of $2|s|\bar g$ with eigenvalue $\frac{1}{|s|}$, which is a linear function.

Therefore, if $i = 1$, 2, or 3, then from scaling considerations there exists ℓ_s such that[5]

$$\left(\int_{\Sigma_s} \left(\operatorname{tr}_{g_{\Sigma_s}} (\nabla_{\Sigma_s}(\nabla_{\Sigma_s}\varphi_i - \bar{\nabla}\ell_s))\right)^2 d\sigma_s\right)^{1/2} = \mathrm{o}(|s|^{-1}) \quad \text{as } |s| \to \infty.$$

Applying this to (8.63), we conclude that

$$(8.64) \qquad \left|\int_{\Sigma_s} (K - \bar{K})\varphi_i d\sigma_s\right| = \mathrm{o}(1) \left(\int_{\Sigma_s} (K - \bar{K})^2 d\sigma_s\right)^{1/2}$$

for $i = 1, 2, 3$. On the other hand, by taking $u = R$ in (8.62), we have
(8.65)

$$\int_{\Sigma_s} |\nabla_{\Sigma_s}R|^2 d\sigma_s - \lambda_4 \int_{\Sigma_s} (R - \bar{R})^2 d\sigma_s \geq -\frac{2.1}{|s|} \sum_{i=1}^{3} \left(\int_{\Sigma_s} (R - \bar{R})\varphi_i d\sigma_s\right)^2.$$

Now, by (8.58), which also implies that $|\bar{K} - \frac{\bar{R}}{2}| \leq \mathrm{O}(|s|^{-2})$, and by (8.64),

$$\left|\int_{\Sigma_s} (R - \bar{R})\varphi_i d\sigma_s\right| = \left|\int_{\Sigma_s} \left(2(K - \bar{K})\varphi_i + \mathrm{O}(|s|^{-2})|\varphi_i|\right) d\sigma_s\right|$$

$$\leq 2\left|\int_{\Sigma_s} (K - \bar{K})\varphi_i d\sigma_s\right| + \mathrm{O}(|s|^{-2})\left(\int \varphi_i^2 d\sigma_s\right)^{\frac{1}{2}}\left(\int d\sigma_s\right)^{\frac{1}{2}}$$

$$(8.66) \qquad \leq \mathrm{o}(1) \left(\int_{\Sigma_s} (K - \bar{K})^2 d\sigma_s\right)^{1/2} + \mathrm{O}(|s|^{-3/2})$$

$$\leq \mathrm{o}(1) \left(\int_{\Sigma_s} \left(|R - \bar{R}| + C|s|^{-2}\right)^2 d\sigma_s\right)^{1/2} + \mathrm{O}(|s|^{-3/2})$$

$$\leq \mathrm{o}(1) \left(\int_{\Sigma_s} (R - \bar{R})^2 d\sigma_s\right)^{1/2} + \mathrm{O}(|s|^{-3/2}).$$

Applying (8.66) to (8.65) yields

$$\int_{\Sigma_s} |\nabla_{\Sigma_s}R|^2 d\sigma_s \geq \frac{3 - \mathrm{o}(1)}{|s|} \int_{\Sigma_s} (R - \bar{R})^2 d\sigma_s$$

$$- \frac{2.1}{|s|} \left(\mathrm{o}(1)\left(\int_{\Sigma_s} (R - \bar{R})^2 d\sigma_s\right)^{1/2} + \mathrm{O}(|s|^{-3/2})\right)^2$$

$$\geq \frac{3 - \mathrm{o}(1)}{|s|} \int_{\Sigma_s} (R - \bar{R})^2 d\sigma_s - \mathrm{O}(|s|^{-4}).$$

The lemma follows. $\qquad\qquad\qquad\qquad\qquad\qquad\qquad\qquad\qquad\qquad\qquad\square$

---

[5]We have the scaling orders $g_{\Sigma_s}^{-1} \sim |s|^{-1}$, $d\sigma_s \sim |s|$, $\nabla \sim 1$, $|\varphi_i| \sim |s|^{-1/2}$, and hence $|\ell_s| \sim |s|^{-1/2}$.

8.3.4. Refined asymptotics for the curvature and its derivatives.

We first obtain a decay estimate for the difference of the scalar curvature from its average on a level surface. Note that $\int_{\Sigma_s}(R - \bar{R})^2 d\sigma_s$ scales like $|s|^{-1}$.

Proposition 8.20. *For our steady GRS, we have*

$$(8.67) \qquad \int_{\Sigma_s}(R - \bar{R}(s))^2 d\sigma_s = O(|s|^{-2}) \quad as \ s \to -\infty.$$

Proof. First observe that by (8.53) and (8.54), we have

$$(8.68) \qquad \lim_{s \to -\infty}\int_{\Sigma_s}(R - \bar{R}(s))^2 d\sigma_s = 0.$$

Let u be a function on $\mathcal{M}^3$. If we consider parametrizations F_s of the Σ_s evolving with velocity $-\frac{1}{|\nabla f|}\nu$, where $\nu = -\frac{\nabla f}{|\nabla f|}$, then $\frac{\partial}{\partial s}d\sigma_s = -\frac{H}{|\nabla f|}d\sigma_s$. From this we obtain

$$-\frac{d}{ds}\int_{\Sigma_s}u\,d\sigma_s = -\int_{\Sigma_s}\left\langle\frac{\nabla f}{|\nabla f|^2}, \nabla u\right\rangle d\sigma_s + \int_{\Sigma_s}u\frac{H}{|\nabla f|}d\sigma_s.$$

In particular,

$$(8.69)$$
$$-\frac{d}{ds}\int_{\Sigma_s}(R - \bar{R}(s))^2 d\sigma_s = 2\int_{\Sigma_s}(R - \bar{R}(s))\left(-\frac{\langle\nabla f, \nabla R\rangle}{|\nabla f|^2} - \frac{d\bar{R}}{ds}(s)\right) d\sigma_s$$
$$+ \int_{\Sigma_s}(R - \bar{R}(s))^2\frac{H}{|\nabla f|}d\sigma_s.$$

Since

$$\int_{\Sigma_s}|\nabla_{\Sigma_s}R|^2 d\sigma_s + \int_{\Sigma_s}(R - \bar{R}(s))\Delta_{\Sigma_s}R\,d\sigma_s = 0$$

and $\int_{\Sigma_s}(R - \bar{R}(s))d\sigma_s = 0$, we may rewrite (8.69) as

$$-\frac{d}{ds}\int_{\Sigma_s}(R - \bar{R}(s))^2 d\sigma_s$$
$$= 2\int_{\Sigma_s}|\nabla_{\Sigma_s}R|^2 d\sigma_s + 2\int_{\Sigma_s}(R - \bar{R}(s))\left(\Delta_{\Sigma_s}R - \frac{\langle\nabla f, \nabla R\rangle}{|\nabla f|^2} + R^2\right) d\sigma_s$$
$$+ \int_{\Sigma_s}(R - \bar{R}(s))^2\left(\frac{H}{|\nabla f|} - 2R - 2\bar{R}(s)\right) d\sigma_s.$$

Applying (8.57) and (8.61) to this, we obtain

$$-\frac{d}{ds}\int_{\Sigma_s}(R-\bar{R}(s))^2 d\sigma_s \geq \int_{\Sigma_s}(R-\bar{R}(s))^2\left(\frac{5}{|s|}+\frac{H}{|\nabla f|}-2R-2\bar{R}(s)\right)d\sigma_s$$
$$-C|s|^{-5/2}\int_{\Sigma_s}|R-\bar{R}(s)|d\sigma_s - C|s|^{-4}$$
$$\geq \frac{1}{|s|}\int_{\Sigma_s}(R-\bar{R}(s))^2 d\sigma_s$$
$$-C|s|^{-5/2}\int_{\Sigma_s}|R-\bar{R}(s)|d\sigma_s - C|s|^{-4},$$

where to obtain the last inequality we used (8.50), $|\nabla f| \to 1$, and (8.29). Since $\int_{\Sigma_s}|R-\bar{R}(s)|d\sigma_s \leq \frac{|s|^{3/2}}{4C}\int_{\Sigma_s}(R-\bar{R}(s))^2 d\sigma_s + C|s|^{-3/2}\,\mathrm{Area}(\Sigma_s)$, we conclude that

$$(8.70)\quad -\frac{d}{ds}\int_{\Sigma_s}(R-\bar{R}(s))^2 d\sigma_s \geq -C|s|^{-4}\,\mathrm{Area}(\Sigma_s) - C|s|^{-4} \geq -C|s|^{-3},$$

where we used (8.53) for the last inequality. The proposition now follows from (8.68) and integrating (8.70). $\qquad\square$

We now can prove the following improvement of (8.54) as well as higher derivative estimates for R.

Proposition 8.21. *For our steady GRS, we have the following estimates on the level sets:*

$$(8.71a)\qquad\qquad \sup_{\Sigma_s}|R-\bar{R}(s)| = \mathrm{O}(|s|^{-5/4}),$$

$$(8.71b)\qquad\qquad \sup_{\Sigma_s}|\nabla_{\Sigma_s}R| = \mathrm{O}(|s|^{-7/4}),$$

$$(8.71c)\qquad\qquad \sup_{\Sigma_s}|\Delta_{\Sigma_s}R| = \mathrm{O}(|s|^{-9/4}).$$

Proof. Let $s \leq -1$ and let $x \in \Sigma_s$. We have $|\mathrm{Rm}| \leq |s|^{-1}$ in $B_{\sqrt{|s|}}(x)$. So, by going to the canonical form, one can deduce from Shi's local derivative estimates (see Lemma 4.22) that

$$(8.72)\qquad\qquad |\nabla^m\mathrm{Rm}|(x) \leq C|s|^{-\frac{m}{2}-1}.$$

Now, by Corollary 12.7 in [**173**], for any Riemannian manifold $(\mathcal{N}^m, h)$ and integers $0 \leq k \leq \ell$ there exists a constant $C > 0$ depending only on ℓ and the metric h such that for any tensor T on $\mathcal{N}^m$ we have the following interpolation estimate:

$$(8.73)\qquad \int_{\mathcal{N}}|\nabla^k T|_h^2 d\mu_h \leq C\left(\int_{\mathcal{N}}|\nabla^\ell T|_h^2 d\mu_h\right)^{\frac{k}{\ell}}\left(\int_{\mathcal{N}}|T|_h^2 d\mu_h\right)^{1-\frac{k}{\ell}}.$$

Since $\frac{1}{2|s|}g_{\Sigma_s} \to g_{\mathbb{S}^2}$ and by the scale invariance of the inequality, we find that there exists C depending only on ℓ (independent of s) such that for any function u on Σ_s,

$$\int_{\Sigma_s} |\nabla^k_{\Sigma_s} u|^2 d\sigma_s \leq C \left(\int_{\Sigma_s} |\nabla^\ell_{\Sigma_s} u|^2 d\sigma_s \right)^{\frac{k}{\ell}} \left(\int_{\Sigma_s} |u - \bar{u}(s)|^2 d\sigma_s \right)^{1-\frac{k}{\ell}},$$

where $\bar{u}(s)$ is the average of u on Σ_s. In particular,

(8.74)
$$\int_{\Sigma_s} |\nabla^k_{\Sigma_s} R|^2 d\sigma_s \leq C \left(\int_{\Sigma_s} |\nabla^\ell_{\Sigma_s} R|^2 d\sigma_s \right)^{\frac{k}{\ell}} \left(\int_{\Sigma_s} |R - \bar{R}(s)|^2 d\sigma_s \right)^{1-\frac{k}{\ell}}$$
$$\leq C|s|^{-2(1-\frac{k}{\ell})} \left(\int_{\Sigma_s} |\nabla^\ell_{\Sigma_s} R|^2 d\sigma_s \right)^{\frac{k}{\ell}},$$

where we used (8.67). By (8.53) and (8.72), we have

$$\int_{\Sigma_s} |\nabla^\ell_{\Sigma_s} R|^2 d\sigma_s \leq C |s|^{-\ell-1}.$$

In particular, taking $\ell = 2k$ for $k \geq 1$, we have the improved estimate

(8.75)
$$\int_{\Sigma_s} |\nabla^k_{\Sigma_s} R|^2 d\sigma_s \leq C|s|^{-1} \left(\int_{\Sigma_s} |\nabla^{2k}_{\Sigma_s} R|^2 d\sigma_s \right)^{\frac{1}{2}} \leq C|s|^{-k-\frac{3}{2}}.$$

Note that by (8.60) we have

$$\int_{\Sigma_s} (R - \bar{R}(s))^2 d\sigma_s \leq 2|s| \int_{\Sigma_s} |\nabla_{\Sigma_s} R|^2 d\sigma_s$$

for $|s|$ sufficiently large. We also have from (8.75) with $k = 1$ that

$$\int_{\Sigma_s} |\nabla_{\Sigma_s} R|^2 d\sigma_s \leq C|s|^{-1} \left(\int_{\Sigma_s} |\nabla^2_{\Sigma_s} R|^2 d\sigma_s \right)^{\frac{1}{2}}.$$

By combining (8.75) for $k = 2$ with the Sobolev inequality ($W^{2,2}$ embeds continuously into L^∞ since $n = 2$), we have

$$\sup_{\Sigma_s} |R - \bar{R}(s)| \leq C \left(\int_{\Sigma_s} \left(|s|^{-1}(R - \bar{R}(s))^2 + |\nabla_{\Sigma_s} R|^2 + |s||\nabla^2_{\Sigma_s} R|^2 \right) d\sigma_s \right)^{1/2}$$
$$\leq C \left(3C|s|^{-1} \left(\int_{\Sigma_s} |\nabla^2_{\Sigma_s} R|^2 d\sigma_s \right)^{\frac{1}{2}} + |s| \int_{\Sigma_s} |\nabla^2_{\Sigma_s} R|^2 d\sigma_s \right)^{1/2}$$
$$\leq C|s|^{1/2} \left(2 \int_{\Sigma_s} |\nabla^2_{\Sigma_s} R|^2 d\sigma_s + C|s|^{-4} \right)^{1/2}$$
$$\leq C|s|^{-5/4}.$$

Finally, by the Sobolev inequality and (8.75), for $m \geq 1$ we have

(8.76)

$$\sup_{\Sigma_s} |\nabla^m_{\Sigma_s} R| \leq C \left(\int_{\Sigma_s} \left(|s|^{-1} |\nabla^m_{\Sigma_s} R|^2 + |\nabla^{m+1}_{\Sigma_s} R|^2 d\mu + |s| |\nabla^{m+1}_{\Sigma_s} R|^2 \right) d\sigma_s \right)^{1/2}$$

$$\leq C|s|^{-\frac{m}{2}-\frac{5}{4}}.$$

$\square$

We have the following improvements of the estimates $|\nabla R| = O(r^{-3/2})$ and $|f| R \to 1$.

Proposition 8.22. *For our steady GRS,*

(8.77) $|\nabla R| = O(r^{-7/4})$ *and* $|f| R - 1 = O(r^{-1/4}).$

Proof. (1) Since

$$|\nabla R|^2 = |\nabla_{\Sigma_s} R|^2 + \frac{\langle \nabla R, \nabla f \rangle^2}{|\nabla f|^2},$$

since $|\langle \nabla R, \nabla f \rangle| \leq C|s|^{-2}$ by (8.39), and since $|\nabla f| \to 1$ by (8.25), we conclude $|\nabla R| = O(r^{-7/4})$ from (8.71b).

(2) We estimate $R^{-1} + f$ by considering its derivative in the ∇f direction:

$$\langle \nabla f, \nabla \left(R^{-1} + f \right) \rangle = |\nabla f|^2 - \frac{\langle \nabla f, \nabla R \rangle}{R^2}$$

$$= |\nabla f|^2 - 1 + \frac{\Delta_{\Sigma_s} R - \langle \nabla f, \nabla R \rangle + R^2}{R^2} - \frac{\Delta_{\Sigma_s} R}{R^2}$$

$$= O(|s|^{-1}) + o(r^{-1/2}) + O(|s|^{-1/4})$$

$$= O(|s|^{-1/4}),$$

by (8.56) and (8.71c). Integrating $\left\langle \frac{\nabla f}{|\nabla f|^2}, \nabla \left(R^{-1} + f \right) \right\rangle \leq C|s|^{-1/4}$ along the integral curves of $\frac{\nabla f}{|\nabla f|^2}$, we obtain $R^{-1} + f = O(|s|^{3/4})$, from which $|f| R - 1 = O(r^{-1/4})$ follows. $\square$

Using the estimates above, we obtain the following estimates for the second fundamental form and Gauss curvature of the level sets.

Proposition 8.23. *For our steady GRS,*

(8.78) $\left| \text{II} - \frac{1}{2|s|} g|_{\Sigma_s} \right| = O(|s|^{-5/4})$ *and* $K = \frac{1}{2|s|} + O(|s|^{-5/4}),$

where K is the intrinsic Gauss curvature of $(\Sigma_s, g|_{\Sigma_s})$.

Proof. By (8.39), (8.49), and (8.77),

$$H = \frac{1}{|\nabla f| \, |f|} + O(r^{-5/4}) = \frac{1}{|f|} + O(r^{-5/4}).$$

The first estimate in (8.78) follows from (8.51). The second estimate in (8.78) follows simply from (8.58) and (8.77). □

8.4. Existence of asymptotically Killing vector fields

In this section we first obtain higher-order estimates for the intrinsic and extrinsic geometry of the level sets of f. We then construct asymptotically Killing vector fields on $(\mathcal{M}^3, g)$, which we shall later perturb to obtain exact Killing vector fields to prove the rotational symmetry of $(\mathcal{M}^3, g)$.

8.4.1. Estimate for the asymptotic rate of roundness of the level sets of the potential function.

Recall that the potential function f is strictly concave, with a unique critical point o, where f attains its maximum value -1. In particular, the level sets $\Sigma_s = \{x \in \mathcal{M}^3 : f(x) = s\}$ are diffeomorphic to $\mathbb{S}^2$ for each $s < -1$. Let $g_s = g|_{\Sigma_s}$ be the induced Riemannian metric on Σ_s. We may consider Σ_s parametrically as evolving embeddings $F_s : \mathbb{S}^2 \to F_s(\mathbb{S}^2) = \Sigma_s \subset \mathcal{M}^3$ defined by

$$(8.79) \qquad \frac{\partial}{\partial s} F_s = \frac{\nabla f}{|\nabla f|^2} = -\frac{1}{|\nabla f|} \nu \quad \text{for } s \in (-\infty, -1),$$

where $\nu := -\frac{\nabla f}{|\nabla f|}$ is the outward normal to $\partial \{f \geq s\}$. Define the rescaled pullback metrics γ_s on $\mathbb{S}^2$ by

$$(8.80) \qquad \gamma_s := \frac{1}{2|s|} F_s^*(g_s).$$

First observe that, by (8.78), $K_{\gamma_s} = 1 + O(|s|^{-1/4})$. So we expect that the γ_s converge in C^m as $s \to -\infty$ to the standard metric on $\mathbb{S}^2$ at a certain rate. This is indeed the case.

Lemma 8.24. *The rescaled metrics γ_s on $\mathbb{S}^2$ satisfy the evolution equation*

$$(8.81) \qquad \frac{\partial}{\partial s} \gamma_s = -\frac{1}{|s| \, |\nabla f|} F_s^* \left(\mathrm{II} - \frac{|\nabla f|}{2|s|} g_s \right).$$

Proof. Firstly,

$$(8.82) \qquad \frac{\partial}{\partial s} \gamma_s = \frac{1}{2\,|s|^2} F_s^*(g_s) + \frac{1}{2|s|} \frac{\partial}{\partial s} (F_s^*(g_s)).$$

Since the normal velocity is $-\frac{1}{|\nabla f|}$, (8.81) follows from the fact that the time-derivative of the metric $F_s^*(g_s)$ of a hypersurface evolving with normal velocity σ is $2\sigma \, \mathrm{II}$; see [190]. For completeness, we now prove this fact.

Let $\phi_s : \mathcal{M}^3 \setminus \{o\} \to \mathcal{M}^3 \setminus \{o\}$ be the 1-parameter group of diffeomorphisms generated by $|\nabla f|^{-2} \nabla f = -|\nabla f|^{-1} \nu$. Then $\phi_s(\Sigma_t) = \Sigma_{t+s}$ and $\phi_s \circ F_t = F_{t+s}$. We have $g_s = g - \nu^\flat \otimes \nu^\flat$, where $\nu^\flat$ is the 1-form metrically dual to ν. We compute that

$$(8.83) \qquad \frac{\partial}{\partial s}(F_s^*(g_s)) = \lim_{\Delta s \to 0} \frac{F_{s+\Delta s}^*(g_{s+\Delta s}) - F_s^*(g_s)}{\Delta s}$$

$$= F_s^* \left(\lim_{\Delta s \to 0} \frac{\phi_{\Delta s}^*(g_{s+\Delta s}) - g_s}{\Delta s} \right)$$

$$= F_s^* \left(\mathcal{L}_{|\nabla f|^{-2}\nabla f} \left(g - \nu^\flat \otimes \nu^\flat \right) \right).$$

Observe, using $(\mathcal{L}_X g)_{ij} = \nabla_i X_j + \nabla_j X_i$ and $d|\nabla f|^2 = -dR$, that

$$\mathcal{L}_{|\nabla f|^{-2}\nabla f} g = \frac{2\nabla^2 f}{|\nabla f|^2} + \frac{dR \otimes df + dR \otimes df}{|\nabla f|^4}.$$

We compute that

$$\mathcal{L}_\nu \nu^\flat = \mathcal{L}_{\frac{\nabla f}{|\nabla f|}} \frac{df}{|\nabla f|}$$

$$= \left(\mathcal{L}_{\frac{\nabla f}{|\nabla f|}} \frac{1}{|\nabla f|} \right) df + \frac{1}{|\nabla f|} \mathcal{L}_{\frac{\nabla f}{|\nabla f|}} df$$

$$= \frac{\nabla f}{|\nabla f|} \left(\frac{1}{|\nabla f|} \right) df + \frac{1}{|\nabla f|^2} \mathcal{L}_{\nabla f} df + d\left(\frac{1}{|\nabla f|} \right) \iota_{\frac{\nabla f}{|\nabla f|}} df$$

$$= \frac{1}{2|\nabla f|^4} \langle \nabla f, \nabla R \rangle df - \frac{1}{2|\nabla f|^2} dR.$$

Thus

$$\mathcal{L}_{-|\nabla f|^{-1}\nu} \nu^\flat = -|\nabla f|^{-1} \mathcal{L}_\nu \nu^\flat - d(|\nabla f|^{-1}) \iota_\nu \nu^\flat$$

$$= -\frac{\langle \nabla f, \nabla R \rangle}{2|\nabla f|^5} df,$$

and hence

$$\mathcal{L}_{-|\nabla f|^{-1}\nu} \left(\nu^\flat \otimes \nu^\flat \right) = -\frac{\langle \nabla f, \nabla R \rangle}{2|\nabla f|^5} \left(df \otimes \nu^\flat + \nu^\flat \otimes df \right)$$

$$= \frac{\langle \nabla f, \nabla R \rangle}{|\nabla f|^4} \nu^\flat \otimes \nu^\flat.$$

So

$$\mathcal{L}_{|\nabla f|^{-2}\nabla f} \left(g - \nu^\flat \otimes \nu^\flat \right) = \frac{2\nabla^2 f}{|\nabla f|^2} + \frac{dR \otimes df + dR \otimes df}{|\nabla f|^4} - \frac{\langle \nabla f, \nabla R \rangle}{|\nabla f|^4} \nu^\flat \otimes \nu^\flat.$$

Now the tangential projection of $\nabla^2 f$ is

$$\left(\nabla^2 f\right)^\top = \nabla^2 f - \left(\nabla^2 f\right)(\nu) \otimes \nu^\flat - \nu^\flat \otimes \left(\nabla^2 f\right)(\nu) + \left(\nabla^2 f\right)(\nu,\nu)\,\nu^\flat \otimes \nu^\flat$$

$$= \nabla^2 f + \frac{1}{2}\frac{dR \otimes df + dR \otimes df}{|\nabla f|^2} - \frac{1}{2}\frac{\langle \nabla f, \nabla R\rangle}{|\nabla f|^2}\nu^\flat \otimes \nu^\flat.$$

By the above, we have

$$(8.84) \qquad \mathcal{L}_{|\nabla f|^{-2}\nabla f}\left(g - \nu^\flat \otimes \nu^\flat\right) = \frac{2\left(\nabla^2 f\right)^\top}{|\nabla f|^2} = -\frac{2\,\mathrm{II}}{|\nabla f|},$$

where our choice of unit normal is $\nu = -\frac{\nabla f}{|\nabla f|}$. Hence we conclude from (8.82), (8.83), and (8.84) that

$$(8.85) \qquad \frac{\partial}{\partial s}\gamma_s = \frac{1}{2\,|s|^2}F_s^*(g_s) - \frac{1}{2|s|}F_s^*\left(\frac{2\,\mathrm{II}}{|\nabla f|}\right). \qquad \square$$

In part, from the estimates for the second fundamental forms of the Σ_s, we obtain the following estimate on the convergence rate of the rescaled pullback metrics γ_s on $\mathbb{S}^2$.

Proposition 8.25. *For each nonnegative integer* m,

$$\left\|\frac{\partial}{\partial s}\gamma_s\right\|_{C^m(\mathbb{S}^2,\gamma_s)} \le O(|s|^{-9/8}).$$

Proof. By (8.81), we have

$$(8.86) \qquad \sup_{\mathbb{S}^2}\left|\frac{\partial}{\partial s}\gamma_s\right|_{\gamma_s} \le C\left|\mathrm{II} - \frac{|\nabla f|}{2|s|}g_s\right|_{g_s} \le O(|s|^{-5/4}).$$

From Perelman's derivative estimates (8.21), the covariant derivatives of the ambient curvature satisfy

$$(8.87) \qquad |\nabla^m \mathrm{Ric}| \le O(|s|^{-1-\frac{m}{2}}) \quad \text{for } m \ge 0.$$

Let D_{g_s} denote the covariant derivative with respect to g_s on Σ_s and let K_{g_s} denote the intrinsic Gauss curvature of g_s.

Claim 8.26.

$$(8.88) \qquad |D_{g_s}^m \mathrm{II}| \le O(|s|^{-1-\frac{m}{2}}) \quad \text{for } m \ge 0.$$

Proof of Claim 8.26. Let α be a k-tensor on $\mathcal{M}$. Then

$$D_{g_s}(\alpha|_{\Sigma_s}) = (\nabla\alpha)|_{\Sigma_s} + \mathrm{II} * \alpha|_{\Sigma_s},$$

where $*$ denotes the product with contractions of tensors. Applying this to $\alpha = \frac{\mathrm{Ric}}{|\nabla f|}$, we have
(8.89)

$$D_{g_s}\mathrm{II} = D_{g_s}\left(\frac{\mathrm{Ric}\,|_{\Sigma_s}}{|\nabla f|}\right) = \frac{1}{2}\frac{(\mathrm{Ric}\otimes\nabla R)\,|_{\Sigma_s}}{|\nabla f|^3} + \frac{(\nabla\mathrm{Ric})|_{\Sigma_s}}{|\nabla f|} + \mathrm{II} * \frac{\mathrm{Ric}\,|_{\Sigma_s}}{|\nabla f|}.$$

Hence

$$|D_{g_s}\mathrm{II}| \leq \mathrm{o}(r^{-3/2}).$$

One can check that for each term in $D_{g_s}^m\mathrm{II}$ (which is obtained from differentiating (8.89) $m-1$ times), each covariant derivative decreases the power of $|s|$ in its bound by $1/2$. Claim 8.26 follows. $\qquad\square$

Claim 8.27.

(8.90) $$\left|D_{g_s}^m K_{g_s}\right|_{g_s} \leq \mathrm{O}(|s|^{-1-\frac{m}{2}}) \quad \textit{for } m \geq 0.$$

Proof of Claim 8.27. Differentiating the Gauss equation $K_{g_s} = \frac{R}{2} - R_{33} + \det(\mathrm{II})$, we have

(8.91) $$dK_{g_s} = \frac{dR}{2} - dR_{33} + d\det(\mathrm{II})$$

$$= \frac{dR}{2} - \nabla\mathrm{Ric}(\cdot, e_3, e_3) + \mathrm{II} * \mathrm{Ric}(\cdot, e_3) + D_{g_s}\mathrm{II} * \mathrm{II},$$

where d denotes the exterior derivative on Σ_s. By (8.87) and (8.52), we have $|dK_{g_s}|_{g_s} \leq \mathrm{O}(|s|^{-\frac{3}{2}})$. Each covariant derivative of (8.91) decreases the power of $|s|$ for its bound by $1/2$. Claim 8.27 follows. $\qquad\square$

Now let D_{γ_s} denote the covariant derivative with respect to γ_s on $\mathbb{S}^2$. From (8.90) we obtain that for each $m \geq 0$,

$$\left|D_{\gamma_s}^m K_{\gamma_s}\right|_{\gamma_s} \leq C(m)$$

is bounded independent of $s \leq -2$.

Claim 8.28. For any constant C,

(8.92) $$\sup_{\{s-C\sqrt{|s|}\leq f\leq s+C\sqrt{|s|}\}} \left|\nabla_g^\ell(\mathcal{L}_{\nabla f/|\nabla f|^2}g)\right|_g \leq \mathrm{O}(|s|^{-1-\frac{\ell}{2}}).$$

Proof of Claim 8.28. We compute that

(8.93) $$(\mathcal{L}_{\nabla f/|\nabla f|^2}g)_{ij} = \nabla_i\left(\frac{\nabla_j f}{|\nabla f|^2}\right) + \nabla_j\left(\frac{\nabla_i f}{|\nabla f|^2}\right)$$

$$= -2\frac{R_{ij}}{|\nabla f|^2} + \frac{\nabla_j f\nabla_i R}{|\nabla f|^4} + \frac{\nabla_i f\nabla_j R}{|\nabla f|^4}.$$

Hence

$$(8.94) \qquad \left| \mathcal{L}_{\nabla f / |\nabla f|^2} g \right|_g \leq O(r^{-1}).$$

One again checks that each covariant derivative of (8.93) decreases the power of r by $1/2$. Claim 8.28 follows. $\qquad \square$

We conclude from (8.92), while using (8.88), that

$$(8.95) \qquad \left\| \frac{\partial}{\partial s} \gamma_s \right\|_{C^m(\mathbb{S}^2, \gamma_s)} \leq O(|s|^{-1}) \quad \text{for } m \geq 0.$$

The proposition now follows from this, (8.86), the interpolation estimate (8.73), and the Sobolev inequality (cf. the proof of Proposition 8.21). $\qquad \square$

Note by (8.29), (8.51), and the Gauss equation that the $(\mathbb{S}^2, \gamma_s)$ converge in the Cheeger–Gromov sense to the unit 2-sphere, with $R_g|_{\Sigma_s} \to 2$ and $R_{\gamma_s} \to 2$ as $s \to -\infty$.

From Proposition 8.25, we have that the γ_s converge to a C^∞ metric $\bar\gamma$ on $\mathbb{S}^2$ with constant Gauss curvature 1. Moreover,

$$\left\| \frac{\partial}{\partial s} \gamma_s \right\|_{C^m(\mathbb{S}^2, \bar\gamma)} \leq O(|s|^{-9/8})$$

for $m \geq 0$, and since $\gamma_{\bar s} - \bar\gamma = \int_{-\infty}^{\bar s} \frac{\partial}{\partial s} \gamma_s ds$ we have

$$(8.96) \qquad \|\gamma_s - \bar\gamma\|_{C^m(\mathbb{S}^2, \bar\gamma)} \leq O(|s|^{-1/8}).$$

8.4.2. Constructing the approximate Killing vector fields.

Let $\bar U_1$, $\bar U_2$, and $\bar U_3$ be Killing vector fields on the unit sphere $(\mathbb{S}^2, \bar\gamma)$ such that (see Exercise 8.9)

$$(8.97) \qquad \sum_{a=1}^3 \bar U_a \otimes \bar U_a = \frac{1}{2} \bar\gamma^{-1} \quad \text{and} \quad [\bar U_a, \bar U_b] = \bar U_c,$$

where $[abc] = [123]$; i.e., $\{a, b, c\}$ is a cyclic permutation of $\{1, 2, 3\}$.

Define pushed-forward vector fields U_1, U_2, and U_3 on $\{f < s_0\} \subset \mathcal{M}^3$ by

$$(8.98) \qquad U_a := (F_s)_* \bar U_a \quad \text{for } s < s_0,$$

where $s_0 \ll -1$, and extend the vector fields U_a smoothly to the compact set $\Omega_{s_0} := \{x \in \mathcal{M}^3 : f(x) \geq s_0\}$ so that the U_a are defined on all of $\mathcal{M}^3$. Note that

$$(8.99) \qquad [U_a, U_b] = U_c \quad \text{on } \{f < s_0\} \subset \mathcal{M}^3 \quad \text{for } [abc] = [123].$$

We have for $m \geq 0$ on $\mathbb{S}^2$ that

$$
\begin{aligned}
\left\|\mathcal{L}_{\bar{U}_a} \gamma_s\right\|_{C^m(\mathbb{S}^2, \bar{\gamma})} &= \left\|\mathcal{L}_{\bar{U}_a}(\gamma_s - \bar{\gamma})\right\|_{C^m(\mathbb{S}^2, \bar{\gamma})} \\
&\leq C \left\|\gamma_s - \bar{\gamma}\right\|_{C^{m+1}(\mathbb{S}^2, \bar{\gamma})} \\
&\leq O(|s|^{-1/8}).
\end{aligned}
\tag{8.100}
$$

Define the map $F : \mathbb{S}^2 \times (-\infty, s_0) \to \mathcal{M}^3 \setminus \Omega_{s_0}$ by $F(x,s) := F_s(x)$. Note that $F(\mathbb{S}^2 \times \{s\}) = \Sigma_s$. Consider $\bar{U}_a$ as vector fields on $\mathbb{S}^2 \times (-\infty, s_0)$ tangent to $\mathbb{S}^2 \times \{s\}$. We have $F_*(\frac{\partial}{\partial s}) = \frac{\nabla f}{|\nabla f|^2}$ and $F_*(\bar{U}_a) = U_a$. Since $[\bar{U}_a, \frac{\partial}{\partial s}] = 0$ on $\mathbb{S}^2 \times (-\infty, s_0)$, we have $[U_a, \frac{\nabla f}{|\nabla f|^2}] = 0$ on $\mathcal{M}^3 \setminus \Omega_{s_0}$. Thus

$$
[U_a, \nabla f] = U_a(|\nabla f|^2) \frac{\nabla f}{|\nabla f|^2}
\tag{8.101}
$$

on $\mathcal{M}^3 \setminus \Omega_{s_0}$. Now, we have that $U_a(|\nabla f|^2) = -U_a(R) \leq o(|s|^{-1})$ by (8.37) and $|U_a| \leq O(|s|^{1/2})$. Hence

$$
|[U_a, \nabla f]| \leq o(|s|^{-1}).
\tag{8.102}
$$

In fact, from (8.71b) we have the improved estimate

$$
|[U_a, \nabla f]| = \left| U_a(|\nabla f|^2) \frac{\nabla f}{|\nabla f|^2} \right| \leq \frac{|U_a| |\nabla_{\Sigma_s} R|}{|\nabla f|} \leq O(|s|^{-5/4}).
$$

Clearly

$$
\|\bar{U}_a\|_{C^\ell(\mathbb{S}^2, \gamma_s)} \leq O(1) \quad \text{for } \ell \geq 0.
$$

Hence, by scaling,

$$
|\nabla_{\Sigma_s}^\ell U_a| \leq O(r^{\frac{1-\ell}{2}}) \quad \text{for } \ell \geq 0.
$$

Since $[U_a, \frac{\nabla f}{|\nabla f|^2}] = 0$, we conclude that

$$
|\nabla^\ell U_a| \leq O(r^{\frac{1-\ell}{2}}) \quad \text{for } \ell \geq 0.
\tag{8.103}
$$

We now check this for $\ell = 1$. If V is tangent to Σ_s, then

$$
|(\nabla - \nabla_{\Sigma_s})_V U_a| \leq |\mathrm{II}| \, |V| \, |U_a| \leq C r^{-\frac{1}{2}} |V|
$$

by (8.88). Moreover,

$$
\left| \nabla_{\frac{\nabla f}{|\nabla f|}} U_a \right| = |\nabla f| \left| \nabla_{\frac{\nabla f}{|\nabla f|^2}} U_a \right| \leq \left| \nabla_{U_a} \left(\frac{\nabla f}{|\nabla f|^2} \right) \right| \leq C |\mathrm{Ric}| \, |U_a| \leq O(r^{-\frac{1}{2}}).
$$

This proves the $\ell = 1$ case of (8.103). We leave the higher derivative of U_a estimates as an exercise (see Exercise 8.16).

8.4.3. Estimates for the approximate Killing vector fields.

We now show that the U_a are approximate Killing vector fields. Recall that $\Delta_f := \Delta - \nabla_{\nabla f}$ acting on vector fields.

Proposition 8.29. *For our steady GRS, the vector fields U_a on $\mathcal{M}^3$ satisfy*

$$(8.104) \qquad |\mathcal{L}_{U_a} g| \leq O(r^{-1/8})$$

and

$$(8.105) \qquad |\Delta_f U_a| \leq O(r^{-9/16})$$

for $a = 1, 2, 3$. Moreover, on Σ_s we have (8.99) and

$$(8.106) \qquad \sum_{a=1}^{3} U_a \otimes U_a = |s| g_s^{-1} + O(|s|^{7/8}).$$

Proof. Let $\{e_1, e_2\}$ be a local orthonormal frame on Σ_s. Then by (2.27) we have

$$
\begin{aligned}
(\mathcal{L}_{U_a} g)(e_i, e_j) &= \langle \nabla_{e_i} U_a, e_j \rangle + \langle \nabla_{e_j} U_a, e_i \rangle \\
&= g_s(\nabla_{e_i}^{g_s} U_a, e_j) + g_s(\nabla_{e_j}^{g_s} U_a, e_i) \\
&= (\mathcal{L}_{U_a} g_s)(e_i, e_j) \\
&= 2(\mathcal{L}_{\bar{U}_a} \gamma_s)(|s|^{1/2} F_s^*(e_i), |s|^{1/2} F_s^*(e_j)).
\end{aligned}
$$

Hence (8.100) implies $|(\mathcal{L}_{U_a} g)(e_i, e_j)| \leq O(|s|^{-1/8})$.

We compute using $\langle U_a, \nabla f \rangle \equiv 0$ and (8.101) that

$$
\begin{aligned}
(\mathcal{L}_{U_a} g)(\nabla f, e_i) &= \langle \nabla_{\nabla f} U_a, e_i \rangle + \langle \nabla_{e_i} U_a, \nabla f \rangle \\
&= \langle \nabla_{U_a} \nabla f, e_i \rangle + \langle [\nabla f, U_a], e_i \rangle - \langle U_a, \nabla_{e_i} \nabla f \rangle \\
&= 0.
\end{aligned}
$$

We also compute using (8.101) again that

$$
\begin{aligned}
(\mathcal{L}_{U_a} g)(\nabla f, \nabla f) &= 2 \langle \nabla_{\nabla f} U_a, \nabla f \rangle \\
&= 2 \langle \nabla_{U_a} \nabla f, \nabla f \rangle + 2 \langle [\nabla f, U_a], \nabla f \rangle \\
&= -U_a(|\nabla f|^2) \\
&= U_a(R) \\
&\leq o(|s|^{-1}).
\end{aligned}
$$

From the above, we conclude (8.104).

Next, observe that (8.103) implies

$$(8.107) \qquad |\nabla^{\ell}(\mathcal{L}_{U_a} g)| \leq O(r^{-\ell/2}).$$

By interpolation, we have:

Claim 8.30.

(8.108) $$|\nabla(\mathcal{L}_{U_a}g)| \leq \mathrm{O}(r^{-9/16}).$$

Proof of Claim 8.30. Let $s \ll -1$ and define

$$A_s := \{s - C\sqrt{s} \leq f \leq s + C\sqrt{s}\}.$$

We shall show that there exists a constant C such that

$$\alpha := \mathcal{L}_{U_a}g$$

satisfies the interpolation inequality

(8.109) $$\sup_{\Sigma_s} |\nabla\alpha|^2 \leq C \sup_{A_s} |\alpha| \max\left\{\sup_{A_s} |\nabla^2\alpha|, |s|^{-1}\right\}.$$

In view of the estimates (8.107) for $\ell = 2$ and (8.104), Claim 8.30 will follow from (8.109).

The following is a standard interpolation inequality argument. Since $\nabla\alpha$ is symmetric in its last two components, there exist $x \in \Sigma_s$ and unit vectors V and W in $T_x\mathcal{M}$ such that

(8.110) $$(\nabla_V \alpha)(W, W) \geq \frac{1}{9}\sup_{\Sigma_s}|\nabla\alpha|.$$

Let $\gamma : [0, t_0] \to A_s$ be the unit speed geodesic with $\gamma'(0) = V$. Parallel transport W along γ. Define

$$\psi(t) := \alpha\left(W\left(\gamma(t)\right), W\left(\gamma(t)\right)\right) \quad \text{for } s \in [0, t_0],$$

so that

$$\psi'(t) = (\nabla_{\gamma'(t)}\alpha)\left(W\left(\gamma(t)\right), W\left(\gamma(t)\right)\right)$$

and in particular $\psi'(0) = (\nabla_V\alpha)(W, W)$. By replacing both α by $-\alpha$ and V by $-V$ if necessary, we may assume that $\psi(0) \geq 0$ while preserving (8.110). By the mean value theorem, there exists $c_0 \in [0, t_0]$ such that

$$\psi(t_0) = \psi(0) + \psi'(0)t_0 + \frac{1}{2}\psi''(c_0)t_0^2.$$

So

(8.111) $$\sup_{A_s}|\alpha| \geq \psi(t_0)$$

$$= \psi(0) + \psi'(0)t_0 + \frac{1}{2}\psi''(c_0)t_0^2$$

$$\geq \frac{1}{9}\sup_{\Sigma_s}|\nabla\alpha|\, t_0 - \frac{1}{2}\sup_{A_s}|\nabla^2\alpha|\, t_0^2.$$

Now taking $t_0 := \frac{1}{6}\frac{\sup_{\Sigma_s}|\nabla\alpha|}{B}$, where $B := \max\left\{\sup_{A_s}|\nabla^2\alpha|, |s|^{-1}\right\}$, we obtain $t_0 \leq O(|s|^{1/2})$, so $\gamma([0, t_0]) \subset A_s$ by choosing C large enough in the definition of A_s. Moreover, (8.111) yields

$$\sup_{A_s}|\alpha| \geq \frac{1}{216}\frac{\sup_{\Sigma_s}|\nabla\alpha|^2}{B}.$$

This completes the proof of Claim 8.30. $\qquad\square$

Since by (1.77)

$$\Delta U_a + \mathrm{Ric}(U_a) = \mathrm{div}(\mathcal{L}_{U_a}g) - \frac{1}{2}\nabla\,\mathrm{tr}(\mathcal{L}_{U_a}g),$$

we obtain

(8.112) $\qquad\qquad |\Delta U_a + \mathrm{Ric}(U_a)| \leq O(r^{-9/16}).$

On the other hand,

(8.113) $\quad |\mathrm{Ric}(U_a) + \nabla_{\nabla f}U_a| = |-\nabla_{U_a}\nabla f + \nabla_{\nabla f}U_a| = |[\nabla f, U_a]| \leq o(|s|^{-1})$

by (8.102). From (8.112) and (8.113), we conclude (8.105).

Finally, by (8.97) and (8.98),

$$\sum_{a=1}^{3} U_a \otimes U_a = \frac{1}{2}(F_s)_*\bar{\gamma}^{-1}.$$

Hence (8.96), with $m = 0$, and $\frac{1}{2}(F_s)_*\gamma_s^{-1} = |s|g_s^{-1}$ yield (8.106). $\qquad\square$

Remark 8.31. As an alternative approach, one may use a localized version of Hamilton's interpolation formula as shown in the proof of Proposition 8.21. Let x be an arbitrary point on the level set $\{f = s\}$ and let ϕ be a standard cutoff function that equals 1 in $B_s(\frac{1}{2}\sqrt{|s|})$ and equals 0 outside $B_s(\sqrt{|s|})$, with $|\nabla\phi| \leq 3|s|^{-1/2}$. We have

$$\int_{\mathcal{M}} \phi^2|\nabla\alpha|^2 = \int_{\mathcal{M}} \phi^2\nabla_i\alpha_{jk}\nabla_i\alpha_{jk}$$

$$= -2\int_{\mathcal{M}} \phi\alpha_{jk}\nabla_i\phi\nabla_i\alpha_{jk} - \int_{\mathcal{M}} \phi^2\alpha_{jk}\Delta\alpha_{jk}$$

$$\leq \frac{1}{2}\int_{\mathcal{M}} \phi^2|\nabla\alpha|^2 + 2\int_{\mathcal{M}} |\nabla\phi|^2|\alpha|^2$$

$$+ \sqrt{3}\left(\int_{\mathcal{M}} \phi^2|\alpha|^2\right)^{\frac{1}{2}}\left(\int_{\mathcal{M}} \phi^2|\nabla^2\alpha|^2\right)^{\frac{1}{2}},$$

where we used $|\Delta\alpha| \leq \sqrt{3}|\nabla^2\alpha|$. By using $|\nabla^\ell\alpha| \leq O(|s|^{-\frac{\ell}{2}})$ and $|\alpha| \leq O(|s|^{-\frac{1}{8}})$, we obtain

$$\int_{B_s(\frac{1}{2}\sqrt{|s|})} |\nabla\alpha|^2 \leq O(|s|^{-\frac{3}{8}}).$$

Iterating this, one may obtain estimates for the L^2 norms of the higher-order derivatives. Furthermore, because of the local cylindrical geometry, one may use a local Sobolev inequality to estimate the L^∞ norm of $|\nabla\alpha|$.

8.5. The f-Laplace equation for vector fields

The key result of this section is Proposition 8.32, which says that if $\Delta_f W = 0$, then $\Delta_{L,f}(\mathcal{L}_W g) = 0$. This proposition, together with a vanishing theorem for solutions of $\Delta_{L,f} h = 0$ to be proved in §8.7, enables us to reduce the proof of the existence of Killing vector fields to solving the equation $\Delta_f W = 0$.

8.5.1. Redux of basic formulas.

We first rederive, in a different way, some formulas from §8.1.2. For any vector field X on a Riemannian manifold $(\mathcal{M}^n, g)$ we have

$$\operatorname{div}(\mathcal{L}_X g)_i = \nabla_j(\nabla_i X_j + \nabla_j X_i)$$
$$= R_{ij} X_j + \nabla_i \nabla_j X_j + \Delta X_i;$$

that is,

$$(8.114) \qquad \operatorname{div}(\mathcal{L}_X g) = \Delta X + \operatorname{Ric}(X) + \nabla \operatorname{div} X.$$

We also observe that on a steady GRS,

$$(8.115) \quad \mathcal{L}_{\nabla f} X = [\nabla f, X] = \nabla_{\nabla f} X - \nabla_X \nabla f = -\Delta_f X + \Delta X + \operatorname{Ric}(X).$$

In particular, if V is a Killing vector field, then

$$(8.116) \qquad \Delta V + \operatorname{Ric}(V) = 0.$$

Furthermore, if we are on a rotationally symmetric steady GRS $(\mathcal{M}^n, g, f)$, where f is preserved by the isometry group, then

$$\Delta V = -\operatorname{Ric}(V) = \nabla^2 f(V) = \nabla_{\nabla f} V,$$

since $[V, \nabla f] = \mathcal{L}_V \nabla f = 0$. That is, $\Delta_f V := \Delta V - \nabla_{\nabla f} V = 0$.

On the other hand, we compute using (2.96) that on any steady GRS,

$$(8.117) \qquad \Delta_f \nabla f = \Delta \nabla f - \nabla_{\nabla f} \nabla f$$
$$= -\operatorname{Ric}(\nabla f) - \nabla^2 f(\nabla f)$$
$$= 0.$$

So, if V is a Killing vector field on a rotationally symmetric steady GRS as above, then for any $\mu \in \mathbb{R}$,

$$(8.118) \qquad \Delta_f(V + \mu \nabla f) = 0.$$

Conversely, this observation opens up the possibility of proving the existence of a Killing vector field by first solving the equation $\Delta_f V = 0$ and then showing there exists $\mu \in \mathbb{R}$ such that $V + \mu \nabla f$ is a Killing vector field.

8.5.2. Link between the f-Laplace equation for vector fields and the f-Lichnerowicz Laplace equation.

Recall that the f-Lichnerowicz Laplacian is defined by $\Delta_{L,f} := \Delta_L - \mathcal{L}_{\nabla f}$, where Δ_L is given by (4.79). Note that by (4.80), on a GRS we have

$$(8.119) \qquad \Delta_{L,f} \operatorname{Ric} = 0.$$

In particular, on a steady GRS, $\Delta_{L,f}(\mathcal{L}_{\nabla f}g) = 0$. Thus, if $\Delta_{L,f}(\mathcal{L}_W g) = 0$, then $\Delta_{L,f}(\mathcal{L}_{W+\mu\nabla f}g) = 0$ for any $\mu \in \mathbb{R}$.

Regarding the aforementioned converse direction of proving the existence of Killing vector fields, we have the following key link between the f-Laplace equation for vector fields and the f-Lichnerowicz Laplace equation for symmetric 2-tensors.

Proposition 8.32. *If on a steady GRS a vector field W satisfies $\Delta_f W = 0$, then*

$$(8.120) \qquad \Delta_{L,f}(\mathcal{L}_W g) = 0.$$

Proof. By a well-known variation formula (see e.g. [**276**, Proposition 2.3.7]), if $\frac{\partial}{\partial s}g_{ij} = v_{ij}$ for a symmetric 2-tensor v_{ij}, then

$$(8.121) \qquad 2\frac{\partial}{\partial s}\operatorname{Ric} = -\Delta_L v - \mathcal{L}_Z g,$$

where $Z := \frac{1}{2}\nabla V - \operatorname{div} v$ and $V = \operatorname{tr} v$. If W is a vector field, then taking $v = \mathcal{L}_W g$ yields

$$(8.122) \qquad \frac{\partial}{\partial s}\operatorname{Ric} = \mathcal{L}_W \operatorname{Ric}$$

by the diffeomorphism invariance of Ric. Hence, by (8.114), (8.121), and (8.122), we have

$$(8.123) \qquad 2\mathcal{L}_W \operatorname{Ric} = -\Delta_L(\mathcal{L}_W g) + \mathcal{L}_{\Delta W + \operatorname{Ric}(W)}g.$$

Since (see Exercise 8.4)

$$(8.124) \qquad \mathcal{L}_X(\mathcal{L}_Y g) = \mathcal{L}_Y(\mathcal{L}_X g) + \mathcal{L}_{[X,Y]}g$$

and $\operatorname{Ric} + \nabla^2 f = 0$, on the other hand we have

$$2\mathcal{L}_W \operatorname{Ric} = -\mathcal{L}_W \mathcal{L}_{\nabla f}g = -\mathcal{L}_{\nabla f}\mathcal{L}_W g - \mathcal{L}_{[W,\nabla f]}g.$$

Hence

$$\Delta_{L,f}(\mathcal{L}_W g) = \Delta_L(\mathcal{L}_W g) - \mathcal{L}_{\nabla f}(\mathcal{L}_W g) = \mathcal{L}_{\Delta W + \mathrm{Ric}(W) + [W, \nabla f]} g.$$

Now

$$\Delta W + \mathrm{Ric}(W) + [W, \nabla f] = \Delta W + (\mathrm{Ric} + \nabla^2 f)(W) - \nabla_{\nabla f} W$$
$$= \Delta_f W,$$

so that $\Delta_{L,f}(\mathcal{L}_W g) = \mathcal{L}_{\Delta_f W} g$. The proposition follows.   □

If V satisfies $\Delta_f V = 0$, then $\Delta_f (V + \mu \nabla f) = 0$, which implies that

$$\Delta_{L,f}(\mathcal{L}_{V + \mu \nabla f} g) = 0.$$

Hence, provided we can prove the vanishing of suitable solutions h to the equation $\Delta_{L,f} h = 0$, there is the possibility of showing that $V + \mu \nabla f$ is a Killing vector field for some $\mu \in \mathbb{R}$.

Definition 8.33. We say that two vector fields V_1 and V_2 defined on the same domain $\Omega \subset \mathcal{M}^3$ are ∇f-**equivalent** on Ω, where f is the potential function, if there exists $\mu \in \mathbb{R}$ such that $V_1 - V_2 = \mu \nabla f$. Clearly, ∇f-equivalence is an equivalence relation.

Finally, we remark that a parabolic version of (8.120) is the following (see e.g. [**111**, Exercise 2.36]). If $(\mathcal{M}^n, g(t))$ is a solution to the Ricci flow, then for any time-dependent vector field $X(t)$ we have

$$\left(\frac{\partial}{\partial t} - \Delta_L \right) (\mathcal{L}_X g) = \mathcal{L}_{(\frac{\partial}{\partial t} - \Delta - \mathrm{Ric}) X} g.$$

In particular, if $\frac{\partial}{\partial t} X = \Delta X + \mathrm{Ric}(X)$, then $\frac{\partial}{\partial t} \mathcal{L}_X g = \Delta_L \mathcal{L}_X g$.

8.6. Solving the inhomogeneous f-Laplace equation

Recall from (8.105) that the approximate Killing vector fields U_a have the property that $|\Delta_f U_a| \leq O(r^{-9/16})$ is small for $a = 1, 2, 3$ in the sense that $-\frac{9}{16}$ is less than the expected decay power $-\frac{1}{2}$ from scaling. For each $a = 1, 2, 3$ we wish to find a small solution of the inhomogeneous equation $\Delta_f V_a = Q_a$, where $Q_a := \Delta_f U_a$, in order to obtain a solution $W_a := U_a - V_a$ of the homogeneous equation $\Delta_f W_a = 0$ close to U_a. Eventually we shall show that W_a differs from a Killing vector field by a multiple of ∇f, the latter of which is a bounded vector field.

8.6.1. Statement of the existence theorem.

Theorem 8.34. *Let Q be a vector field on $\mathcal{M}^3$ satisfying $|Q| \leq O(r^{-\frac{1}{2}-2\varepsilon})$, where $\varepsilon > 0$. Then there exists a vector field V solving*

$$\Delta_f V = Q \tag{8.125}$$

and satisfying $|V| \leq O(r^{\frac{1}{2}-\varepsilon})$ and $|\nabla V| \leq O(r^{-\varepsilon})$.

We will prove the theorem below.

A significance of the condition $|Q| \leq O(r^{-\frac{1}{2}-2\varepsilon})$, where $\varepsilon > 0$, is that it implies in the limit that Q vanishes after suitable rescaling at infinity. In view of (8.105), the theorem is applicable to $Q = \Delta_f U_a$ with $\varepsilon = \frac{1}{32}$. Note that for the Bryant soliton, Killing vector fields have norms of order $r^{1/2}$.

To prove Theorem 8.34, we solve the Dirichlet problem for (8.125) on domains comprising an exhaustion of $\mathcal{M}^3$. By studying the linearization at infinity of the equation, we obtain weighted estimates for the solutions independent of the domain of the exhaustion. This enables us to take a limit to obtain a global solution of (8.125).

8.6.2. Estimates for solutions on superlevel sets of the potential function.

We begin with the following estimate for a solution V on a superlevel set (a.k.a. upper level set) of f, proved using the elliptic maximum principle and a power of $|f|$ as a barrier.

Lemma 8.35. *Let Q be a vector field on $\mathcal{M}^3$ satisfying $|Q| \leq O(r^{-\frac{1}{2}-2\varepsilon})$, where $\varepsilon > 0$. Then there exists C with the following property. If $\Delta_f V = Q$ on the superlevel set $\{f \geq s\}$, where $s < -1$, then*

$$\sup_{\{f \geq s\}} |V| \leq \sup_{\{f=s\}} |V| + C|s|^{\frac{1}{2}-2\varepsilon}. \tag{8.126}$$

Proof. We compute using (4.118) that

$$\frac{1}{2}\Delta_f |V|^2 = |\nabla V|^2 + \langle \Delta_f V, V \rangle$$
$$\geq |\nabla|V||^2 - |Q|\,|V|.$$

Hence

$$\Delta_f |V| = \Delta_f(|V|^2)^{1/2} \tag{8.127}$$
$$= \frac{1}{2|V|}\Delta_f |V|^2 - \frac{1}{4|V|^3}|\nabla|V|^2|^2$$
$$\geq -|Q|.$$

Using (2.56), we compute for $p > 0$ that

$$\Delta_f \left(|f|^p \right) = p \, |f|^{p-1} \, \Delta_f \, |f| + p(p-1) \, |f|^{p-2} \, |\nabla f|^2$$
$$= p \, |f|^{p-1} + p(p-1) \, |f|^{p-2} \, |\nabla f|^2.$$

Since $|f| \geq 1$, this implies for $p < 1$ that

$$(8.128) \qquad\qquad \Delta_f \left(|f|^p \right) \geq p^2 \, |f|^{p-1}.$$

Since $|Q| \leq Cr^{-\frac{1}{2} - 2\varepsilon} \leq C \, |f|^{-\frac{1}{2} - 2\varepsilon}$, by (8.127) and (8.128) we have

$$\Delta_f \left(|V| + \left(\frac{1}{2} - 2\varepsilon \right)^{-2} C \, |f|^{\frac{1}{2} - 2\varepsilon} \right) \geq 0.$$

Applying the elliptic maximum principle to this yields

$$\sup_{\{f \geq s\}} \left(|V| + \left(\frac{1}{2} - 2\varepsilon \right)^{-2} C \, |f|^{\frac{1}{2} - 2\varepsilon} \right)$$
$$\leq \sup_{\{f = s\}} \left(|V| + \left(\frac{1}{2} - 2\varepsilon \right)^{-2} C \, |f|^{\frac{1}{2} - 2\varepsilon} \right).$$

The lemma follows. $\qquad\qquad\qquad\qquad\qquad\qquad\qquad\qquad\qquad\qquad \square$

8.6.3. Estimates for the linearized equation.

It is useful to consider the canonical form $(\mathcal{M}^3, g(t), f(t))$ of our steady GRS, which satisfies $g(t) = \varphi_t^* g$ and $f(t) = f \circ \varphi_t$, where φ_t is the 1-parameter group of diffeomorphisms generated by ∇f. Given $\Delta_{g,f} V = Q$, we define $V(t) = \varphi_t^* V = (\varphi_{-t})_* V$ and $Q(t) = \varphi_t^* Q$. Then $\Delta_{g(t), f(t)} V(t) = Q(t)$. On the other hand, for $V = V(t)$, $g = g(t)$, $f = f(t)$, $\nabla = \nabla^{g(t)}$, and $\mathrm{Ric}_g = \mathrm{Ric}_{g(t)}$, we have

$$\frac{\partial}{\partial t} V = \mathcal{L}_{\nabla f} V = [\nabla f, V] = \nabla_{\nabla f} V - \nabla_V \nabla f = \nabla_{\nabla f} V + \mathrm{Ric}_g(\nabla f).$$

Thus,

$$(8.129) \qquad\qquad \frac{\partial}{\partial t} V = \Delta V + \mathrm{Ric}_g(\nabla f) - Q.$$

Given two vector fields $\bar{V}_1$ and $\bar{V}_2$ on $\mathbb{S}^2 \times \mathbb{R}$, we say that they are $\frac{\partial}{\partial z}$-**equivalent** if there exists $c \in \mathbb{R}$ such that $\bar{V}_1 - \bar{V}_2 = c \frac{\partial}{\partial z}$.

In view of the dimension reduction result of Corollary 8.14 and the quantitatively improved estimates of Propositions 8.21, 8.22, and 8.23, we consider the following, which is the equation for vector fields linearized (rescaled) at spatial infinity.[6]

---

[6]By this we mean that our steady soliton is asymptotically cylindrical at spatial infinity and the linearization of (8.129) becomes equation (8.130) below.

Lemma 8.36. *Let $\bar{g}(t) := 2\,(1-t)\,g_{\mathbb{S}^2} + dz^2$, $t \in (-\infty, 1)$, be the round cylinder solution to the Ricci flow on $\mathbb{S}^2 \times \mathbb{R}$. Suppose that a 1-parameter family of vector fields $\bar{V}(t)$ on $\mathbb{S}^2 \times \mathbb{R}$, each invariant under translation along $\mathbb{R}$, solves*

$$(8.130) \qquad \frac{\partial}{\partial t}\bar{V} = \Delta_{\bar{g}(t)}\bar{V} + \mathrm{Ric}_{\bar{g}(t)}(\bar{V}) \quad \text{for } t \in (0,1),$$

where $\mathrm{Ric}_{\bar{g}(t)}$ is considered to be a $(1,1)$-tensor and where $\bar{V}(t)$ satisfies $\sup_{\mathbb{S}^2 \times \mathbb{R}} |\bar{V}(t)|_{\bar{g}(t)} \leq 1$ for $t \in (0, \frac{1}{2}]$. Then there exists a constant C such that

$$(8.131) \qquad \inf_{\lambda \in \mathbb{R}} \left(\sup_{\mathbb{S}^2 \times \mathbb{R}} \left| \bar{V}(t) - \lambda \frac{\partial}{\partial z} \right|_{\bar{g}(t)} \right) \leq C\sqrt{1-t} \quad \text{for } t \in \left[\tfrac{1}{2}, 1\right).$$

That is, the minimal L^∞ norm on any $\mathbb{S}^2 \times \{z\}$ of vector fields in the $\frac{\partial}{\partial z}$-equivalence class of $\bar{V}(t)$ is bounded by $C\sqrt{1-t}$.

Remark 8.37. Note that for any $\lambda \in \mathbb{R}$, $\lambda \frac{\partial}{\partial z}$ is a solution to (8.130) with each term in (8.130) vanishing.

Proof of Lemma 8.36. Observe that (8.130) and $\frac{\partial}{\partial t}\bar{g}(t) = -2\,\mathrm{Ric}_{\bar{g}(t)}$ imply that

$$\frac{\partial}{\partial t}|\bar{V}(t)|^2_{\bar{g}(t)} = \bar{g}(t)(\Delta_{\bar{g}(t)}V, V(t))$$
$$= \Delta_{g(t)}|\bar{V}(t)|^2_{\bar{g}(t)} - 2|\nabla_{g(t)}\bar{V}(t)|^2_{\bar{g}(t)}.$$

By applying the maximum principle to this equation, we obtain

$$(8.132) \qquad |\bar{V}(t)|_{\bar{g}(t)} \leq C \quad \text{on } (\mathbb{S}^2 \times \mathbb{R}) \times (0,1)$$

for some constant C. Since $\bar{V}(t)$ is invariant under translation along $\mathbb{R}$, we may write it as

$$\bar{V}(t) = \bar{U}(t) + v(t)\frac{\partial}{\partial z},$$

where the $\bar{U}(t)$ are vector fields on $\mathbb{S}^2$ and the $v(t)$ are functions on $\mathbb{S}^2$. Since $\bar{g}(t)|_{\mathbb{S}^2} := 2\,(1-t)\,g_{\mathbb{S}^2}$ and $\mathrm{Ric}_{\bar{g}(t)} = \frac{1}{2(1-t)}\bar{g}(t)|_{\mathbb{S}^2}$, we have that (8.130) is equivalent to

$$(8.133a) \qquad \frac{\partial \bar{U}}{\partial t}(t) = \frac{1}{2\,(1-t)}\left(\Delta_{\mathbb{S}^2}\bar{U}(t) + \bar{U}(t)\right),$$

$$(8.133b) \qquad \frac{\partial v}{\partial t}(t) = \frac{1}{2\,(1-t)}\Delta_{\mathbb{S}^2}v(t).$$

Now since $|\bar{V}(t)|_{\bar{g}(t)} \leq 1$ for $t \in (0, \frac{1}{2}]$, we have

$$\sup_{\mathbb{S}^2} |\bar{U}(t)|_{g_{\mathbb{S}^2}} \leq \frac{1}{\sqrt{2(1-t)}} \leq 1,$$

$$\sup_{\mathbb{S}^2} |v(t)| \leq 1,$$

for $t \in (0, \frac{1}{2}]$. We do not use the estimates from (8.132) up to time 1 since we need to improve them by considering the spectrum of $\Delta_{\mathbb{S}^2}$.

Let $\tau = -\frac{1}{2} \ln(1-t)$. Then $\frac{\partial v}{\partial \tau}(\tau) = \Delta_{\mathbb{S}^2} v(\tau)$, $\tau \in (0, \infty)$. We have from (8.133b) that

$$\frac{\partial}{\partial \tau} \left(e^{2\tau} v(\tau) \right) = \Delta_{\mathbb{S}^2} \left(e^{2\tau} v(\tau) \right) + 2 \left(e^{2\tau} v(\tau) \right).$$

By using the linear static solutions to the equation $\frac{\partial w}{\partial \tau} = \Delta_{\mathbb{S}^2} w + 2w$ as barriers and the Aleksandrov reflection method (see [**107**]), we find that there exists a constant such that (see Exercise 8.20 for details)

$$(8.134) \qquad |\nabla(e^{2\tau} v(x, \tau))| \leq \frac{C}{\pi} \quad \text{for all } x \in \mathbb{S}^2 \text{ and } \tau \in \left[\frac{1}{2} \ln 2, \infty \right).$$

Integrating this estimate over geodesics in $\mathbb{S}^2$ at each time, we conclude that[7]

$$(8.135) \qquad |v(x, \tau) - v_{\text{avg}}| \leq C e^{-2\tau},$$

where $v_{\text{avg}} \in \mathbb{R}$ denotes the average of $v(\tau)$ on $\mathbb{S}^2$ (which is independent of τ since v is a solution to the heat equation); that is,

$$(8.136) \qquad |v(x, \tau) - v_{\text{avg}}| \leq C(1-t) \quad \text{for all } x \in \mathbb{S}^2 \text{ and } t \in \left[\frac{1}{2}, 1 \right).$$

Next, we observe that from (8.133a) and $\frac{\partial \tau}{\partial t} = \frac{1}{2(1-t)}$ we have

$$\frac{\partial \bar{U}}{\partial \tau}(\tau) = \Delta_{\mathbb{S}^2} \bar{U}(\tau) + \bar{U}(\tau).$$

Let $\lambda_1 \leq \lambda_2 \leq \cdots$ denote the eigenvalues of $\Delta_{\mathbb{S}^2}$ acting on vector fields (or equivalently, 1-forms) with associated eigenvector fields $\varphi_1, \varphi_2, \ldots$ normalized by $\int_{\mathbb{S}^2} \langle \varphi_i, \varphi_j \rangle \, dA_{\mathbb{S}^2} = \delta_{ij}$. Then we may express the solution $\bar{U}(\tau)$

---

[7] A more general fact, which can be proved by an eigenfunction expansion and $W^{k,2}$-estimates (see Exercise 8.21 for the case of $\mathbb{S}^2$) is that for any closed Riemannian manifold, any solution to $\frac{\partial w}{\partial \tau} = \Delta_g w$ satisfies $|w(x, \tau) - w_{\text{avg}}| \leq C e^{-\lambda_1 \tau}$, where w_{avg} is the (spatial) average of w and where λ_1 is the first nonzero eigenvalue of Δ_g.

as

$$\bar{U}(\tau) = \sum_{i=1}^{\infty} e^{(1-\lambda_i)\tau} a_i \varphi_i,$$

where $a_i = \int_{\mathbb{S}^2} \langle \bar{U}(0), \varphi_i \rangle \, d\mu_{\mathbb{S}^2}$. We have $\lambda_1(\Delta) = 1$ for the rough Laplacian acting on 1-forms.[8] We claim that this implies

$$(8.137) \qquad |\bar{U}(x,\tau)|_{g_{\mathbb{S}^2}} \leq C \quad \text{for } x \in \mathbb{S}^2 \text{ and } \tau \in \left[\tfrac{1}{2}\ln 2, \infty\right),$$

which in turn implies that

$$(8.138) \qquad |\bar{U}(x,t)|_{\bar{g}(t)} \leq C\sqrt{1-t} \quad \text{for } x \in \mathbb{S}^2 \text{ and } t \in (0,1).$$

Lemma 8.36 will then follow from (8.136) and (8.138).

To prove (8.137), we first observe that there exists a constant C such that[9]

$$\int_{\mathbb{S}^2} |\bar{U}(x,\tau)|^2_{g_{\mathbb{S}^2}} d\mu_{\mathbb{S}^2} = \sum_{i=1}^{\infty} e^{2(1-\lambda_i)\tau} a_i^2 \leq C \quad \text{for } x \in \mathbb{S}^2 \text{ and } \tau \in \left[\tfrac{1}{2}\ln 2, \infty\right).$$

Next, on $\mathbb{S}^2 \times \left[\tfrac{1}{2}\ln 2, \infty\right)$,

$$\int_{\mathbb{S}^2} |\nabla \bar{U}(x,\tau)|^2_{g_{\mathbb{S}^2}} d\mu_{\mathbb{S}^2} = -\int_{\mathbb{S}^2} \bar{U}(x,\tau) \Delta_{\mathbb{S}^2} \bar{U}(\tau) d\mu_{\mathbb{S}^2}$$
$$= \sum_{i=1}^{\infty} \lambda_i e^{2(1-\lambda_i)\tau} a_i^2$$
$$\leq C,$$

since for $\tau \in \left[\tfrac{1}{2}\ln 2, \infty\right)$ the function $\lambda \mapsto \lambda e^{2(1-\lambda)\tau}$ is uniformly bounded above for $\lambda \in [0, \infty)$.

---

[8]The lowest eigenvalue $\lambda_1(\Delta_d, n, p)$ of the Hodge Laplacian $\Delta_d = -(d\delta + \delta d)$ acting on p-forms on $\mathbb{S}^n$ is $\min\{p(n-p+1), (p+1)(n-p)\}$. In particular, $\lambda_1(\Delta_d, 2, 1) = 2$. On the other hand, we have the Bochner formula $\Delta\alpha - \Delta_d\alpha = \mathrm{Ric}_{\mathbb{S}^2}(\alpha) = \alpha$ for a 1-form α.

[9]We immediately get $\|\bar{U}(\tau)\|_{L^2} \leq C$. The L^∞-bound follows from Moser iteration [**231**].

Secondly, since $R^{\mathbb{S}^2}_{ijk\ell} = (g_{\mathbb{S}^2})_{i\ell}(g_{\mathbb{S}^2})_{jk} - (g_{\mathbb{S}^2})_{ik}(g_{\mathbb{S}^2})_{j\ell}$ and $R^{\mathbb{S}^2}_{ij} = (g_{\mathbb{S}^2})_{ij}$, we have

$$
\begin{aligned}
\int_{\mathbb{S}^2} \left|\nabla^2 \bar{U}\right|^2 d\mu &= \int_{\mathbb{S}^2} \nabla_i \nabla_j \bar{U}_k \nabla_i \nabla_j \bar{U}_k d\mu \\
&= \int_{\mathbb{S}^2} \left(\nabla_i \nabla_j \bar{U}_k \nabla_j \nabla_i \bar{U}_k - \nabla_i \nabla_j \bar{U}_k R^{\mathbb{S}^2}_{ijk\ell} \bar{U}_\ell \right) d\mu \\
&= \int_{\mathbb{S}^2} \left(\bar{U}_j (\nabla_i \nabla_j \bar{U}_i - \nabla_j \nabla_i \bar{U}_i) - \nabla_j \bar{U}_k \nabla_i \nabla_j \nabla_i \bar{U}_k \right) d\mu \\
&= \int_{\mathbb{S}^2} \bar{U}_j R^{\mathbb{S}^2}_{jk} \bar{U}_k d\mu - \int_{\mathbb{S}^2} \nabla_j \bar{U}_k \nabla_j \nabla_i \nabla_i \bar{U}_k d\mu \\
&\quad + \int_{\mathbb{S}^2} \nabla_j \bar{U}_k \left(-R^{\mathbb{S}^2}_{j\ell} \nabla_\ell \bar{U}_k + R^{\mathbb{S}^2}_{ijk\ell} \nabla_i \bar{U}_\ell \right) d\mu \\
&= \int_{\mathbb{S}^2} |\bar{U}|^2 d\mu + \int_{\mathbb{S}^2} |\Delta \bar{U}|^2 d\mu - \int_{\mathbb{S}^2} |\nabla \bar{U}|^2 d\mu \\
&\quad + \int_{\mathbb{S}^2} (\operatorname{div} \bar{U})^2 d\mu - \int_{\mathbb{S}^2} \nabla_j \bar{U}_i \nabla_i \bar{U}_j d\mu \\
&= \int_{\mathbb{S}^2} \left(|\bar{U}|^2 + |\Delta \bar{U}|^2 - \frac{1}{2} \left| \nabla_i \bar{U}_j + \nabla_j \bar{U}_i - \operatorname{div} \bar{U}(g_{\mathbb{S}^2})_{ij} \right|^2 \right) d\mu.
\end{aligned}
$$

Hence, on $\mathbb{S}^2 \times [\frac{1}{2}\ln 2, \infty)$,

$$
\int_{\mathbb{S}^2} \left|\nabla^2 \bar{U}(x,\tau)\right|^2 d\mu \leq \sum_{i=1}^\infty (1 + \lambda_i^2)\, e^{2(1-\lambda_i)\tau} a_i^2 \leq C.
$$

Since $\dim \mathbb{S}^2 = 2$, we can now apply the Sobolev inequality on $\mathbb{S}^2$ to obtain (8.137). This completes the proof of Lemma 8.36. $\qquad\square$

We can replace part of the above argument with the following. We have

$$
\int_{\mathbb{S}^2} |\bar{U}(x,\tau)|^2_{g_{\mathbb{S}^2}} d\mu_{\mathbb{S}^2} \leq C \quad \text{for } x \in \mathbb{S}^2 \text{ and } \tau \in \left[\tfrac{1}{4}\ln 2, \infty\right).
$$

Notice here that the start time is $\frac{1}{4}\ln 2$ instead of $\frac{1}{2}\ln 2$ to give extra time to apply the mean value inequality. By the evolution equation for $\bar{U}$, we have

$$
\begin{aligned}
\frac{\partial}{\partial \tau} |\bar{U}|^2 &= 2 \left\langle \bar{U}, \frac{\partial \bar{U}}{\partial \tau} \right\rangle \\
&= 2 \left\langle \bar{U}, \Delta_{\mathbb{S}^2} \bar{U} \right\rangle + 2|\bar{U}|^2 \\
&= \Delta_{\mathbb{S}^2} |\bar{U}|^2 - 2|\nabla \bar{U}|^2 + 2|\bar{U}|^2 \\
&\leq \Delta_{\mathbb{S}^2} |\bar{U}|^2 + 2|\bar{U}|^2.
\end{aligned}
$$

Hence $|\bar{U}|^2$ is a subsolution to a linear heat equation. Let $r_0 := \sqrt{\frac{1}{4}\ln 2}$. We have for any $\tau \in [\frac{1}{2}, \infty)$ that $[\tau - r_0^2, \tau] \subset [\frac{1}{4}\ln 2, \infty)$. Hence, by applying the mean value inequality for parabolic equations on the parabolic cylinder $B_{r_0}^{g_{S^2}}(x) \times [\tau - r_0^2, \tau]$ for $x \in \mathbb{S}^2$ (see Theorem 25.2 in [**103**]), we have that

$$|\bar{U}(x,\tau)| \leq C \quad \text{for } x \in \mathbb{S}^2, \ \tau \in \left[\tfrac{1}{2}, \infty\right).$$

8.6.4. Existence of global solutions to the f-Laplace equation.

We begin the construction of a global solution to $\Delta_f V = Q$ by considering the Dirichlet problems on the superlevel sets of f comprising an exhaustion of $\mathcal{M}^3$. Let $s_k \to -\infty$ and let $V^{(k)}$ be the unique vector field such that

(8.139a) $$\Delta_f V^{(k)} = Q \quad \text{in } \{f \geq s_k\},$$

(8.139b) $$V^{(k)} = 0 \quad \text{on } \{f = s_k\}.$$

Recall that since f has a unique critical point, each domain $\{f \geq s_k\}$ has a boundary $\{f = s_k\}$ consisting of a smooth 2-sphere.

Ideally, to obtain the global solution, we wish to prove uniform (distance) weighted estimates for $V^{(k)}$ so that we can pass to a subsequential limit as $k \to \infty$. However, in view of the ambiguity indicated by (8.117) and (8.118) (see also Exercise 8.18) as well as Lemma 8.36 to be applied to any limit, we define

(8.140) $$A^{(k)}(s) := \inf_{\lambda \in \mathbb{R}} \sup_{\{f=s\}} |V^{(k)} + \lambda \nabla f| \quad \text{for } s \geq s_k.$$

That is, $A^{(k)}(s)$ is the minimal L^∞ norm on $\{f = s\}$ of vector fields in the ∇f-equivalence class of $V^{(k)}$. We wish to obtain uniform weighted estimates for $A^{(k)}(s)$. If one makes an analogy between this equivalence and gauge equivalence for Yang–Mills connections, this brings to mind Uhlenbeck's work on removable singularities [**280**] as only a rough analogy.

Modulo the aforementioned ambiguity, the following estimate shall imply the decay of solutions inside a superlevel set. Its proof, by contradiction, ultimately relies on the linearized estimate of Lemma 8.36. Because of this, the estimate holds at large enough interior distances.

Lemma 8.38. *Let C be as in Lemma 8.36 and let $\tau \in (0, \tfrac{1}{2})$ be sufficiently small so that $\tau^{-\varepsilon} > 2C$. Then there exist $s_0 < -1$ and k_0 such that for all $s \in [s_k, s_0]$ and $k \geq k_0$, we have*

(8.141) $$2\tau^{-\frac{1}{2}+\varepsilon} A^{(k)}(\tau s) \leq A^{(k)}(s) + |s|^{\frac{1}{2}-\varepsilon},$$

which is equivalent to

$$2|\tau s|^{-\frac{1}{2}+\varepsilon} A^{(k)}(\tau s) \leq |s|^{-\frac{1}{2}+\varepsilon} A^{(k)}(s) + 1$$

and to

$$A^{(k)}(\tau s) \leq \frac{1}{2} \tau^{\frac{1}{2}-\varepsilon} A^{(k)}(s) + \frac{1}{2}|\tau s|^{\frac{1}{2}-\varepsilon}.$$

Proof. Suppose that the lemma were false. Then there would exist $r_k \in [s_k, s_0]$ with $|r_k| \to \infty$ such that the "contradiction hypothesis"

$$(8.142) \qquad 2\tau^{-\frac{1}{2}+\varepsilon} A^{(k)}(\tau r_k) \geq A^{(k)}(r_k) + |r_k|^{\frac{1}{2}-\varepsilon}$$

would hold for all k. For each k, choose $\lambda_k \in \mathbb{R}$ so that

$$\sup_{\{f=r_k\}} |V^{(k)} + \lambda_k \nabla f| = A^{(k)}(r_k).$$

The optimized vector fields $V^{(k)} + \lambda_k \nabla f$ on $\{f \geq s_k\}$ have the best bounds on $\{f = r_k\}$ among all the $V^{(k)} + \lambda \nabla f$ for $\lambda \in \mathbb{R}$. By Lemma 8.35,

$$\sup_{\{f \geq r_k\}} |V^{(k)} + \lambda_k \nabla f| \leq \sup_{\{f=r_k\}} |V^{(k)} + \lambda_k \nabla f| + C|r_k|^{\frac{1}{2}-2\varepsilon}$$

$$\leq A^{(k)}(r_k) + |r_k|^{\frac{1}{2}-\varepsilon}$$

for k sufficiently large. Rescale the optimized vector fields by defining

$$(8.143) \qquad \tilde{V}^{(k)} := (A^{(k)}(r_k) + |r_k|^{\frac{1}{2}-\varepsilon})^{-1}(V^{(k)} + \lambda_k \nabla f),$$

so that

$$(8.144) \qquad \sup_{\{f \geq r_k\}} |\tilde{V}^{(k)}|_g \leq 1.$$

Note that

$$(8.145) \qquad \Delta_f \tilde{V}^{(k)} = (A^{(k)}(r_k) + |r_k|^{\frac{1}{2}-\varepsilon})^{-1} Q.$$

By this and (8.115), we have that $\tilde{V}^{(k)}$ satisfies

$$(8.146) \qquad \mathcal{L}_{\nabla f} \tilde{V}^{(k)} = \Delta \tilde{V}^{(k)} + \mathrm{Ric}(\tilde{V}^{(k)}) - (A^{(k)}(r_k) + |r_k|^{\frac{1}{2}-\varepsilon})^{-1} Q.$$

Recall that $\{\varphi_t\}$ is the 1-parameter group of diffeomorphisms generated by $\nabla_g f$. Define the sequence of rescaled solutions to the Ricci flow (because we shall choose the base point on $\{f = r_k\}$) by

$$(8.147) \qquad \hat{g}^{(k)}(t) = |r_k|^{-1} \varphi^*_{|r_k|t}(g), \quad -\infty < t < \infty,$$

and corresponding rescaled time-dependent vector fields

$$(8.148) \qquad \widehat{V}^{(k)}(t) = |r_k|^{1/2} \varphi^*_{|r_k|t}(\tilde{V}^{(k)}).$$

Note that

$$(8.149) \qquad |\widehat{V}^{(k)}(t)|_{\hat{g}^{(k)}(t)}(x) = |\tilde{V}^{(k)}|_g(\varphi_{|r_k|t}(x)).$$

Then (8.145) and (8.146) imply that

$$(8.150) \qquad \frac{\partial}{\partial t}\widehat{V}^{(k)}(t) = |r_k|^{1/2} \mathcal{L}_{\hat{X}^{(k)}(t)}\widehat{V}^{(k)}(t)$$

$$= \Delta_{\hat{g}^{(k)}(t)}\widehat{V}^{(k)}(t) + \mathrm{Ric}_{\hat{g}^{(k)}(t)}(\widehat{V}^{(k)}(t)) - \widehat{Q}^{(k)}(t),$$

where

(8.151)
$$\hat{X}^{(k)}(t) := |r_k|^{1/2}\varphi^*_{|r_k|t}(\nabla_g f)$$

and

(8.152)
$$\widehat{Q}^{(k)}(t) := |r_k|^{3/2}(A^{(k)}(r_k) + |r_k|^{\frac{1}{2}-\varepsilon})^{-1}\varphi^*_{|r_k|t}(Q).$$

Let $\delta \in (0, \frac{1}{2})$. We claim that, by (8.144),

(8.153)
$$\sup_{t\in[\delta,1-\delta]}\left(\sup_{\{r_k-\delta^{-1}\sqrt{|r_k|}\leq f\leq r_k+\delta^{-1}\sqrt{|r_k|}\}}|\widehat{V}^{(k)}(t)|_{\hat{g}^{(k)}(t)}\right) \leq 1$$

for k sufficiently large. Moreover,
(8.154)
$$\limsup_{k\to\infty}\left(\sup_{t\in[\delta,1-\delta]}\left(\sup_{\{r_k-\delta^{-1}\sqrt{|r_k|}\leq f\leq r_k+\delta^{-1}\sqrt{|r_k|}\}}|\widehat{Q}^{(k)}(t)|_{\hat{g}^{(k)}(t)}\right)\right) = 0.$$

Indeed, via pulling back by the diffeomorphism $\varphi_{|r_k|t}$ and rescaling by $|r_k|^{-1}$, the metric $\hat{g}^{(k)}(t)$ on the set $\{r_k - \delta^{-1}\sqrt{|r_k|} \leq f \leq r_k + \delta^{-1}\sqrt{|r_k|}\}$ corresponds to the metric g on the set

$$\{(1-t)r_k - \delta^{-1}\sqrt{|r_k|} \lesssim f \lesssim (1-t)r_k + \delta^{-1}\sqrt{|r_k|}\} \subset \{f \geq r_k\},$$

where $\lesssim$ denotes "less than approximately". This is why (8.144) and (8.149) give (8.153) and why $|Q| \leq O(r^{-\frac{1}{2}-2\varepsilon})$ gives (8.154).

Choose p_k such that $f(p_k) = r_k$. Since $p_k \to \infty$, we have by Corollary 8.14 that $(\mathcal{M}^3, \hat{g}^{(k)}(t), p_k)$ subconverges in the pointed C^∞ Cheeger–Gromov sense to $(\mathbb{S}^2 \times \mathbb{R}, \bar{g}(t), p_\infty)$, $t \in (-\infty, 1)$. By the time-dependent version of Lemma 8.17 we have that $\hat{X}^{(k)}(t)$ subconverges to the vector field $\frac{\partial}{\partial z}$ on $\mathbb{S}^2 \times \mathbb{R}$ for each t. In particular, $|r_k|^{1/2}\nabla_g f$ subconverges to $\frac{\partial}{\partial z}$. This implies that under the above Cheeger–Gromov convergence, for $z \in \mathbb{R}$ the sequences of diffeomorphisms $\varphi_{|r_k|^{1/2}z}$ of $\mathcal{M}^3$ converge as $k \to \infty$ to the isometries $\psi_z : \mathbb{S}^2 \times \mathbb{R} \to \mathbb{S}^2 \times \mathbb{R}$ of $\bar{g}(t)$ generated by the vector field $\frac{\partial}{\partial z}$.

By (8.150), (8.153), and (8.154), the vector fields $\widehat{V}^{(k)}(t)$ subconverge in C^0_{loc} to a vector field $\bar{V}(t)$ on $\mathbb{S}^2 \times \mathbb{R}$ satisfying

(8.155)
$$\frac{\partial}{\partial t}\bar{V}(t) = \Delta_{\bar{g}(t)}\bar{V}(t) + \text{Ric}_{\bar{g}(t)}(\bar{V}(t)).$$

By (8.148), we have

$$\widehat{V}^{(k)}(t + |r_k|^{-1/2}z) = |r_k|^{1/2}\varphi^*_{|r_k|t+|r_k|^{1/2}z}(\tilde{V}^{(k)}) = \varphi^*_{|r_k|^{1/2}z}\widehat{V}^{(k)}(t).$$

Taking the limit as $k \to \infty$, we obtain for $z \in \mathbb{R}$,

$$\bar{V}(t) = \psi_z^* \bar{V}(t),$$

since $\varphi_{|r_k|^{1/2}z} \to \psi_z$ under the Cheeger–Gromov convergence. By (8.144),

$$|\bar{V}(t)|_{\bar{g}(t)} \leq 1 \quad \text{for } t \in \left(0, \tfrac{1}{2}\right].$$

Hence, by (8.155) and Lemma 8.36,

$$(8.156) \qquad \inf_{\lambda \in \mathbb{R}} \left(\sup_{\mathbb{S}^2 \times \mathbb{R}} \left| \bar{V}(t) - \lambda \frac{\partial}{\partial z} \right|_{\bar{g}(t)} \right) \leq C\, (1 - t)^{1/2} \quad \text{for } t \in \left[\tfrac{1}{2}, 1\right).$$

On the other hand, by (8.147) and (8.148) and since $\varphi_t^{-1} = \varphi_{-t}$, we have

(8.157)

$$\inf_{\lambda \in \mathbb{R}} \left(\sup_{\varphi_{|r_k|(\tau-1)}(\{f=\tau r_k\})} \left| \widehat{V}^{(k)}(1 - \tau) + \lambda |r_k|^{1/2} \varphi_{|r_k|(1-\tau)}^* (\nabla_g f) \right|_{\hat{g}^{(k)}(1-\tau)} \right)$$

$$= \inf_{\lambda \in \mathbb{R}} \left(\sup_{\{f=\tau r_k\}} \left| \tilde{V}^{(k)} + \lambda \nabla f \right|_g \right)$$

$$= (A^{(k)}(r_k) + |r_k|^{\frac{1}{2}-\varepsilon})^{-1} \inf_{\lambda \in \mathbb{R}} \left(\sup_{\{f=\tau r_k\}} |V^{(k)} + \lambda \nabla f|_g \right)$$

$$= \frac{A^{(k)}(\tau r_k)}{A^{(k)}(r_k) + |r_k|^{\frac{1}{2}-\varepsilon}}$$

$$\geq \frac{1}{2} \tau^{\frac{1}{2}-\varepsilon},$$

where we used (8.143) for the second equality and we used the contradiction hypothesis (8.142) for the last inequality. Taking the limit of this as $k \to \infty$ while using the subconvergence in (8.151), we conclude that

$$(8.158) \qquad \inf_{\lambda \in \mathbb{R}} \left(\sup_{\mathbb{S}^2 \times \mathbb{R}} \left| \bar{V}(1 - \tau) - \lambda \frac{\partial}{\partial z} \right|_{\bar{g}(1-\tau)} \right) \geq \frac{1}{2} \tau^{\frac{1}{2}-\varepsilon}.$$

Since $\tau^{-\varepsilon} > 2C$, we have $\frac{1}{2} \tau^{\frac{1}{2}-\varepsilon} > C\tau^{\frac{1}{2}}$. This and $t = 1 - \tau \in (\frac{1}{2}, 1)$ imply that (8.158) contradicts (8.156). The proof of the lemma is complete.   □

By iterating the above estimate, essentially using the dyadic-type intervals $\left[\tau^j s_k, \tau^{j+1} s_k\right]$, we obtain the following decay estimate.

Proposition 8.39. *There exists $\lambda_k \in \mathbb{R}$ such that*

$$(8.159) \qquad \sup_{k \geq 1} \left(\sup_{\{f \geq s_k\}} |f|^{-\frac{1}{2}+\varepsilon} |V^{(k)} + \lambda_k \nabla f| \right) < \infty.$$

That is, there exists a constant C such that for each $k \geq 1$ there exists $W^{(k)}$ in the ∇f-equivalence class of $V^{(k)}$ such that $|W^{(k)}| \leq C|f|^{\frac{1}{2}-\varepsilon}$ on $\{f \geq s_k\}$.

Proof. Let C be as in Lemma 8.36 and let $\tau \in (0, \frac{1}{2})$ satisfy $\tau^{-\varepsilon} > 2C$ as in Lemma 8.38, so that we have (8.141). Since $V^{(k)}$ satisfies (8.139), by Lemma 8.35 we have

$$\sup_{\{s_k \leq s \leq s_0\}} A^{(k)}(s) \leq \sup_{\{f \geq s_k\}} |V^{(k)}| \leq C|s_k|^{\frac{1}{2}-2\varepsilon}.$$

Now, iterating (8.141) from Lemma 8.38 yields for $k \geq k_0$ and all $j \geq 1$ that

$$(8.160) \qquad 2^j |\tau^j s|^{-\frac{1}{2}+\varepsilon} A^{(k)}\left(\tau^j s\right) \leq |s|^{-\frac{1}{2}+\varepsilon} A^{(k)}(s) + 2^j - 1$$

for $s \in [s_k, \tau s_k]$ with $\tau^j s \leq s_0$. Indeed, for $j = 1$ this is just (8.141). By induction, if (8.160) holds for some $j \geq 1$, then

$$\begin{aligned}
2^{j+1} |\tau^{j+1} s|^{-\frac{1}{2}+\varepsilon} A^{(k)}\left(\tau^{j+1} s\right) &= 2 \cdot 2^j |\tau^j (\tau s)|^{-\frac{1}{2}+\varepsilon} A^{(k)}\left(\tau^j (\tau s)\right) \\
&\leq 2 \left(|\tau s|^{-\frac{1}{2}+\varepsilon} A^{(k)}(\tau s) + 2^j - 1 \right) \\
&\leq |s|^{-\frac{1}{2}+\varepsilon} (A^{(k)}(s) + |s|^{\frac{1}{2}-\varepsilon}) + 2^{j+1} - 2 \\
&= |s|^{-\frac{1}{2}+\varepsilon} A^{(k)}(s) + 2^{j+1} - 1,
\end{aligned}$$

where we used (8.141) again in the third line. This proves (8.160).

Rewriting (8.160), we have

$$|\tau^j s|^{-\frac{1}{2}+\varepsilon} A^{(k)}\left(\tau^j s\right) \leq 2^{-j} |s|^{-\frac{1}{2}+\varepsilon} A^{(k)}(s) + 1 \leq 2^{-j} C + 1$$

for all $k \geq k_0$, $j \geq 1$, and $s \in [s_k, \tau s_k]$ with $\tau^j s \leq s_0$. We deduce the uniform weighted estimate

$$(8.161) \qquad \sup_{k \geq k_0} \left(\sup_{s_k \leq s \leq s_0} |s|^{-\frac{1}{2}+\varepsilon} A^{(k)}(s) \right) < \infty.$$

Note that we still have work to do since the norm-minimizing λ depends on s.

Let $\bar{s}_1 < s_0$ be such that $\sup_{\{f=\bar{s}_1\}} |\nabla f| \geq \frac{1}{2}$. By definition, there exist $\lambda_k \in \mathbb{R}$ such that

$$A^{(k)}(\bar{s}_1) = \sup_{\{f=\bar{s}_1\}} |V^{(k)} + \lambda_k \nabla f|.$$

By Lemma 8.35, for $\lambda \in \mathbb{R}$ and $s \in [s_k, \bar{s}_1]$ we have

$$\sup_{\{f=\bar{s}_1\}} |V^{(k)} + \lambda \nabla f| \leq \sup_{\{f \geq s\}} |V^{(k)} + \lambda \nabla f|$$

$$\leq \sup_{\{f=s\}} |V^{(k)} + \lambda \nabla f| + C|s|^{\frac{1}{2}-2\varepsilon}.$$

Now

$$\sup_{\{f=s\}} |V^{(k)} + \lambda_k \nabla f| \leq \sup_{\{f=s\}} |V^{(k)} + \lambda \nabla f| + |\lambda - \lambda_k|$$

$$\leq \sup_{\{f=s\}} |V^{(k)} + \lambda \nabla f| + 2 \sup_{\{f=\bar{s}_1\}} |(\lambda - \lambda_k) \nabla f|$$

$$\leq \sup_{\{f=s\}} |V^{(k)} + \lambda \nabla f| + 2 \sup_{\{f=\bar{s}_1\}} |V^{(k)} + \lambda_k \nabla f|$$

$$+ 2 \sup_{\{f=\bar{s}_1\}} |V^{(k)} + \lambda \nabla f|.$$

Combining the above two displays, we obtain for any $\lambda \in \mathbb{R}$ and $s \in [s_k, \bar{s}_1]$,

$$\sup_{\{f=s\}} |V^{(k)} + \lambda_k \nabla f| \leq 3 \sup_{\{f=s\}} |V^{(k)} + \lambda \nabla f| + 2A^{(k)}(\bar{s}_1) + 2C|s|^{\frac{1}{2}-2\varepsilon}.$$

By taking the infimum over all $\lambda \in \mathbb{R}$, we have

$$\sup_{\{f=s\}} |V^{(k)} + \lambda_k \nabla f| \leq 3A^{(k)}(s) + 2A^{(k)}(\bar{s}_1) + 2C|s|^{\frac{1}{2}-2\varepsilon}.$$

Hence (8.161) implies a uniform weighted estimate on the exhaustion of $\{f \leq \bar{s}_1\}$:

$$(8.162) \qquad \sup_{k \geq k_0} \left(\sup_{\{s_k \leq s \leq \bar{s}_1\}} \left(\sup_{\{f=s\}} |s|^{-\frac{1}{2}+\varepsilon} \left| V^{(k)} + \lambda_k \nabla f \right| \right) \right) < \infty;$$

that is,

$$(8.163) \qquad \sup_{k \geq k_0} \left(\sup_{\{s_k \leq f \leq \bar{s}_1\}} |f|^{-\frac{1}{2}+\varepsilon} \left| V^{(k)} + \lambda_k \nabla f \right| \right) < \infty.$$

On the other hand, from (8.163) and applying Lemma 8.35 to $V^{(k)} + \lambda_k \nabla f$, we conclude that

$$\sup_{\{f \geq \bar{s}_1\}} \left| V^{(k)} + \lambda_k \nabla f \right| \leq \sup_{\{f=\bar{s}_1\}} \left| V^{(k)} + \lambda_k \nabla f \right| + C|\bar{s}_1|^{\frac{1}{2}-2\varepsilon}$$

$$\leq C|\bar{s}_1|^{\frac{1}{2}-\varepsilon} + C|\bar{s}_1|^{\frac{1}{2}-2\varepsilon}$$

independent of $k \geq k_0$. That is, we have a uniform estimate on the compact set $\{f \geq \bar{s}_1\}$:

$$(8.164) \qquad \sup_{k \geq k_0} \left(\sup_{\{f \geq \bar{s}_1\}} |V^{(k)} + \lambda_k \nabla f| \right) < \infty.$$

The proposition follows from (8.163) and (8.164). $\qquad\qquad\qquad \Box$

With the above uniform weighted estimates, we can conclude this section with the following:

Proof of Theorem 8.34. (Solving the inhomogeneous Δ_f equation on $\mathcal{M}$.) Let λ_k be as in Proposition 8.39. By (8.117) and (8.139a),

$$(8.165) \qquad \Delta_f(V^{(k)} + \lambda_k \nabla f) = Q \quad \text{in } \{f \geq s_k\}.$$

By (8.159), there exists a constant C such that for all $k \geq 1$ and for all $x \in \{f \geq s_k\}$,

$$|V^{(k)} + \lambda_k \nabla f|(x) \leq C |f(x)|^{\frac{1}{2} - \varepsilon}.$$

By this uniform weighted bound and by standard interior estimates for elliptic systems applied to (8.165), there exists a subsequence such that $V^{(k)} + \lambda_k \nabla f$ converges in each C^m on compact sets to a C^∞ vector field V on $\mathcal{M}$ satisfying

$$\Delta_f V = Q, \quad |V| \leq O\left(r^{\frac{1}{2} - \varepsilon}\right).$$

We shall now show that $|\nabla V| \leq O(r^{-\varepsilon})$. Let $\rho_i \to \infty$ and define

$$\hat{g}^{(i)}(t) := \rho_i^{-1} \varphi_{\rho_i t}^*(g) = \rho_i^{-1} g(\rho_i t) \quad \text{for } t \in \left[-\tfrac{1}{2}, 0 \right].$$

Define

$$\widehat{V}^{(i)}(t) := \varphi_{\rho_i t}^*(V) \quad \text{and} \quad \widehat{Q}^{(i)}(t) := \rho_i \varphi_{\rho_i t}^*(Q) \quad \text{for } t \in \left[-\tfrac{1}{2}, 0 \right].$$

Then

$$(8.166) \qquad \frac{\partial}{\partial t} \widehat{V}^{(i)}(t) = \Delta_{\hat{g}^{(i)}(t)} \widehat{V}^{(i)}(t) + \mathrm{Ric}_{\hat{g}^{(i)}(t)}(\widehat{V}^{(i)}(t)) - \hat{Q}^{(i)}(t).$$

By applying standard interior estimates for parabolic systems to (8.166), we have

$$\sup_{|f| = \rho_i} |\nabla \widehat{V}^{(i)}(0)|_{\hat{g}^{(i)}(0)} \leq C \sup_{t \in [-\frac{1}{2}, 0]} \left(\sup_{\rho_i - C\sqrt{\rho_i} \leq |f| \leq \rho_i + C\sqrt{\rho_i}} |\widehat{V}^{(i)}(t)|_{\hat{g}^{(i)}(t)} \right)$$

$$+ C \sup_{t \in [-\frac{1}{2}, 0]} \left(\sup_{\rho_i - C\sqrt{\rho_i} \leq |f| \leq \rho_i + C\sqrt{\rho_i}} |\widehat{Q}^{(i)}(t)|_{\hat{g}^{(i)}(t)} \right).$$

Since $|V| \leq O(r^{\frac{1}{2}-\varepsilon})$, where $\varepsilon > 0$, we have from $\widehat{V}^{(i)}(t) = \varphi_{\rho_i t}^*(V)$ and $\hat{g}^{(i)}(t) = \rho_i^{-1}\varphi_{\rho_i t}^*(g)$ that

$$\sup_{t\in[-\frac{1}{2},0]}\left(\sup_{\rho_i-C\sqrt{\rho_i}\leq|f|\leq\rho_i+C\sqrt{\rho_i}}|\widehat{V}^{(i)}(t)|_{\hat{g}^{(i)}(t)}\right) \leq O(\rho_i^{-\varepsilon}).$$

Since $|Q| \leq O(r^{-\frac{1}{2}-2\varepsilon})$, we have from $\widehat{Q}^{(i)}(t) = \rho_i\varphi_{\rho_i t}^*(Q)$ that

$$\sup_{t\in[-\frac{1}{2},0]}\left(\sup_{\rho_i-C\sqrt{\rho_i}\leq|f|\leq\rho_i+C\sqrt{\rho_i}}|\widehat{Q}^{(i)}(t)|_{\hat{g}^{(i)}(t)}\right) \leq O(\rho_i^{-2\varepsilon}).$$

We conclude that

$$\sup_{|f|=\rho_i} |\nabla V|_g = \sup_{|f|=\rho_i} |\nabla\widehat{V}^{(i)}(0)|_{\hat{g}^{(i)}(0)} \leq O(\rho_i^{-\varepsilon}).$$

Since $\rho_i \to \infty$ is arbitrary, this implies $|\nabla V| \leq O(r^{-\varepsilon})$. The proof of the theorem is complete. $\qquad\square$

8.7. Uniqueness theorem for the f-Lichnerowicz Laplace equation

The main result of this section is Proposition 8.40, which may be viewed as a Liouville-type theorem for the f-Lichnerowicz Laplace equation.

Let $(\mathcal{M}^3, g(t), f(t))$, $t \in (-\infty, \infty)$, be the canonical form associated to the steady GRS $(\mathcal{M}^3, g, f)$ given by Proposition 4.19. We have

$$(8.167) \qquad g(t) = \varphi_t^* g \quad \text{and} \quad f(t) = f \circ \varphi_t,$$

where $\varphi_0 = \mathrm{id}$ and $\frac{\partial}{\partial t}\varphi_t(x) = (\nabla_g f)(\varphi_t(x))$ for $t \in (-\infty, \infty)$. Moreover, $g(t)$ is a solution to the Ricci flow evolving by diffeomorphisms:

$$(8.168) \qquad \frac{\partial}{\partial t}g(t) = -2\operatorname{Ric}_{g(t)} = \nabla_{g(t)}^2 f(t).$$

8.7.1. Uniqueness theorem for the linearized GRS equation.

Let $(\mathcal{M}^3, g, f)$ be a steady GRS. Let $\Delta_{L,g}$ be the Lichnerowicz Laplacian with respect to g, defined by (4.79). The linearized GRS equation is

$$(8.169) \qquad \Delta_{L,g,f}h := \Delta_{L,g}h - \mathcal{L}_{\nabla_g f}h = 0.$$

Note that, by (4.80), $a\operatorname{Ric} = -\frac{a}{2}\mathcal{L}_{\nabla f}g$ is a solution of (8.169) for any $a \in \mathbb{R}$. Hence h is a solution of (8.169) if and only if $h+a\operatorname{Ric}$ is a solution of (8.169). If two solutions h_1 and h_2 of (8.169) differ by a constant multiple of Ric, then we say that h_1 and h_2 are Ric-**equivalent**. Since this condition is the same as h_1 and h_2 differing by a constant multiple of $\mathcal{L}_{\nabla f}g$, we may also say that h_1 and h_2 are $\mathcal{L}_{\nabla f}g$-**equivalent**.

Note also that if φ is a diffeomorphism, then

$$(8.170) \quad \Delta_{L,\varphi^*g}(\varphi^*h) = \varphi^*(\Delta_{L,g}h) \quad \text{and} \quad \mathcal{L}_{\nabla_{\varphi^*g}f\circ\varphi}\varphi^*h = \varphi^*(\mathcal{L}_{\nabla_g f}h).$$

Hence, if h satisfies $\Delta_{L,g,f}h = 0$, then φ^*h satisfies $\Delta_{L,\varphi^*g,f\circ\varphi}(\varphi^*h) = 0$.

In this section we shall prove the following uniqueness theorem for solutions to the linearized GRS equation.

Proposition 8.40. *Let* $(\mathcal{M}^3, g, f)$ *be a nonflat noncompact steady GRS which is κ-noncollapsed at below scales for some $\kappa > 0$. If h is a solution to (8.169) with $|h|(x) \le Cr(x)^{-\varepsilon}$ for some $\varepsilon > 0$, then there exists $b \in \mathbb{R}$ such that*

$$h = b\operatorname{Ric}_g = -\frac{b}{2}\mathcal{L}_{\nabla f}g.$$

That is, h is Ric-equivalent to the zero tensor, which is the same as saying that h is $\mathcal{L}_{\nabla f}g$-equivalent to the zero tensor.

The remainder of this subsection is devoted to the proof of the proposition.

Let h be a solution to (8.169). Let $(\mathcal{M}^3, g(t), f(t), \varphi_t)$, $t \in (-\infty, \infty)$, be the canonical form associated to $(\mathcal{M}^3, g, f)$. Define $h(t) := \varphi_t^*h$. Then, using (8.170), we see that

$$\frac{\partial h}{\partial t}(t) = \mathcal{L}_{\nabla_{g(t)}f(t)}h(t) = \Delta_{L,g(t)}h(t);$$

in particular,

$$\Delta_{L,g(t),f(t)}h(t) = 0.$$

The following is a result of Greg Anderson and the author [3]:

Lemma 8.41. *If $(\mathcal{M}^n, g(t), h(t))$ is a solution to the linearized Ricci flow $\frac{\partial}{\partial t}g = -2\operatorname{Ric}$ and $\frac{\partial}{\partial t}h = \Delta_L h$ with positive scalar curvature, then*

$$(8.171) \quad \frac{\partial}{\partial t}\left(\frac{|h|^2}{R^2}\right) = \Delta\left(\frac{|h|^2}{R^2}\right) + \frac{2}{R}\left\langle \nabla R, \nabla\left(\frac{|h|^2}{R^2}\right)\right\rangle$$

$$- \frac{2}{R^4}|R\nabla_i h_{jk} - \nabla_i R h_{jk}|^2 + 4P,$$

where

$$P := \frac{1}{R^3}\left(RR_{ijk\ell}h_{i\ell}h_{jk} - |h|^2|\operatorname{Ric}|^2\right).$$

Moreover, when $n = 3$, by (8.43) we have

(8.172)
$$P = \frac{1}{R^3}\left(2R\left\langle \mathrm{Ric}, h\right\rangle H - 2R\left\langle \mathrm{Ric}, h^2\right\rangle + \frac{1}{2}R^2\left(|h|^2 - H^2\right) - |h|^2\left|\mathrm{Ric}\right|^2\right),$$

where $H := \mathrm{tr}\, h$. Moreover, $P \le 0$. Hence, for $n = 3$,

(8.173)
$$\frac{\partial}{\partial t}\left(\frac{|h|^2}{R^2}\right) \le \Delta\left(\frac{|h|^2}{R^2}\right) + \frac{2}{R}\left\langle \nabla R, \nabla\left(\frac{|h|^2}{R^2}\right)\right\rangle.$$

Remark 8.42. A simplified proof that $P \le 0$ is given in Bamler and Kleiner [**29**], where the main result of their paper is a proof of the uniqueness of 3-dimensional Ricci flow surgery, as conjectured by Perelman. The applications of their proof of the generalized Smale conjecture on the diffeomorphism groups of 3-dimensional spherical space forms and prime 3-manifolds are in [**26, 27, 30**].

Let h be a solution to (8.169) on a steady GRS $\left(\mathcal{M}^3, g, f\right)$ with $R > 0$. Then, by (8.173) and $\frac{\partial}{\partial t}\frac{|h|^2}{R^2} = \left\langle \nabla f, \nabla\frac{|h|^2}{R^2}\right\rangle$ for $(g(t), f(t), h(t))$ evaluated at $t = 0$, we see that the triple (g, f, h) satisfies

(8.174)
$$\Delta_f\frac{|h|^2}{R^2} + \left\langle\frac{2\nabla R}{R}, \nabla\frac{|h|^2}{R^2}\right\rangle \ge 0.$$

From applying the elliptic maximum principle (Lemma B.1) to this inequality, we immediately obtain:

Lemma 8.43. *Let Ω be a bounded domain in $\mathcal{M}^3$. If h is a solution to (8.169) on a steady GRS $\left(\mathcal{M}^3, g, f\right)$ with $R > 0$, then*

$$\sup_{\Omega}\frac{|h|}{R} \le \sup_{\partial\Omega}\frac{|h|}{R}.$$

In particular, for any $s < -1$,

(8.175)
$$\sup_{\{f \ge s\}}\frac{|h|}{R} = \sup_{\{f = s\}}\frac{|h|}{R}.$$

Hence there exists a constant B such that

(8.176)
$$\sup_{\{f \ge s\}}|f|\,|h| \le B\,|s|\,\sup_{\{f = s\}}|h|.$$

We now give the following:

Proof of Proposition 8.40. Define

(8.177)
$$A(s) := \inf_{a \in \mathbb{R}}\,\sup_{\{f = s\}}|h - a\,\mathrm{Ric}|.$$

Since $\{f = s\}$ is compact, the infimum in (8.177) is attained for each s. Observe also that $A(s) \le \sup_{\{f=s\}}|h|$. We have applied a similar device to (8.140) for vector fields. In particular, $A(s)$ is the minimal L^∞ norm on $\{f = s\}$ of symmetric 2-tensors in the Ric-equivalence class of h.

CASE 1 (Fast decay): *There exists $s_i \to -\infty$ such that $|s_i| A(s_i) \to 0$.* Since the infimum is attained, for each i there exists $a_i \in \mathbb{R}$ such that

$$\sup_{\{f=s_i\}} |h - a_i \operatorname{Ric}_g| = A(s_i).$$

On the other hand, since $h - a_i \operatorname{Ric}_g$ is a solution to (8.169), by Lemma 8.43 we have

$$\sup_{\{f \geq s_i\}} \frac{|h - a_i \operatorname{Ric}_g|}{R} \leq \sup_{\{f = s_i\}} \frac{|f| \, |h - a_i \operatorname{Ric}_g|}{|f| \, R} \leq |s_i| A(s_i) \sup_{\{f = s_i\}} \frac{1}{|f| \, R}.$$

Since $\sup_{\{f=s_i\}} \frac{1}{|f|R} \to 1$ by Lemma 8.13 and since $|s_i| A(s_i) \to 0$, we conclude that

$$\sup_{\{f \geq s_i\}} \frac{|h - a_i \operatorname{Ric}_g|}{R} \to 0 \quad \text{as } i \to \infty.$$

Since $\operatorname{Ric}_g \not\equiv 0$, it is easy to see from this that $\{a_i\}$ converges to some $b \in \mathbb{R}$ and that

$$h = b \operatorname{Ric}_g \quad \text{on } \mathcal{M}^3.$$

CASE 2: *There exists $c > 0$ such that $|s| A(s) \geq c$ for all sufficiently large $|s|$.* In particular, $A(s) > 0$ for $s \ll -1$. The idea is to obtain a contradiction to the elliptic maximum principle estimate of Lemma 8.43 to rule out this case.

Let B be as in (8.176) and let C be as in estimate (8.186) below for the linearized equation, so that

$$\sup_{\{f \geq s\}} |f| \, |h| \leq B |s| \sup_{\{f=s\}} |h|,$$

$$\inf_{\lambda \in \mathbb{R}} \left(\sup_{\mathbb{S}^2 \times \mathbb{R}} \left| \bar{h}(t) - \lambda \operatorname{Ric}_{\bar{g}(t)} \right|_{\bar{g}(t)} \right) \leq C \quad \text{for } t \in \left[\frac{1}{2}, 1 \right].$$

Choose $\tau \in (0, \frac{1}{2})$ so that $\tau^{-\varepsilon} > 2BC$.

Since $|h|(x) \leq Cr(x)^{-\varepsilon}$ by hypothesis and since $|f|(x) \leq Cr(x)$, there exists C such that

$$|s|^\varepsilon A(s) \leq |s|^\varepsilon \sup_{\{f=s\}} |h| \leq C.$$

Hence there exist $r_k \to -\infty$ such that

$$|r_k|^\varepsilon A(r_k) \leq 2 |\tau r_k|^\varepsilon A(\tau r_k);$$

that is,

$$A(r_k) \leq 2\tau^\varepsilon A(\tau r_k).$$

By definition (8.177), there exist $\lambda_k \in \mathbb{R}$ such that

$$(8.178) \qquad A(r_k) = \sup_{\{f = r_k\}} |h - \lambda_k \operatorname{Ric}|.$$

Define

$$\tilde{h}^{(k)} := \frac{1}{A(r_k)} (h - \lambda_k \operatorname{Ric}).$$

Then $\sup_{\{f = r_k\}} |\tilde{h}^{(k)}| = 1$. By (8.175),

$$\sup_{\{f \geq r_k\}} \frac{|\tilde{h}^{(k)}|}{R} = \sup_{\{f = r_k\}} \frac{|\tilde{h}^{(k)}|}{R} \leq \sup_{\{f = r_k\}} R^{-1} \leq O(|r_k|)$$

and

$$\Delta_L \tilde{h}^{(k)} - \mathcal{L}_{\nabla f} \tilde{h}^{(k)} = 0.$$

Since $\tilde{h}^{(k)}$ is a solution to (8.169), by (8.176) we have for $r \geq r_k$,

$$(8.179) \qquad |r| \sup_{\{f = r\}} |\tilde{h}^{(k)}| \leq B |r_k| \sup_{\{f = r_k\}} |\tilde{h}^{(k)}|$$
$$= \frac{B |r_k|}{A(r_k)} \sup_{\{f = r_k\}} |h - \lambda_k \operatorname{Ric}|$$
$$= B |r_k|$$

by (8.178).

Define

$$\hat{g}^{(k)}(t) := |r_k|^{-1} \varphi_{|r_k| t}^* (g) = |r_k|^{-1} g(r_k t)$$

and

$$\hat{h}^{(k)}(t) := |r_k|^{-1} \varphi_{|r_k| t}^* (\tilde{h}^{(k)})$$
$$= \frac{1}{|r_k| A(r_k)} \varphi_{|r_k| t}^* (h - \lambda_k \operatorname{Ric}).$$

Then

$$\frac{\partial}{\partial t} \hat{h}^{(k)}(t) = \Delta_{L, \hat{g}^{(k)}(t)} \hat{h}^{(k)}(t).$$

By (8.179) we have that for any $\delta \in (0, \tfrac{1}{2})$,

$$\limsup_{k \to \infty} \left(\sup_{t \in [\delta, 1-\delta]} \left(\sup_{\{r_k - \delta^{-1} |r_k|^{1/2} \leq f \leq r_k + \delta^{-1} |r_k|^{1/2}\}} |\hat{h}^{(k)}(t)|_{\hat{g}^{(k)}(t)} \right) \right) < \infty.$$

Choosing any point $p_k \in \Sigma_{r_k}$, by Perelman's compactness theorem (Theorem 8.6) there exists a subsequence such that $(\mathcal{M}^3, \hat{g}^{(k)}(t), p_k)$ converges in the pointed C^∞ Cheeger–Gromov sense to a standard cylinder solution $(\mathbb{S}^2 \times \mathbb{R}, \bar{g}(t), \bar{p})$, $t \in (0, 1)$. By Lemma 8.17, the vector fields $|r_k|^{1/2} \nabla f$ subconverge to $\frac{\partial}{\partial z}$.

Passing to a subsequence, we have that the $\hat{h}^{(k)}(t)$ converge to a 1-parameter family of symmetric 2-tensors $\bar{h}(t)$ (with respect to the Cheeger–Gromov convergence of $\hat{g}^{(k)}(t)$ to $\bar{g}(t)$), $t \in (-\infty, 1)$, on the cylinder satisfying

$$\frac{\partial}{\partial t}\bar{h}(t) = \Delta_{L,\bar{g}(t)}\bar{h}(t).$$

Let $\psi_s : \mathbb{S}^2 \times \mathbb{R} \to \mathbb{S}^2 \times \mathbb{R}$ be the 1-parameter group of isometries $\psi_s : \mathbb{S}^2 \times \mathbb{R} \to \mathbb{S}^2 \times \mathbb{R}$ of $\bar{g}(t)$ generated by $\frac{\partial}{\partial z}$; i.e., $\psi_s(x, z) = (x, z + s)$. Then, with respect to the Cheeger–Gromov convergence of $\hat{g}^{(k)}(t)$ to $\bar{g}(t)$, we have that the $\varphi_{|r_k|^{1/2}s}$ converge to the ψ_s uniformly in any C^m norm on compact sets. Note that

$$\begin{aligned}
\varphi^*_{\sqrt{|r_k|}s}(\hat{h}^{(k)}(t)) &= |r_k|^{-1}\varphi^*_{|r_k|t+|r_k|^{1/2}s}(\tilde{h}^{(k)}) \\
&= |r_k|^{-1}\varphi^*_{|r_k|\left(t+\frac{s}{|r_k|^{1/2}}\right)}(\tilde{h}^{(k)}) \\
&= \hat{h}^{(k)}\left(t + \frac{s}{|r_k|^{1/2}}\right).
\end{aligned}$$

Hence, taking the limit as $k \to \infty$, we have that the left-hand side converges to $\psi_s(\bar{h}(t))$, whereas the right-hand side converges to $\bar{h}(t)$. We conclude that for all s, the 2-tensor $\bar{h}(t)$ is translation invariant along the $\mathbb{R}$ factor of $\mathbb{S}^2 \times \mathbb{R}$; i.e.,

$$\psi_s^*(\bar{h}(t)) = \bar{h}(t).$$

We claim that by (8.179), we have for $t \in (0, \frac{1}{2}]$,

(8.180) $$(1 - t)\left|\bar{h}(t)\right|_{\bar{g}(t)} \leq B.$$

To see this, let $r \in [r_k - \delta^{-1}\sqrt{|r_k|}, r_k + \delta^{-1}\sqrt{|r_k|}]$ and $t \in [\delta, \frac{1}{2}]$. We have that

(8.181) $$\sup_{\{f=r\}} (1 - t)|\hat{h}^{(k)}(t)|_{\hat{g}^{(k)}(t)} \leq (1 - t) \sup_{\varphi_{-|r_k|t}(\{f=r\})} |\tilde{h}^{(k)}|_g.$$

On the other hand,

$$\begin{aligned}
f|_{\varphi_{-|r_k|t}(\{f=r\})} &\leq r + \int_0^{|r_k|t} \sup |\nabla f| \\
&\leq r_k + \delta^{-1}\sqrt{|r_k|} + |r_k|t \\
&= -(1 - t)|r_k| + \delta^{-1}\sqrt{|r_k|},
\end{aligned}$$

so that (8.181) implies

$$\sup_{\{f=r\}} (1-t)|\hat{h}^{(k)}(t)|_{\hat{g}^{(k)}(t)} \leq \frac{1}{|r_k|} \sup_{\varphi_{-|r_k|t}(\{f=r\})} |f| |\tilde{h}^{(k)}|_g$$

$$+ \frac{1}{\delta\sqrt{|r_k|}} \sup_{\varphi_{-|r_k|t}(\{f=r\})} |\tilde{h}^{(k)}|_g.$$

By (8.179) and taking $k \to \infty$, we obtain (8.180).

Now, from estimate (8.186) in Lemma 8.44 below, we have

$$(8.182) \qquad \inf_{\lambda \in \mathbb{R}} \left(\sup_{\mathbb{S}^2 \times \mathbb{R}} \left| \bar{h}(t) - \lambda \operatorname{Ric}_{\bar{g}(t)} \right|_{\bar{g}(t)} \right) \leq BC \quad \text{for } t \in \left[\tfrac{1}{2}, 1\right).$$

On the other hand (compare with (8.157)),

$$\inf_{\lambda \in \mathbb{R}} \left(\sup_{\varphi_{-|r_k|(1-\tau)}(\{f=\tau r_k\})} |\hat{h}^{(k)}(1-\tau) - \lambda \operatorname{Ric}_{\hat{g}^{(k)}(1-\tau)}|_{\hat{g}^{(k)}(1-\tau)} \right)$$

$$= \inf_{\lambda \in \mathbb{R}} \left(\sup_{\{f=\tau r_k\}} |\tilde{h}^{(k)} - \lambda \operatorname{Ric}_g|_g \right)$$

$$= \frac{1}{A(r_k)} \inf_{\lambda \in \mathbb{R}} \left(\sup_{\{f=\tau r_k\}} |h - \lambda \operatorname{Ric}_g|_g \right)$$

$$= \frac{A(\tau r_k)}{A(r_k)}$$

$$\geq \frac{1}{2}\tau^{-\varepsilon}.$$

Taking the limit of this as $k \to \infty$, we conclude that

$$\inf_{\lambda \in \mathbb{R}} \left(\sup_{\mathbb{S}^2 \times \mathbb{R}} \left| \bar{h}(1-\tau) - \lambda \operatorname{Ric}_{\bar{g}(1-\tau)} \right|_{\bar{g}(1-\tau)} \right) \geq \frac{1}{2}\tau^{-\varepsilon}.$$

Since $\frac{1}{2}\tau^{-\varepsilon} > BC$, this contradicts (8.182). Since this contradiction rules out Case 2, Proposition 8.40 is proved. $\square$

8.7.2. Linearized Ricci flow on $\mathbb{S}^2 \times \mathbb{R}$.

In this subsection we consider the equation for symmetric 2-tensors which is the linearization of the steady Ricci soliton equation at spatial infinity via blow down. Recall that the standard product solution to the Ricci flow on $\mathbb{S}^2 \times \mathbb{R}$, forming a singularity at $t = 1$, is given by

$$(8.183) \qquad \bar{g}(t) = 2(1-t) g_{\mathbb{S}^2} + dz \otimes dz,$$

where $g_{\mathbb{S}^2}$ denotes the standard metric on $\mathbb{S}^2$. Note that $\operatorname{Ric}_{\bar{g}(t)} = g_{\mathbb{S}^2}$ and $R_{\bar{g}(t)} = \frac{1}{1-t}$.

We now state and prove the estimate for the linearized Ricci flow on $\mathbb{S}^2 \times \mathbb{R}$ that we used in the previous subsection.

Lemma 8.44. *Let $\bar{h}(t)$, $t \in (0,1)$, be a solution to*

$$\frac{\partial}{\partial t}\bar{h}(t) = \Delta_{L,\bar{g}(t)}\bar{h}(t),$$

where $(\mathbb{S}^2 \times \mathbb{R}, \bar{g}(t))$, $t \in (0,1)$, is the shrinking cylinder. Suppose that $\bar{h}(t)$ is translation invariant along the $\mathbb{R}$ factor of $\mathbb{S}^2 \times \mathbb{R}$; i.e.,

(8.184)
$$\psi_s^* \left(\bar{h}(t)\right) = \bar{h}(t),$$

where the family $\psi_s : \mathbb{S}^2 \times \mathbb{R} \to \mathbb{S}^2 \times \mathbb{R}$ is generated by $\frac{\partial}{\partial z}$; i.e., $\psi_s(x,z) = (x, z+s)$. If

(8.185)
$$\left|\bar{h}(t)\right|_{\bar{g}(t)} \le \frac{1}{1-t} \quad \text{for } t \in \left(0, \tfrac{1}{2}\right],$$

then there exists a constant C such that

(8.186)
$$\inf_{\lambda \in \mathbb{R}} \left(\sup_{\mathbb{S}^2 \times \mathbb{R}} \left|\bar{h}(t) - \lambda \operatorname{Ric}_{\bar{g}(t)}\right|_{\bar{g}(t)}\right) \le C \quad \text{for } t \in \left[\tfrac{1}{2}, 1\right).$$

Proof. By definition (4.79) and $(\operatorname{Rm}_{g_{\mathbb{S}^2}})_{kij\ell} = g_{k\ell}g_{ij} - g_{kj}g_{i\ell}$, the Lichnerowicz Laplacian on $\mathbb{S}^2$ is given by

(8.187)
$$\Delta_{L,g_{\mathbb{S}^2}} h = \Delta_{\mathbb{S}^2} h - 4\overset{\circ}{h},$$

where $\overset{\circ}{h} := h - \frac{1}{2}(\operatorname{tr}_{g_{\mathbb{S}^2}} h)g_{\mathbb{S}^2}$ denotes the trace-free part of a symmetric 2-tensor h on $\mathbb{S}^2$. By the translation invariance of (8.184), we may express the solution $\bar{h}(t)$, $t \in (0,1)$, as

(8.188) $\bar{h}(\theta, z, t) = \alpha(\theta, t) + \beta(\theta, t) \otimes dz + dz \otimes \beta(\theta, t) + \gamma(\theta, t)\, dz \otimes dz,$

where $\alpha(\cdot, t)$ is a symmetric 2-tensor on $\mathbb{S}^2$, $\beta(\cdot, t)$ is a 1-form on $\mathbb{S}^2$, and $\gamma(\cdot, t)$ is a function on $\mathbb{S}^2$. By (4.79) and (8.183), we have that

$$\Delta_{L,\bar{g}(t)}\bar{h} = \frac{1}{2(1-t)}(\Delta_{\mathbb{S}^2}\bar{h} - 4\overset{\circ}{\alpha}),$$

where we used that $\frac{\partial^2 \bar{h}}{\partial z^2} = 0$. Since $\Delta_{\mathbb{S}^2}$ preserves the decomposition (8.188), the equation $\frac{\partial}{\partial t}\bar{h} = \Delta_{L,\bar{g}(t)}\bar{h}$ is equivalent to the following system for (α, β, γ):

(8.189a)
$$\frac{\partial \alpha}{\partial t} = \frac{1}{2(1-t)}(\Delta_{\mathbb{S}^2}\alpha - 4\overset{\circ}{\alpha}),$$

(8.189b)
$$\frac{\partial \beta}{\partial t} = \frac{1}{2(1-t)}\Delta_{\mathbb{S}^2}\beta,$$

(8.189c)
$$\frac{\partial \gamma}{\partial t} = \frac{1}{2(1-t)}\Delta_{\mathbb{S}^2}\gamma.$$

By hypothesis (8.185) we have that

$$\frac{1}{(1-t)^2} \geq \left|\bar{h}(t)\right|^2_{\bar{g}(t)} = \frac{1}{(2(1-t))^2}\left|\alpha(t)\right|^2_{g_{\mathbb{S}^2}} + \frac{1}{2(1-t)}\left|\beta(t)\right|^2_{g_{\mathbb{S}^2}} + \gamma^2(t),$$

so that on $\mathbb{S}^2 \times (0, \frac{1}{2}]$ we have

(8.190a) $$\left|\alpha(t)\right|_{g_{\mathbb{S}^2}} \leq 2,$$

(8.190b) $$\left|\beta(t)\right|_{g_{\mathbb{S}^2}} \leq \frac{\sqrt{2}}{\sqrt{1-t}},$$

(8.190c) $$\left|\gamma(t)\right| \leq \frac{1}{1-t}.$$

Defining $\tau(t) = -\frac{1}{2}\ln(1-t)$, we have

(8.191) $$\frac{\partial \alpha}{\partial \tau} = \Delta_{\mathbb{S}^2}\alpha - 4\overset{\circ}{\alpha}.$$

Tracing this, we have that $A := \operatorname{tr}_{g_{\mathbb{S}^2}}\alpha$ satisfies

(8.192) $$\frac{\partial A}{\partial \tau} = \Delta_{\mathbb{S}^2}A.$$

To estimate $\alpha(t)$ we consider the lowest two eigenvalues and corresponding eigenspaces of the linear elliptic operator $L(\alpha) := \Delta_{\mathbb{S}^2}\alpha - 4\overset{\circ}{\alpha}$ on the right-hand side of (8.189a). Suppose that $\alpha \not\equiv 0$ satisfies

(8.193) $$(\Delta_{\mathbb{S}^2}\alpha - 4\overset{\circ}{\alpha}) + \mu\alpha = 0, \quad \mu \in \mathbb{R}.$$

Taking the trace-free part of this equation, we have that

$$\Delta_{\mathbb{S}^2}\overset{\circ}{\alpha} + (\mu - 4)\overset{\circ}{\alpha} = 0.$$

Hence, if $\overset{\circ}{\alpha} \not\equiv 0$, then $\mu \geq 4$. We also have that the trace A satisfies

(8.194) $$\Delta_{\mathbb{S}^2}A + \mu A = 0.$$

If $A \equiv 0$, then $\overset{\circ}{\alpha} \not\equiv 0$, so that $\mu \geq 4$. Now assume that $A \not\equiv 0$. Since the kth eigenvalue of Δ acting on functions of $\mathbb{S}^2$ is $k(k-1)$ for $k \geq 1$, (8.194) implies that $\mu \geq 0$ and that, unless A is constant, we have $\mu \geq 2$. Moreover, unless A is constant or a linear function on $\mathbb{R}^3$ restricted to $\mathbb{S}^2$, we have $\mu \geq 6$. Thus the first eigenvalue of L is 0. If $\Delta_{\mathbb{S}^2}\alpha - 4\overset{\circ}{\alpha} = 0$, then

$$0 = \int_{\mathbb{S}^2}\langle\alpha, \Delta_{\mathbb{S}^2}\alpha - 4\overset{\circ}{\alpha}\rangle d\mu_{\mathbb{S}^2} = -\int_{\mathbb{S}^2}(|\nabla\alpha|^2 + 4|\overset{\circ}{\alpha}|^2)d\mu_{\mathbb{S}^2}.$$

Hence, the associated eigenspace is

$$E_0 := \mathbb{R}g_{\mathbb{S}^2} := \{sg_{\mathbb{S}^2} : s \in \mathbb{R}\}.$$

We also see that the second eigenvalue of L is 2, with the associated eigenspace Vg, where V is the space of linear functions on $\mathbb{R}^3$ restricted to $\mathbb{S}^2$.

In general, if a tensor δ satisfies $\frac{\partial}{\partial t}\delta = \Delta_g \delta$ on a compact Riemannian manifold $(\mathcal{M}^n, g)$, where δ is perpendicular in L^2 to the eigenspaces with eigenvalue $\lambda < \ell_0$, where $\ell_0 \in \mathbb{R}$, then $|\delta|_g \leq Ce^{-\ell_0 t}$ (with our sign convention for eigenvalues of Δ_g). This follows from the eigentensor expansion of a solution to the heat equation.

Thus for any solution $\alpha(\tau)$ of (8.191), its projection onto the subspace perpendicular to the first eigenspace E_0 with respect to the L^2 metric on the space of symmetric 2-tensors satisfies the inequalities

(8.195)

$$\inf_{\lambda \in \mathbb{R}} \sup_{x \in \mathbb{S}^2} |\alpha(x,t) - \lambda g_{\mathbb{S}^2}(x)|_{g_{\mathbb{S}^2}} \leq \sup_{x \in \mathbb{S}^2} \left| \alpha(x,t) - \frac{1}{2} \frac{\int_{\mathbb{S}^2} A(t)\, d\mu_{\mathbb{S}^2}}{\int_{\mathbb{S}^2} d\mu_{\mathbb{S}^2}} g_{\mathbb{S}^2}(x) \right|_{g_{\mathbb{S}^2}}$$

$$\leq C_1 e^{-2\tau(t)}$$

$$= C_1(1-t)$$

for each $t \in [0,1)$. Note that by (8.192), we have that $\int_{\mathbb{S}^2} A(t)\, d\mu_{\mathbb{S}^2} / \int_{\mathbb{S}^2} d\mu_{\mathbb{S}^2}$ is independent of t.

Next, we estimate $\beta(t)$. Here, (8.189b) implies that $\frac{\partial \beta}{\partial \tau} = \Delta_{\mathbb{S}^2}\beta$. Recall that the lowest nontrivial eigenvalue for the Hodge–de Rham Laplacian Δ_d acting on 1-forms on $(\mathbb{S}^2, \bar{g})$ is equal to 2. Indeed, for closed 1-forms the eigenvalues are $\lambda_k^c = k(k+1)$ for $k \geq 1$ and for co-closed 1-forms the eigenvalues are also $\lambda_k^{cc} = k(k+1)$ for $k \geq 1$. Since $\Delta_d = \Delta_{\mathbb{S}^2} - 1$, where $\Delta_{\mathbb{S}^2}$ is the rough Laplacian, the lowest nontrivial eigenvalue for $\Delta_{\mathbb{S}^2}$ acting on 1-forms is 1. Hence we have

(8.196) $$|\beta|_{g_{\mathbb{S}^2}}(x,t) \leq C_2(1-t)^{1/2} \quad \text{on } \mathbb{S}^2 \times [0,1).$$

Finally, we estimate $\gamma(t)$. Since $\frac{\partial \gamma}{\partial \tau} = \Delta_{\mathbb{S}^2}\gamma$ by (8.189c), from the parabolic maximum principle we have that

(8.197) $$|\gamma|(x,t) \leq C_3 \quad \text{on } \mathbb{S}^2 \times [0,1).$$

By combining (8.195), (8.196), and (8.197), we conclude that

$$\inf_{\lambda \in \mathbb{R}} \sup_{\mathbb{S}^2 \times \mathbb{R}} \left| \bar{h}(t) - \lambda \operatorname{Ric}_{\bar{g}(t)} \right|_{\bar{g}(t)}^2 \leq (2(1-t))^{-2} \inf_{\bar{\lambda} \in \mathbb{R}} \sup_{x \in \mathbb{S}^2} \left| \alpha(x,t) - \bar{\lambda} g_{\mathbb{S}^2}(x) \right|_{g_{\mathbb{S}^2}}^2$$

$$+ 2(1-t) \sup_{x \in \mathbb{S}^2} |\beta|_{g_{\mathbb{S}^2}}^2(x,t) + \sup_{x \in \mathbb{S}^2} |\gamma|^2(x,t)$$

$$\leq 4C_1^2 + 2C_2^2 + C_3^2. \qquad \square$$

8.8. Completion of the proof of the main theorem

In this section, by combining the results proved in the previous sections, we complete the proof of Theorem 8.1.

Recall that we defined diffeomorphisms $\Phi_s : \mathbb{S}^2 \to \Sigma_s$ by (8.79) and vector fields U_1, U_2, U_3 on $\mathcal{M}^3$ by (8.98) to satisfy $U_a := (\Phi_s)_* \bar{U}_a$ for $s < s_0$, where $s_0 \ll -1$. Here, $\bar{U}_1, \bar{U}_2, \bar{U}_3$ are Killing vector fields on the unit sphere $(\mathbb{S}^2, \bar{\gamma})$ satisfying (8.97). Note by (8.96) that the metric $\gamma_s = \frac{1}{2|s|}\Phi_s^*(g_s)$, where $g_s := g|_{\Sigma_s}$ is the induced Riemannian metric on Σ_s, satisfies for $m \geq 0$,

$$\|\gamma_s - \bar{\gamma}\|_{C^m(\mathbb{S}^2, \bar{\gamma})} \leq O(|s|^{-1/8}).$$

By Proposition 8.29, the vector fields U_a on $\mathcal{M}^3$ are approximately Killing vector fields in the sense that

$$|\mathcal{L}_{U_a}g| \leq O(r^{-1/8}) \quad \text{and} \quad |\Delta_f U_a| \leq O(r^{-9/16})$$

for $a = 1, 2, 3$, where here and henceforth the norms are with respect to g unless otherwise indicated. Moreover, by (8.99) and (8.106) they satisfy for $[abc] = [123]$, i.e., for $\{a, b, c\}$ a cyclic permutation of $\{1, 2, 3\}$,

$$(8.198) \quad [U_a, U_b] = U_c \text{ on } \{f < s_0\} \quad \text{and} \quad \sum_{a=1}^3 U_a \otimes U_a = |s|g_s^{-1} + O(|s|^{7/8}),$$

respectively.

We may apply Theorem 8.34 to the vector field $Q := \Delta_f U_a$ with $\varepsilon = \frac{1}{32}$ to obtain that there exist vector fields V_1, V_2, V_3 on $\mathcal{M}$ such that $\Delta_f V_a = \Delta_f U_a$, $|V_a| \leq O(r^{\frac{15}{32}})$, and $|\nabla V_a| \leq O(r^{-\frac{1}{32}})$. Define $W_a := U_a - V_a$ for $a = 1, 2, 3$. Then
(8.199)
$$\Delta_f W_a = 0, \quad |W_a - U_a| \leq O(r^{\frac{15}{32}}), \quad \text{and} \quad |\nabla(W_a - U_a)| \leq O(r^{-\frac{1}{32}}).$$

Moreover, $|\mathcal{L}_{U_a}g| \leq O(r^{-\frac{1}{8}})$ and $|\nabla V_a| \leq O(r^{-\frac{1}{32}})$ imply that $|\mathcal{L}_{W_a}g| \leq O(r^{-\frac{1}{32}})$. Since $\Delta_f W_a = 0$ and by Proposition 8.32, we have that on $\mathcal{M}$,

$$\Delta_{L,f}(\mathcal{L}_{W_a}g) = 0 \quad \text{for } a = 1, 2, 3.$$

Since $\Delta_{L,f}(\mathcal{L}_{W_a}g) = 0$ and $|\mathcal{L}_{W_a}g| \leq O(r^{-\frac{1}{32}})$, by Proposition 8.40 there exists $b_a \in \mathbb{R}$ such that on $\mathcal{M}^3$,

$$\mathcal{L}_{W_a}g = b_a \operatorname{Ric} = -\frac{b_a}{2}\mathcal{L}_{\nabla f}g.$$

Hence, the vector fields

$$(8.200) \qquad\qquad \widehat{U}_a := W_a + \frac{b_a}{2}\nabla f$$

are Killing vector fields on $(\mathcal{M}^3, g)$ with

$$|\widehat{U}_a - U_a| \le \mathrm{O}(r^{\frac{15}{32}}) \quad \text{and} \quad |\nabla \widehat{U}_a - \nabla U_a| \le \mathrm{O}(r^{-\frac{1}{32}}) \quad \text{for } a = 1, 2, 3.$$

Furthermore, by (8.198), (8.199), and the formula $[X, Y] = \nabla_X Y - \nabla_Y X$, we have

$$(8.201) \qquad |[\widehat{U}_a, \widehat{U}_b] - \widehat{U}_c| \le \mathrm{O}(r^{\frac{15}{32}}) \quad \text{for } [abc] = [123].$$

We claim that for $a = 1, 2, 3$, $\langle \widehat{U}_a, \nabla f \rangle = 0$, $[\widehat{U}_a, \nabla f] = 0$, and as symmetric 2-tensors of Σ_s,

$$(8.202) \qquad \sum_{a=1}^{3} \widehat{U}_a \otimes \widehat{U}_a = |s| g_s^{-1} + \mathrm{O}(|s|^{1-\varepsilon})$$

for some $\varepsilon > 0$. This is all because of the following. We compute that

$$\nabla^2 \langle \widehat{U}_a, \nabla f \rangle = \nabla^2 (\mathcal{L}_{\widehat{U}_a} f) = \mathcal{L}_{\widehat{U}_a}(\nabla^2 f) = \frac{1}{2} \mathcal{L}_{\widehat{U}_a}(\mathcal{L}_{\nabla f} g).$$

We can now argue as follows.[10] Since the metric is invariant under the flow generated by $\widehat{U}_a$, the Ricci tensor is also invariant under the flow generated by $\widehat{U}_a$. In other words, $\mathcal{L}_{\widehat{U}_a}(\nabla^2 f) = -\mathcal{L}_{\widehat{U}_a}(\mathrm{Ric}) = 0$, where in the last step we have used the fact that $\widehat{U}_a$ is a Killing vector field. From this, we can deduce that the Hessian of the function $\langle \widehat{U}_a, \nabla f \rangle$ vanishes. Therefore, since the manifold does not split, the function $\langle \widehat{U}_a, \nabla f \rangle$ is constant. Since the function $\langle \widehat{U}_a, \nabla f \rangle$ vanishes at the tip, the function $\langle \widehat{U}_a, \nabla f \rangle$ must vanish identically.

An alternative (and perhaps more elegant) argument is as follows. The identity $\Delta_f W_a = 0$ together with $\Delta_f(\nabla f) = 0$ implies $\Delta_f \widehat{U}_a = 0$. On the other hand, $\Delta \widehat{U}_a + \mathrm{Ric}(\widehat{U}_a) = 0$ since $\widehat{U}_a$ is a Killing vector field. Subtracting the two identities gives $[\widehat{U}_a, \nabla f] = 0$. This finally implies that

$$(8.203) \qquad \nabla(\langle \widehat{U}_a, \nabla f \rangle) = \nabla(\mathcal{L}_{\widehat{U}_a}(f)) = \mathcal{L}_{\widehat{U}_a}(\nabla f) = [\widehat{U}_a, \nabla f] = 0.$$

Since the function $\langle \widehat{U}_a, \nabla f \rangle$ vanishes at the tip, the function $\langle \widehat{U}_a, \nabla f \rangle$ must vanish identically.

Finally, it is easy to verify (8.202) from the properties of W_a and from (8.198).

Fix s and consider Σ_s. Since $\widehat{U}_a$ is a Killing vector field of $(\mathcal{M}^3, g)$ which is tangent to Σ_s, we have that $\widehat{U}_a^s := \widehat{U}_a|_{\Sigma_s}$ is a Killing vector field of (Σ_s, g_s) for $a = 1, 2, 3$. By (8.201), for s sufficiently large, $\{\widehat{U}_1^s, \widehat{U}_2^s, \widehat{U}_3^s\}$ generates a Lie subalgebra $\mathfrak{h}$ of the Lie algebra $\mathfrak{g}$ of all Killing vector fields

---

[10]We are greatly indebted to Simon Brendle for personally communicating to the author all of the arguments in this paragraph and the next, which corrects an earlier version of our exposition of his theorem in this book.

on (Σ_s, g_s), where $\dim \mathfrak{h} \geq 3$. On the other hand, since Σ_s is 2-dimensional (diffeomorphic to $\mathbb{S}^2$), $\dim \mathfrak{g} \leq 3$. Hence $\mathfrak{h} = \mathfrak{g}$ has dimension 3 and consequently $[\hat{U}_a^s, \hat{U}_b^s] = \sum_{c=1}^3 c_{ab}^c(s)\hat{U}_c^s$, where $c_{ab}^c(s) \in \mathbb{R}$. This implies that $[\hat{U}_a, \hat{U}_b] = \sum_{c=1}^3 \bar{c}_{ab}^c \hat{U}_c$, where $\bar{c}_{ab}^c \in \mathbb{R}$. Hence $c_{ab}^c(s) \equiv \bar{c}_{ab}^c$ is independent of s. Now (8.201) and $|\hat{U}_c| = \mathrm{O}(r^{1/2})$ imply that $[\hat{U}_a^s, \hat{U}_b^s] = \hat{U}_c^s$ on Σ_s for $[abc] = [123]$. Therefore $\mathfrak{h} = \mathfrak{g}$ is isomorphic to $\mathfrak{so}(3)$ and each (Σ_s, g_s) is a round 2-sphere for all $s < -1$. We conclude that

$$(8.204) \qquad [\hat{U}_1, \hat{U}_2] = \hat{U}_3, \quad [\hat{U}_2, \hat{U}_3] = \hat{U}_1, \quad [\hat{U}_3, \hat{U}_1] = \hat{U}_2 \quad \text{on } \mathcal{M}.$$

That is, the Lie algebra generated by $\hat{U}_1, \hat{U}_2, \hat{U}_3$ is 3-dimensional and isomorphic to $\mathfrak{so}(3)$. Corresponding to this Lie subalgebra of the Lie algebra of all Killing vector fields is a Lie subgroup of the isometry group of $(\mathcal{M}^3, g)$ which is isomorphic to $SO(3)$. We conclude that g is rotationally symmetric. This completes the proof of Theorem 8.1.

8.9. Notes and commentary

The main result of this chapter, Theorem 8.1, is due to Brendle [42].

Brendle generalized his theorem and proved Perelman's conjecture on the classification of noncompact 3-dimensional ancient κ-solutions to the Ricci flow. For further work on the classification and asymptotics in the compact case and in higher-dimensional cases, see Brendle [45–47], Bamler and Kleiner [28], Brendle, Daskalopoulos, and Šešum [50], Angenent, Brendle, Daskalopoulos, and Šešum [7], Brendle and Naff [51], and Brendle, Daskalopoulos, Naff, and Šešum [49].

Earlier work on 3-dimensional steady GRS was done by Bryant [54] (discussed in Chapter 6 of this book), Ivey [192], Hamilton [179], Chu [116], Guo [167], and Brendle [41]. A higher-dimensional result has been given by Brendle [43].

Brendle's technique has been applied to expanding GRS, in the Riemannian case by Chodosh [97] and in the Kähler case by Chodosh and Fong [98].

For related work on translating self-similar solutions to the mean curvature flow, see Haslhofer [183].

8.10. Exercises

Exercise 8.1. Prove formula (8.3): $\nabla_j \nabla_k V_i = R_{kij\ell} V_\ell$.
HINT: Use the first Bianchi identity.

Exercise 8.2. Prove (8.10): $\Delta_f V - \nabla \langle \nabla f, V \rangle = 0$.
HINT:

$$(\nabla_{\nabla f} V)_i = \nabla_j f \nabla_j V_i = -\nabla_j f \nabla_i V_j.$$

Exercise 8.3. Show that if V is a Killing vector field on a steady GRS $(\mathcal{M}^n, g, f)$, then

(8.205) $$\nabla \langle V, \nabla f \rangle = [V, \nabla f].$$

Exercise 8.4. Prove the following generalization of (8.124): For any tensor T, we have

$$\mathcal{L}_X(\mathcal{L}_Y T) = \mathcal{L}_Y(\mathcal{L}_X T) + \mathcal{L}_{[X,Y]} T.$$

HINT: If T is a vector field, then this is the Jacobi identity. Use (2.25).

Exercise 8.5. Show for any vector field V on a Riemannian manifold $(\mathcal{M}^n, g)$ that

$$g_{k\ell} \delta_{\mathcal{L}_V g}(\Gamma_{ij}^\ell) = \nabla_i \nabla_j V_k - R_{jki\ell} V_\ell,$$

where $\delta_{\mathcal{L}_V g}(\Gamma_{ij}^\ell)$ denotes the variation of the Christoffel symbols in the direction of the symmetric 2-tensor $\mathcal{L}_V g$. This formula provides an alternate proof of (8.3).

Exercise 8.6. We consider the linearization of the nonlinear operator which takes (g, f) to $\mathrm{Ric} + \nabla^2 f$. Show that if $\delta g = v$ and $\delta f = h$, where v is a symmetric 2-tensor and h is a function, then

$$\delta(\mathrm{Ric} + \nabla^2 f) = -\frac{1}{2}\left(\Delta_{L,f} v + \mathcal{L}_Y g\right),$$

where $\Delta_{L,f}$ is defined by (8.12) and where

$$Y := \nabla\left(\frac{V}{2} - h\right) - \mathrm{div}_f v \quad \text{and} \quad \mathrm{div}_f v := \mathrm{div}\, v - v(\nabla f).$$

HINT: Use (8.121) and recall that

$$(\mathcal{L}_{\nabla f} v)_{ij} = (\nabla_{\nabla f} v)_{ij} + \nabla_i \nabla_k f v_{kj} + v_{ik} \nabla_j \nabla_k f.$$

Exercise 8.7. By scaling the Ricci flow solution by $R(x, 0)$ in Theorem 8.11, prove by contradiction that for any 3-dimensional ancient κ-solution $(\mathcal{M}^3, g(t))$, $t \in (-\infty, 0]$, any $p \in \mathcal{M}^3$, and any sequence $x_i \to \infty$, we have $R(x_i, 0) \, d_{g(0)}^2(x_i, p) \to \infty$.

Exercise 8.8. Prove the estimates for the potential function in (8.28). HINT: Let $\varepsilon \in (0, \mathrm{inj}(o))$, where $\mathrm{inj}(o)$ is the injectivity radius of o. Observe that $\sup_{\partial B_\varepsilon(o)} \langle \nabla f, \nabla r \rangle =: a < 0$. Let $\gamma : [0, L] \to \mathcal{M}^3$ be a unit speed minimal geodesic emanating from o. Show that $-1 \leq (f \circ \gamma)'(s) \leq a$ for $s \geq \varepsilon$.

Exercise 8.9. Consider the unit 2-sphere $(\mathbb{S}^2, g_{\mathbb{S}^2})$. Write the metric in local spherical coordinates $(\phi, \theta) \in (0, \pi) \times (0, 2\pi)$ as

$$g_{\mathbb{S}^2} = d\phi^2 + \sin^2 \phi \, d\theta^2.$$

The nonzero Christoffel symbols are $\Gamma_{\theta\theta}^\phi = -\sin\phi \cos\phi$ and $\Gamma_{\theta\phi}^\theta = \Gamma_{\phi\theta}^\theta = \cot\phi$.

(1) Show that the vector fields

$$U_1 := \frac{\partial}{\partial\theta},$$

$$U_2 := \cos\theta\frac{\partial}{\partial\phi} - \cot\phi\sin\theta\frac{\partial}{\partial\theta},$$

$$U_3 := -\sin\theta\frac{\partial}{\partial\phi} - \cot\phi\cos\theta\frac{\partial}{\partial\theta}$$

are Killing vector fields.

(2) Show that $[U_1, U_2] = U_3$, $[U_2, U_3] = U_1$, and $[U_3, U_1] = U_2$. Conclude that the Lie algebra of the isometry group of $(\mathbb{S}^2, g_{\mathbb{S}^2})$ is isomorphic to $\mathfrak{so}(3)$.

(3) Show that

$$\sum_{a=1}^{3} U_a \otimes U_a = \frac{\partial}{\partial\phi} \otimes \frac{\partial}{\partial\phi} + \csc^2\phi\frac{\partial}{\partial\theta} \otimes \frac{\partial}{\partial\theta} = g_{\mathbb{S}^2}.$$

(4) Show that $\|U_1\|_{L^2}^2 = \frac{8}{3}\pi$, $\|U_2\|_{L^2}^2 = \|U_3\|_{L^2}^2 = \frac{3}{2}\pi^2$, and $\langle U_a, U_b \rangle_{L^2} = 0$ for $a \neq b$.

Exercise 8.10. Give a heuristic explanation for why the rescaled (blowdown) limit of the Bryant soliton at spatial infinity, which is the round cylinder solution $\bar{g}(t) := 2(1-t)g_{\mathbb{S}^2} + dz^2$ to the Ricci flow on $\mathbb{S}^2 \times \mathbb{R}$, forms a singularity at time $t = 1$.

HINT: How does Example 8.16 explain why rescaling near spatial infinity the diffeomorphism flow of the Bryant soliton corresponds to the shrinking in time of $\mathbb{S}^2$ in the cylinder limit? That is, how does the rescaling of flowing in space correspond to flowing in time?

Exercise 8.11. With regards to the limit solution $(\mathbb{S}^2 \times \mathbb{R}, \bar{g}(t))$, explain how the space scale is like $\sqrt{r}$ and the time scale is like r.

Exercise 8.12. Prove the time-dependent version of Lemma 8.17, i.e., the subconvergence of (8.151) to the vector field $\frac{\partial}{\partial z}$ on $\mathbb{S}^2 \times \mathbb{R}$. Use this to show that the sequence of diffeomorphisms $\varphi_{|r_k|^{1/2}z}$ of $\mathcal{M}^3$ converges as $k \to \infty$ to the isometry ψ_z of $\mathbb{S}^2 \times \mathbb{R}$.

Exercise 8.13. Prove, using (8.87), that (8.88) is true: $\left|D_{g_s}^m \mathrm{II}\right| \le O(|s|^{-1-\frac{m}{2}})$ for $m \geq 0$. To be rigorous, one must also use induction.

Exercise 8.14. Prove the estimates (8.90) and (8.92).

Exercise 8.15. Prove (8.99).

HINT : Consider the map $F : \mathbb{S}^2 \times (-\infty, s_0) \to \{f < s_0\}$ defined by $\Phi(y, s) = \Phi_s(y)$ for $y \in \mathbb{S}^2$ and $s < s_0$. Define each $\bar{U}_a$ as a vector field on $\mathbb{S}^2 \times (-\infty, s_0)$ and push forward by F.

Exercise 8.16. Prove (8.103): $|\nabla^\ell U_a| \leq O(r^{\frac{1-\ell}{2}})$ for $\ell \geq 0$.

Exercise 8.17. Following the proof of (8.109), while using (8.86) and (8.95), complete the proof of Proposition 8.25.

Exercise 8.18. Let $(\mathcal{M}^n, g, f)$ be the Bryant soliton. Suppose that V is a vector field satisfying $\Delta_f V = 0$. Does there exist $\mu \in \mathbb{R}$ such that $V - \mu \nabla f$ is a Killing vector field? This is related to the discussion in the paragraph containing (8.117) and (8.118).

Exercise 8.19. Show for any 1-form W that

(8.206)
$$\nabla_i (\Delta W_j + R_{jk} W_k) = \Delta (\nabla_i W_j) + 2R_{kijl} \nabla_k W_\ell - R_{ik} \nabla_k W_j + R_{jk} \nabla_i W_k$$
$$+ (\nabla_i R_{jk} - \nabla_j R_{ik} + \nabla_k R_{ij}) W_k.$$

By symmetrizing in i and j the above formula, give another proof of (8.123).

Exercise 8.20. Prove estimate (8.134) as follows. Let $w : \mathbb{S}^n \times [0, T) \to \mathbb{R}$ be a solution to
$$\frac{\partial w}{\partial t} = \Delta_{\mathbb{S}^n} w + nw.$$
Let $\nu \in \mathbb{S}^n$ and define the hemispheres $\mathbb{S}^n_\pm := \{x \in \mathbb{S}^n : \pm \langle x, \nu \rangle \geq 0\}$ and the equator
$$\mathcal{E} := \{x \in \mathbb{S}^n : \langle x, \nu \rangle = 0\} = \mathbb{S}^n_+ \cap \mathbb{S}^n_-.$$
Define the reflection about the equator $R_\nu : \mathbb{S}^n \to \mathbb{S}^n$ by $R_\nu(x) = x - 2 \langle x, \nu \rangle \nu$.

(1) Define $\ell : \mathbb{S}^n \to \mathbb{R}$ by $\ell(x) = \langle x, \nu \rangle$. Prove that there exists a constant C such that
$$w - w \circ R_\nu \leq C\ell \quad \text{on } \mathbb{S}^n_+ \times [0, T).$$

HINT: Prove that there exists C such that this is true initially and then apply the parabolic maximum principle.

(2) Derive from (1) that
$$\nu(w) \leq \frac{C}{2} \quad \text{on } \mathcal{E} \times [0, T).$$

Conclude that $|\nabla w| \leq \frac{C}{2}$ on $\mathbb{S}^n \times [0, T)$.

Exercise 8.21. Give another proof of (8.135) by obtaining a $W^{2,2}$-estimate and then applying the Sobolev inequality.

Exercise 8.22. Prove (8.153) and (8.154).

Exercise 8.23. Prove (8.151).

Exercise 8.24. Verify (8.204).

Exercise 8.25. Try estimating $V^{(k)}$ of (8.139) without considering $A^{(k)}(s)$ of (8.140), i.e., effectively taking $\lambda = 0$ in (8.140). What goes wrong?

Exercise 8.26. Try estimating h of (8.169) without considering $A(s)$ of (8.177), i.e., effectively taking $a = 0$ in (8.177). What goes wrong?

Geometric Preliminaries

A.1. The Kazdan–Warner identity

We recall the proof of the **Kazdan–Warner identity** [199] for conformal vector fields on the sphere. This result can be used in the classification of closed 2-dimensional Ricci solitons; see Theorem 3.5(1).

Theorem A.1 (Kazdan and Warner). *For any conformal vector field Y on a Riemannian 2 sphere $(\mathbb{S}^2, g)$, i.e., $\mathcal{L}_Y g = \operatorname{div}(Y)g$, we have*

$$(A.1) \qquad \int_{\mathbb{S}^2} \langle \nabla_g R, Y \rangle_g \, d\mu_g = 0.$$

Proof of Theorem A.1. By the uniformization theorem, there exists a function u on $\mathbb{S}^2$ such that $g = e^{2u}\bar{g}$, where $\bar{g}$ has constant Gauss curvature $K_{\bar{g}} \equiv 1$, so $\Delta_{\bar{g}} u + e^{2u}K_g = 1$. Let ℓ be the restriction of a linear function on $\mathbb{R}^3$ to $\mathbb{S}^2$ (considering $\mathbb{S}^2 \subset \mathbb{R}^3$ as the unit sphere), so that $\Delta_{\bar{g}}\ell + 2\ell = 0$. (The eigenspace associated to the eigenvalue 2 of $-\Delta_{\bar{g}}$ consists of the linear functions and the next smallest eigenvalue is 6.) In the following, ∇, Δ, and $\langle \cdot, \cdot \rangle$ are all with respect to $\bar{g}$. We compute, integrating by parts twice while using $\nabla^2 \ell + \ell\bar{g} = 0$ and $\Delta_{\bar{g}}\ell + 2\ell = 0$, that

$$
\int_{\mathbb{S}^2} \langle \nabla u, \nabla \ell \rangle \, \Delta u \, dA_{\bar{g}} = -\int_{\mathbb{S}^2} (\langle \nabla^2 \ell, \nabla u \otimes \nabla u \rangle + \langle \nabla^2 u, \nabla u \otimes \nabla \ell \rangle) \, dA_{\bar{g}}
$$
$$
= \int_{\mathbb{S}^2} \left(\ell |\nabla u|^2 - \frac{1}{2} \langle \nabla |\nabla u|^2, \nabla \ell \rangle \right) dA_{\bar{g}}
$$
$$
= 0.
$$

Integrating by parts and again using $\Delta_{\bar{g}}\ell = -2\ell$, we obtain

$$
\int_{\mathbb{S}^2} \langle \nabla u, \nabla \ell \rangle \, \mathrm{e}^{2u} K_g \, dA_{\bar{g}} = \frac{1}{2} \int_{\mathbb{S}^2} \langle \nabla \mathrm{e}^{2u}, \nabla \ell \rangle \, K_g \, dA_{\bar{g}}
$$

$$
= \int_{\mathbb{S}^2} \left(\ell K_g - \frac{1}{2} \langle \nabla K_g, \nabla \ell \rangle \right) \mathrm{e}^{2u} \, dA_{\bar{g}}.
$$

Hence

$$
\int_{\mathbb{S}^2} \left(\ell K_g - \frac{1}{2} \langle \nabla K_g, \nabla \ell \rangle \right) \mathrm{e}^{2u} \, dA_{\bar{g}} = \int_{\mathbb{S}^2} \langle \nabla u, \nabla \ell \rangle \, (\mathrm{e}^{2u} K_g + \Delta u) \, dA_{\bar{g}}
$$

$$
= \int_{\mathbb{S}^2} \langle \nabla u, \nabla \ell \rangle \, dA_{\bar{g}}
$$

$$
= - \int_{\mathbb{S}^2} \ell \Delta u \, dA_{\bar{g}}
$$

$$
= \int_{\mathbb{S}^2} \ell (\mathrm{e}^{2u} K_g - 1) \, dA_{\bar{g}}.
$$

Since $\int_{\mathbb{S}^2} \ell \, dA_{\bar{g}} = 0$, we conclude that

$$
(A.2) \qquad \int_{\mathbb{S}^2} \langle \nabla_g K_g, \nabla_{\bar{g}} \ell \rangle_g \, dA_g = \int_{\mathbb{S}^2} \langle \nabla_{\bar{g}} K_g, \nabla_{\bar{g}} \ell \rangle_{\bar{g}} \, \mathrm{e}^{2u} \, dA_{\bar{g}} = 0.
$$

The group of conformal transformations of $\mathbb{S}^2$ is the Möbius group $\mathrm{PSL}(2, \mathbb{C})$, which has complex dimension 3. Hence, the vector space of conformal vector fields of g (which is the same as that of $\bar{g}$) is 6 real dimensional, of which 3 dimensions are spanned by Killing vector fields of $\bar{g}$ and the other 3 dimensions are spanned by the $\nabla_{\bar{g}} \ell$. If X is a Killing vector field of $\bar{g}$, then

$$
\int_{\mathbb{S}^2} \langle \nabla_g K_g, X \rangle_g \, dA_g = \int_{\mathbb{S}^2} \langle \nabla_{\bar{g}} K_g, X \rangle_{\bar{g}} \, \mathrm{e}^{2u} \, dA_{\bar{g}}
$$

$$
= -2 \int_{\mathbb{S}^2} \langle \nabla_{\bar{g}} u, X \rangle_{\bar{g}} \, K_g \, \mathrm{e}^{2u} \, dA_{\bar{g}}
$$

$$
= 2 \int_{\mathbb{S}^2} \langle \nabla_{\bar{g}} u, X \rangle_{\bar{g}} \, (\Delta_{\bar{g}} u - 1) \, dA_{\bar{g}}
$$

$$
= -2 \int_{\mathbb{S}^2} (\langle \nabla_{\bar{g}}^2 u, \nabla_{\bar{g}} u \otimes X \rangle_{\bar{g}} + \langle \nabla_{\bar{g}} X, \nabla_{\bar{g}} u \otimes \nabla_{\bar{g}} u \rangle_{\bar{g}}) \, dA_{\bar{g}}
$$

$$
= - \int_{\mathbb{S}^2} \langle \nabla_{\bar{g}} |\nabla_{\bar{g}} u|^2, X \rangle_{\bar{g}} \, dA_{\bar{g}}
$$

$$
= 0
$$

by integrating by parts and by using $\nabla_i^{\bar{g}} X_j + \nabla_j^{\bar{g}} X_i = 0$ and $\operatorname{div}_{\bar{g}} X = 0$. The theorem follows. $\square$

A.2. Splitting and dimension reduction

We give a proof of Theorem 1.19. Recall that the statement is:

Theorem A.2. *Let $(\mathcal{M}^n, g)$ be a complete Riemannian manifold with non-negative sectional curvature and let $o \in \mathcal{M}^n$. Suppose there exist sequences $x_i \in \mathcal{M}^n$ and $r_i > 0$ with $r_i^{-1} d(o, x_i) \to \infty$ and such that $(\mathcal{M}^n, g_i, x_i)$, where $g_i = r_i^{-2} g$, converges in the pointed C^∞ Cheeger–Gromov sense to a complete limit $(\mathcal{M}_\infty^n, g_\infty, x_\infty)$. Then $(\mathcal{M}_\infty^n, g_\infty)$ is the product of a line with a C^∞ Riemannian manifold with nonnegative sectional curvature.*

Proof. Since $d_{g_i}(o, x_i) \to \infty$, the limit manifold $\mathcal{M}_\infty^n$ is noncompact. According to the Toponogov splitting theorem [**275**], it suffices to show that $(\mathcal{M}_\infty^n, g_\infty)$ contains a geodesic line. By passing to a subsequence, we may assume that $d(x_{i+1}, o) \geq 10 d(x_i, o)$ and that

$$\delta_i := \angle x_i o x_{i+1} \to 0$$

(the angle formed by the tangent vectors at o to choices of minimal geodesics from o to x_i and x_{i+1}); this limit for δ_i is possible as a consequence of the compactness of the unit sphere in $T_o \mathcal{M}$. Since the sectional curvatures of g are nonnegative, by the Toponogov comparison theorem (see [**84**, Theorem 2.2]) for the hinge $x_i o x_{i+1}$, we have that

$$
\begin{aligned}
\text{(A.3)} \quad d^2(x_i, x_{i+1}) &\leq d^2(o, x_i) + d^2(o, x_{i+1}) - 2 d(o, x_i) d(o, x_{i+1}) \cos \delta_i \\
&\leq d^2(o, x_i) + d^2(o, x_{i+1}) - 2 d(o, x_i) d(o, x_{i+1}) \\
&\quad + 2 \delta_i d(o, x_i) d(o, x_{i+1}) \\
&\leq (1 + \mathrm{O}(\delta_i)) \left(d(o, x_i) - d(o, x_{i+1}) \right)^2,
\end{aligned}
$$

since $\cos \delta_i \geq 1 - \delta_i$ and $d(x_{i+1}, o) \geq 10 d(x_i, o)$.

Applying the Toponogov comparison theorem for the triangle $\triangle x_i o x_{i+1}$, we have that the comparison angle in Euclidean space satisfies

$$
\begin{aligned}
\bar{\angle} \bar{o} \bar{x}_i \bar{x}_{i+1} &= \cos^{-1} \left(\frac{d^2(o, x_i) + d^2(x_i, x_{i+1}) - d^2(o, x_{i+1})}{2 d(o, x_i) d(x_i, x_{i+1})} \right) \\
&\leq \angle o x_i x_{i+1},
\end{aligned}
$$

where $\bar{\angle}$ denotes the Euclidean angle. Observe that by squaring the triangle inequality we have

$$d^2(o, x_{i+1}) \leq d^2(o, x_i) + d^2(x_i, x_{i+1}) + 2 d(o, x_i) d(x_i, x_{i+1}),$$

and hence

$$\frac{d^2(o, x_i) + d^2(x_i, x_{i+1}) - d^2(o, x_{i+1})}{2 d(o, x_i) d(x_i, x_{i+1})} \geq -1.$$

Furthermore, by (A.3) we compute that

$$\frac{d^2(o, x_i) + d^2(x_i, x_{i+1}) - d^2(o, x_{i+1})}{2d(o, x_i)d(x_i, x_{i+1})}$$

$$\leq \frac{1}{2d(o, x_i)d(x_i, x_{i+1})} \Big(d^2(o, x_i) - d^2(o, x_{i+1}) + d^2(o, x_i)$$

$$+ d^2(o, x_{i+1}) - 2d(o, x_i)d(o, x_{i+1}) + 2\delta_i d(o, x_i)d(o, x_{i+1}) \Big)$$

$$= -\frac{d(o, x_{i+1}) - d(o, x_i)}{d(x_i, x_{i+1})} + 2\delta_i \frac{d(o, x_{i+1})}{d(x_i, x_{i+1})}.$$

Since $d(x_{i+1}, o) \geq 10d(x_i, o)$ we have that

$$\frac{d(o, x_{i+1})}{d(x_i, x_{i+1})} \leq \frac{d(o, x_{i+1})}{d(o, x_{i+1}) - d(o, x_i)} \leq \frac{10}{9}.$$

Inequality (A.3) also implies that

$$\frac{d(o, x_{i+1}) - d(o, x_i)}{d(x_i, x_{i+1})} \geq 1 - \mathrm{O}(\delta_i).$$

Let $\triangle \bar{x}_i \bar{o} \bar{x}_{i+1}$ be the comparison triangle in Euclidean space (the triangle with the same side lengths as $\triangle x_i o x_{i+1}$). It then follows from the above inequalities that

$$\cos \bar{\angle} \bar{o} \bar{x}_i \bar{x}_{i+1} \leq -1 + \mathrm{O}(\delta_i) \to -1$$

and hence

$$\bar{\angle} \bar{o} \bar{x}_i \bar{x}_{i+1} \to \pi.$$

For points x and y, let $\overline{xy}$ denote a choice of minimal geodesic from x to y. Observe that for any $L_1, L_2 \in (0, \infty)$ we can choose $p_i \in \overline{ox_i}$ and $q_i \in \overline{x_i x_{i+1}}$ such that $d_{g_i}(x_i, p_i) = L_1$ and $d_{g_i}(x_i, q_i) = L_2$ for i sufficiently large, since $d_{g_i}(x_{i+1}, x_i) \geq d_{g_i}(o, x_i) \to \infty$. By the Toponogov monotonicity principle,[1] we have

$$\bar{\angle} \bar{p}_i \bar{x}_i \bar{q}_i \geq \bar{\angle} \bar{o} \bar{x}_i \bar{x}_{i+1} \to \pi,$$

and thus

$$\frac{d_{g_i}^2(x_i, p_i) + d_{g_i}^2(x_i, q_i) - d_{g_i}^2(p_i, q_i)}{2d_{g_i}(x_i, p_i)d_{g_i}(x_i, q_i)} \to -1 \quad \text{as } i \to \infty.$$

This implies that

$$\frac{L_1 + L_2}{d_{g_i}(p_i, q_i)} \to 1 \quad \text{as } i \to \infty$$

for any L_1 and L_2. It then follows that the concatenation of $\overline{ox_i}$ and $\overline{x_i x_{i+1}}$ will subconverge to a geodesic line in $(\mathcal{M}_\infty^n, g_\infty)$. $\square$

---

[1] Given minimal geodesics $\alpha(s)$ and $\beta(t)$ such that $\alpha(0) = \beta(0) = x_i$, the Euclidean comparison angle $\bar{\angle} \overline{\alpha(s)} \bar{x}_i \overline{\beta(t)}$ is nonincreasing in both s and t. See Cheeger and Ebin [84].

A.3. Uhlenbeck's trick

In this section we provide a background discussion of Hamilton's use of Uhlenbeck's trick for Ricci flow. See the exercises in §5.4.2 for some of the applications we discuss in this book. A more detailed account is given in [**108**].

Uhlenbeck's trick is a device, employed by Hamilton [**174, 177**], to simplify the calculation of the evolution of the Riemann curvture tensor and its related tensors such as the Ricci tensor. For simplicity, we discuss Uhlenbeck's trick in the special case of a parallelizable manifold, i.e., a manifold whose tangent bundle is isomorphic to a trivial vector bundle. (It is a well-known fact that any closed orientable 3-dimensional manifold is parallelizable.)

Let $(\mathcal{M}^n, g(t))$, $t \in [0, T]$, be a solution to the Ricci flow on a parallelizable manifold. Then there exists a global frame field $\{F_i\}_{i=1}^n$. By the Gram–Schmidt process, we may then construct a globally defined frame field $\{e_i^0\}_{i=1}^n$ on $\mathcal{M}^n$ that is orthonormal with respect to the metric $g(0)$.

We may extend this orthonormal frame field in time so that it remains orthonormal with respect to $g(t)$ by defining $\{e_i(t)\}_{i=1}^n$ to be the unique solution to the ODE:

$$(A.4) \qquad \frac{\partial}{\partial t} e_i(t) = \mathrm{Ric}_{g(t)}(e_i(t))$$

with $e_i(0) = e_i^0$ for $1 \leq i \leq n$. By an easy calculation, we see that

$$\frac{\partial}{\partial t} \big(g(t)(e_i(t), e_j(t)) \big) = 0.$$

Thus, $\{e_i(t)\}_{i=1}^n$ remains orthonormal with respect to $g(t)$; see Exercise 5.7.

To effectively fix the fiber metrics, we define the vector bundle isomorphisms $\iota_t : E := \mathcal{M}^n \times \mathbb{R}^n \to TM$, $t \in [0, T)$, by

$$(A.5) \qquad \iota_t(x, V^1, \ldots, V^n) = V^1 e_1(x, t) + \cdots + V^n e_n(x, t)$$

for $x \in \mathcal{M}^n$ and $(V^1, \ldots, V^n) \in \mathbb{R}^n$. Then the pullbacks by ι_t of the fiber metrics $g(x, t)$ on $T_x M$ are all equal to the Euclidean metric g_{Euc} on $\mathbb{R}^n$, which in particular is independent of time.

Now, in local coordinates, the evolution of the Riemann curvature tensor Rm_t is given by

$$(A.6) \qquad \frac{\partial}{\partial t} R_{ijk\ell} = \Delta R_{ijk\ell} + 2 \left(B_{ijk\ell} - B_{ij\ell k} + B_{ikj\ell} - B_{i\ell jk} \right)$$
$$- \left(R_{ip} R_{pjk\ell} + R_{jp} R_{ipk\ell} + R_{kp} R_{ijp\ell} + R_{\ell p} R_{ijkp} \right),$$

where

$$(A.7) \qquad B_{ijk\ell} := -g^{pr} g^{qs} R_{ipjq} R_{kr\ell s} = -R_{pijq} R_{q\ell kp}.$$

Let $D_t := \iota_t^* \nabla_{g(t)}$ be the pullback of the Levi-Civita connection of $g(t)$ on $T\mathcal{M}$ to E. Let D_t also denote the associated connections on the tensor product bundles of E and E^*. Let $\Delta_{D_t} := \operatorname{tr}_{g(t)} D_t^2$ be the Laplacian acting on these vector bundles. The pullbacks of the Riemann curvature tensors $\iota_t \operatorname{Rm}_t$ satisfy the simpler looking evolution equation

$$(A.8) \qquad \frac{\partial}{\partial t} \iota_t \operatorname{Rm}_t = \Delta_D \operatorname{Rm}_t + \operatorname{Rm}_t^2 + \operatorname{Rm}_t^\#.$$

This is Uhlenbeck's trick. Here, Rm^2 denotes the square of Rm as a fiberwise self-adjoint linear map from $\Lambda^2 T^*\mathcal{M}$ to itself. To describe $\operatorname{Rm}_t^\#$, recall that the Lie algebra structure on each $\Lambda^2 T_x^*\mathcal{M}$ is defined by

$$[U, V]_{ij} := g^{k\ell} (U_{ik} V_{\ell j} - V_{ik} U_{\ell j})$$

for $U, V \in \Lambda^2 T_x^*\mathcal{M}$ and $\Lambda^2 T_x^*\mathcal{M} \cong \mathfrak{so}(n)$. Choose a basis $\{\varphi^\alpha\}$ of $\Lambda^2 T_x^*\mathcal{M}$ and let $C_\gamma^{\alpha\beta}$ denote the structure constants defined by

$$\left[\varphi^\alpha, \varphi^\beta\right] := \sum_\gamma C_\gamma^{\alpha\beta} \varphi^\gamma.$$

We then define the **Lie algebra square** $\operatorname{Rm}^\# : \Lambda^2 T_x^*\mathcal{M} \to \Lambda^2 T_x^*\mathcal{M}$ by

$$(A.9) \qquad (\operatorname{Rm}^\#)_{\alpha\beta} := C_\alpha^{\gamma\delta} C_\beta^{\varepsilon\zeta} \operatorname{Rm}_{\gamma\varepsilon} \operatorname{Rm}_{\delta\zeta}.$$

See [**108**] for more expository details.

Analytic Preliminaries

B.1. Elliptic maximum principles

Lemma B.1. *Let* $(\mathcal{M}^n, g)$ *be a complete Riemannian manifold and let* Ω *be a bounded open domain in* $\mathcal{M}^n$. *Suppose that* $u : \bar{\Omega} \to \mathbb{R}$ *satisfies*

$$(B.1) \qquad\qquad \Delta_g u + \langle X, \nabla u \rangle + cu \geq 0,$$

where X *is a vector field and* $c : \mathcal{M}^n \to (-\infty, 0]$. *Then:*

(1) (Weak maximum principle)

$$\sup_{\bar{\Omega}} u \leq \sup_{\partial\Omega} u_+,$$

where $u_+ := \max\{u, 0\}$.

(2) (Strong maximum principle) *Let* Ω *be connected. If* $x \in \text{int}(\mathcal{M}^n)$ *satisfies* $u(x) = \sup_{\bar{\Omega}} u_+$, *then* u *is constant on* $\bar{\Omega}$.

(3) *If* $\mathcal{M}^n$ *is closed and* $\Omega = \mathcal{M}^n$ *and* c *is not identically zero, then* $u \leq 0$ *on* $\mathcal{M}^n$.

For the parabolic maximum principle, see Lemma 1.21.

B.2. Types of singular solutions to the Ricci flow

Recall that for any Riemannian metric g_0 on a closed smooth manifold $\mathcal{M}^n$, there exists a unique solution to the Ricci flow $g(t)$ with $g(0) = g_0$ and which is defined on a maximal time interval $[0, T)$, where $T \in (0, \infty]$. If $T < \infty$, then we say that the solution forms a **finite-time singularity** at time T. If $T = \infty$, then we say that the solution is an **immortal solution**.

For a finite-time singular solution $g(t)$, $t \in [0, T)$, we say that the solution is **Type I** if

$$
(B.2) \qquad\qquad \sup_{\mathcal{M}^n \times [0,T)} (T - t)|\mathrm{Rm}| < \infty.
$$

Otherwise, for a finite-time singularity we say that the solution is **Type II** (a.k.a. **Type IIa**). An example of a Type II singular solution is the degenerate neckpinch; see §1.4.2 for a heuristic description of this.

For example, if g_0 is an Einstein metric with positive scalar curvature, or more generally a shrinking gradient Ricci soliton, then the associated solution to the Ricci flow forms a finite-time singularity and is a Type I singular solution.

For an immortal solution $g(t)$, $t \in [0, \infty)$, we say that the solution is **Type III** if

$$
(B.3) \qquad\qquad \sup_{\mathcal{M}^n \times [0,\infty)} t|\mathrm{Rm}| < \infty.
$$

Otherwise, for an immortal solution we say that the solution is **Type IIb**. There are many homogeneous examples of Type III immortal solutions of the Ricci flow.

If a solution $g(t)$ to the Ricci flow exists on the time interval $(-\infty, 0)$, then we say that it is an **ancient solution**. We say that an ancient solution is **Type I** if

$$
(B.4) \qquad\qquad \sup_{\mathcal{M}^n \times (-\infty,0)} |t||\mathrm{Rm}| < \infty.
$$

Otherwise, for an ancient solution we say that the ancient solution is **Type II**. For example, the canonical form of a shrinking gradient Ricci soliton with bounded curvature is a Type I ancient solution. Examples of Type II ancient solutions are the canonical forms of nonflat steady gradient Ricci solitons. These solutions are in fact **eternal**; that is, they exist on the time interval $(-\infty, \infty)$.

B.3. The real analyticity of Ricci solitons

Given a manifold $\mathcal{M}^n$, a **real-analytic structure** on $\mathcal{M}^n$ is a **maximal atlas** $\{(U_\alpha, \{x_\alpha^i\}_{i=1}^n)\}_{\alpha \in \Lambda}$ such that the transition functions are real analytic; see e.g. Kobayashi and Nomizu [**202**]. Coordinates in this atlas are called **real-analytic coordinates**. We say that a Riemannian metric g on a manifold $\mathcal{M}^n$ with a real-analytic structure is a **real-analytic metric** if, with respect to a subatlas of real-analytic coordinates, the components g_{ij} of the metric are real-analytic functions. Given a smooth manifold $\mathcal{M}^n$, the completion of the atlas of **harmonic coordinates** (i.e., coordinates $\{x^i\}$ that are harmonic functions: $\Delta x^i = 0$) is a real-analytic structure. In fact, any two real-analytic structures on $\mathcal{M}^n$ are equivalent by the work of Whitney [**286**], Morrey [**230**], and Grauert [**159**].[1]

It was shown in DeTurck and Kazdan [**139**] that a manifold $(\mathcal{M}^n, g)$ has optimal regularity in an atlas of g-harmonic coordinates. In particular, if g is real analytic with respect to any coordinate system, it is real analytic in harmonic coordinates. In the same reference, it is proven that every Einstein manifold is real analytic.

For Ricci solitons, we have the following statement proven by Ivey [**196**].

Theorem B.2. *If $(\mathcal{M}^n, g, X, \lambda)$ is a Ricci soliton, where g is $C^{2,\alpha}$ and X is $C^{1,\alpha}$, then both g and X are real analytic. In particular, for a $C^{2,\alpha}$ GRS, f is real analytic.*

We also have the following variant.

Theorem B.3. *If $(\mathcal{M}^n, g, X, \lambda)$ is a Ricci soliton, where g is C^2 and X is C^2, then both g and X are real analytic. In particular, if a GRS has $g \in C^2$ and $f \in C^3$, then g and f are real analytic.*

In the special case of a *gradient* Ricci soliton $(\mathcal{M}^n, g, f, \lambda)$, a proof of Theorem B.2 may be given as follows. By Proposition 4.19 (which holds when g is $C^{2,\alpha}$), we may extend the Ricci soliton structure to a complete solution $g(t)$ to the Ricci flow with $g(0) = g$ and which is defined for $\lambda t < 1$. Hence, by the results of Bando [**31**] and Kotschwar [**205**], g is real analytic. Since g and $\nabla^2 f = -\operatorname{Ric} + \frac{\lambda}{2} g$ are both real analytic, we have that f is real analytic. Therefore $|\nabla f|^2$ is real analytic.

---

[1]By equivalence we mean that there is a diffeomorphism ϕ from $\mathcal{M}^n$ with one real-analytic structure to $\mathcal{M}^n$ with the other real-analytic structure where both ϕ and ϕ^{-1} are real analytic.

Bibliography

[1] P. Albin, C. L. Aldana, and F. Rochon, *Ricci flow and the determinant of the Laplacian on non-compact surfaces*, Comm. Partial Differential Equations **38** (2013), no. 4, 711–749, DOI 10.1080/03605302.2012.721853. MR3040681

[2] S. Alexakis, D. Chen, and G. Fournodavlos, *Singular Ricci solitons and their stability under the Ricci flow*, Comm. Partial Differential Equations **40** (2015), no. 12, 2123–2172, DOI 10.1080/03605302.2015.1081609. MR3421757

[3] G. Anderson and B. Chow, *A pinching estimate for solutions of the linearized Ricci flow system on 3-manifolds*, Calc. Var. Partial Differential Equations **23** (2005), no. 1, 1–12, DOI 10.1007/s00526-003-0212-2. MR2133658

[4] B. Andrews and P. Bryan, *Curvature bounds by isoperimetric comparison for normalized Ricci flow on the two-sphere*, Calc. Var. Partial Differential Equations **39** (2010), no. 3-4, 419–428, DOI 10.1007/s00526-010-0315-5. MR2729306

[5] B. Andrews, B. Chow, C. Guenther, and M. Langford, *Extrinsic geometric flows*, Graduate Studies in Mathematics, vol. 206, American Mathematical Society, Providence, RI, [2020] ©2020, DOI 10.1090/gsm/206. MR4249616

[6] S. Angenent, P. Daskalopoulos, and N. Sesum, *Uniqueness of two-convex closed ancient solutions to the mean curvature flow*, Ann. of Math. (2) **192** (2020), no. 2, 353–436, DOI 10.4007/annals.2020.192.2.2. MR4151080

[7] S. Angenent, S. Brendle, P. Daskalopoulos, and N. Šešum, *Unique asymptotics of compact ancient solutions to three-dimensional Ricci flow*, Comm. Pure Appl. Math. **75** (2022), no. 5, 1032–1073, DOI 10.1002/cpa.21955. MR4400906

[8] S. B. Angenent, J. Isenberg, and D. Knopf, *Formal matched asymptotics for degenerate Ricci flow neckpinches*, Nonlinearity **24** (2011), no. 8, 2265–2280, DOI 10.1088/0951-7715/24/8/007. MR2819450

[9] S. B. Angenent, J. Isenberg, and D. Knopf, *Degenerate neckpinches in Ricci flow*, J. Reine Angew. Math. **709** (2015), 81–117, DOI 10.1515/crelle-2013-0105. MR3430876

[10] A. Appleton, *A family of non-collapsed steady Ricci solitons in even dimensions greater or equal to four*, arXiv:1708.00161.

[11] A. J. Appleton, *Singularities in U(2)-Invariant 4d Ricci Flow*, ProQuest LLC, Ann Arbor, MI, 2019. Thesis (Ph.D.)–University of California, Berkeley. MR4051493

[12] P. Baird, *A class of three-dimensional Ricci solitons*, Geom. Topol. **13** (2009), no. 2, 979–1015, DOI 10.2140/gt.2009.13.979. MR2470968

[13] P. Baird and L. Danielo, *Three-dimensional Ricci solitons which project to surfaces*, J. Reine Angew. Math. **608** (2007), 65–91, DOI 10.1515/CRELLE.2007.053. MR2339469

[14] R. H. Bamler, *Compactness theory of the space of super Ricci flows*, arXiv:2008.09298v2.

[15] R. H. Bamler, *Entropy and heat kernel bounds on a Ricci flow background*, arXiv:2008.07093v3.

[16] R. H. Bamler, *Structure theory of non-collapsed limits of Ricci flows*, arXiv:2009.03243v2.

[17] R. H. Bamler, *Long-time behavior of 3-dimensional Ricci flow: introduction*, Geom. Topol. **22** (2018), no. 2, 757–774, DOI 10.2140/gt.2018.22.757. MR3748679

[18] R. H. Bamler, *Long-time behavior of 3-dimensional Ricci flow: A: Generalizations of Perelman's long-time estimates*, Geom. Topol. **22** (2018), no. 2, 775–844, DOI 10.2140/gt.2018.22.775. MR3748680

[19] R. H. Bamler, *Long-time behavior of 3-dimensional Ricci flow: B: Evolution of the minimal area of simplicial complexes under Ricci flow*, Geom. Topol. **22** (2018), no. 2, 845–892, DOI 10.2140/gt.2018.22.845. MR3748681

[20] R. H. Bamler, *Long-time behavior of 3-dimensional Ricci flow: C: 3-manifold topology and combinatorics of simplicial complexes in 3-manifolds*, Geom. Topol. **22** (2018), no. 2, 893–948, DOI 10.2140/gt.2018.22.893. MR3748682

[21] R. H. Bamler, *Long-time behavior of 3-dimensional Ricci flow: D: Proof of the main results*, Geom. Topol. **22** (2018), no. 2, 949–1068, DOI 10.2140/gt.2018.22.949. MR3748683

[22] R. H. Bamler, *Recent developments in Ricci flows*, Notices Amer. Math. Soc. **68** (2021), no. 9, 1486–1498, DOI 10.1090/noti2343. MR4323820

[23] R. H. Bamler, P.-Y. Chan, Z. Ma, and Y. Zhang, *An optimal volume growth estimate for noncollapsed steady gradient Ricci solitons*, Peking Mathematical Journal (2023), https://doi.org/10.1007/s42543-023-00060-w.

[24] R. H. Bamler, B. Chow, Y. Deng, Z. Ma, and Y. Zhang, *Four-dimensional steady gradient Ricci solitons with 3-cylindrical tangent flows at infinity*, Adv. Math. **401** (2022), Paper No. 108285, 21, DOI 10.1016/j.aim.2022.108285. MR4392220

[25] R. H. Bamler, C. Cifarelli, R. J. Conlon, and A. Deruelle, *A new complete two-dimensional shrinking gradient Kähler-Ricci soliton*, arXiv:2206.10785.

[26] R. H. Bamler and B. Kleiner, *Diffeomorphism groups of prime 3-manifolds*, arXiv:2108.03302.

[27] R. H. Bamler and B. Kleiner, *Ricci flow and contractibility of spaces of metrics*, arXiv:1909.08710.

[28] R. H. Bamler and B. Kleiner, *On the rotational symmetry of 3-dimensional κ-solutions*, J. Reine Angew. Math. **779** (2021), 37–55, DOI 10.1515/crelle-2021-0037. MR4319063

[29] R. H. Bamler and B. Kleiner, *Uniqueness and stability of Ricci flow through singularities*, Acta Math. **228** (2022), no. 1, 1–215, DOI 10.4310/acta.2022.v228.n1.a1. MR4448680

[30] R. H. Bamler and B. Kleiner, *Ricci flow and diffeomorphism groups of 3-manifolds*, J. Amer. Math. Soc. **36** (2023), no. 2, 563–589, DOI 10.1090/jams/1003. MR4536904

[31] S. Bando, *Real analyticity of solutions of Hamilton's equation*, Math. Z. **195** (1987), no. 1, 93–97, DOI 10.1007/BF01161602. MR888130

[32] J. Bartz, M. Struwe, and R. Ye, *A new approach to the Ricci flow on S^2*, Ann. Scuola Norm. Sup. Pisa Cl. Sci. (4) **21** (1994), no. 3, 475–482. MR1310637

[33] J. Bernstein and T. Mettler, *Two-dimensional gradient Ricci solitons revisited*, Int. Math. Res. Not. IMRN **1** (2015), 78–98, DOI 10.1093/imrn/rnt177. MR3340295

[34] A. L. Besse, *Einstein manifolds*, reprint of the 1987 edition, Classics in Mathematics, Springer-Verlag, Berlin, 2008. MR2371700

[35] L. Bessières, G. Besson, S. Maillot, M. Boileau, and J. Porti, *Geometrisation of 3-manifolds*, EMS Tracts in Mathematics, vol. 13, European Mathematical Society (EMS), Zürich, 2010, DOI 10.4171/082. MR2683385

[36] A. Betancourt de la Parra, A. S. Dancer, and M. Y. Wang, *A Hamiltonian approach to the cohomogeneity one Ricci soliton equations and explicit examples of non-Kähler solitons*, J. Math. Phys. **57** (2016), no. 12, 122501, 17, DOI 10.1063/1.4972216. MR3584802

[37] C. Böhm and B. Wilking, *Manifolds with positive curvature operators are space forms*, Ann. of Math. (2) **167** (2008), no. 3, 1079–1097, DOI 10.4007/annals.2008.167.1079. MR2415394

[38] J.-M. Bony, *Principe du maximum, inégalite de Harnack et unicité du problème de Cauchy pour les opérateurs elliptiques dégénérés* (French, with English summary), Ann. Inst. Fourier (Grenoble) **19** (1969), no. fasc. 1, 277–304 xii. MR262881

[39] T. Bourni, M. Langford, and G. Tinaglia, *Convex ancient solutions to mean curvature flow*, Differential geometry in the large, London Math. Soc. Lecture Note Ser., vol. 463, Cambridge Univ. Press, Cambridge, 2021, pp. 50–74. MR4420785

[40] S. Brendle, *Ricci flow and the sphere theorem*, Graduate Studies in Mathematics, vol. 111, American Mathematical Society, Providence, RI, 2010, DOI 10.1090/gsm/111. MR2583938

[41] S. Brendle, *Uniqueness of gradient Ricci solitons*, Math. Res. Lett. **18** (2011), no. 3, 531–538, DOI 10.4310/MRL.2011.v18.n3.a13. MR2802586

[42] S. Brendle, *Rotational symmetry of self-similar solutions to the Ricci flow*, Invent. Math. **194** (2013), no. 3, 731–764, DOI 10.1007/s00222-013-0457-0. MR3127066

[43] S. Brendle, *Rotational symmetry of Ricci solitons in higher dimensions*, J. Differential Geom. **97** (2014), no. 2, 191–214. MR3231974

[44] S. Brendle, *Ricci flow with surgery in higher dimensions*, Ann. of Math. (2) **187** (2018), no. 1, 263–299, DOI 10.4007/annals.2018.187.1.6. MR3739233

[45] S. Brendle, *Ancient solutions to the Ricci flow in dimension 3*, Acta Math. **225** (2020), no. 1, 1–102, DOI 10.4310/acta.2020.v225.n1.a1. MR4176064

[46] S. Brendle, *Ancient solutions to the Ricci flow in dimension 3*, Acta Math. **225** (2020), no. 1, 1–102, DOI 10.4310/acta.2020.v225.n1.a1. MR4176064

[47] S. Brendle, *Singularity models in the three-dimensional Ricci flow*, Recent progress in mathematics, KIAS Springer Ser. Math., vol. 1, Springer, Singapore, [2022] ©2022, pp. 87–118, DOI 10.1007/978-981-19-3708-8_3. MR4510955

[48] S. Brendle and K. Choi, *Uniqueness of convex ancient solutions to mean curvature flow in higher dimensions*, Geom. Topol. **25** (2021), no. 5, 2195–2234, DOI 10.2140/gt.2021.25.2195. MR4310889

[49] S. Brendle, P. Daskalopoulos, K. Naff, and N. Sesum, *Uniqueness of compact ancient solutions to the higher-dimensional Ricci flow*, J. Reine Angew. Math. **795** (2023), 85–138, DOI 10.1515/crelle-2022-0075. MR4542417

[50] S. Brendle, P. Daskalopoulos, and N. Sesum, *Uniqueness of compact ancient solutions to three-dimensional Ricci flow*, Invent. Math. **226** (2021), no. 2, 579–651, DOI 10.1007/s00222-021-01054-0. MR4323639

[51] S. Brendle and K. Naff, *Rotational symmetry of ancient solutions to the Ricci flow in higher dimensions*, Geom. Topol. **27** (2023), no. 1, 153–226, DOI 10.2140/gt.2023.27.153. MR4584263

[52] S. Brendle and R. Schoen, *Manifolds with 1/4-pinched curvature are space forms*, J. Amer. Math. Soc. **22** (2009), no. 1, 287–307, DOI 10.1090/S0894-0347-08-00613-9. MR2449060

[53] S. Brendle and R. M. Schoen, *Classification of manifolds with weakly 1/4-pinched curvatures*, Acta Math. **200** (2008), no. 1, 1–13, DOI 10.1007/s11511-008-0022-7. MR2386107

[54] R. L. Bryant, *Ricci flow solitons in dimension three with SO(3)-symmetries*, http://www.math.duke.edu/bryant/3DRotSymRicciSolitons.pdf.

[55] R. L. Bryant, *Gradient Kähler Ricci solitons* (English, with English and French summaries), Géométrie différentielle, physique mathématique, mathématiques et société. I, Astérisque **321** (2008), 51–97. MR2521644

[56] D. Burago, Y. Burago, and S. Ivanov, *A course in metric geometry*, Graduate Studies in Mathematics, vol. 33, American Mathematical Society, Providence, RI, 2001, DOI 10.1090/gsm/033. MR1835418

[57] R. Buzano and M. Rupflin, *Smooth long-time existence of harmonic Ricci flow on surfaces*, J. Lond. Math. Soc. (2) **95** (2017), no. 1, 277–304, DOI 10.1112/jlms.12005. MR3653093

[58] E. Cabezas-Rivas and P. M. Topping, *The canonical expanding soliton and Harnack inequalities for Ricci flow*, Trans. Amer. Math. Soc. **364** (2012), no. 6, 3001–3021, DOI 10.1090/S0002-9947-2012-05391-8. MR2888237

[59] E. Cabezas-Rivas and P. M. Topping, *The canonical shrinking soliton associated to a Ricci flow*, Calc. Var. Partial Differential Equations **43** (2012), no. 1-2, 173–184, DOI 10.1007/s00526-011-0407-x. MR2886114

[60] H.-D. Cao, *Existence of gradient Kähler-Ricci solitons*, Elliptic and parabolic methods in geometry (Minneapolis, MN, 1994), A K Peters, Wellesley, MA, 1996, pp. 1–16. MR1417944

[61] H.-D. Cao, *Recent progress on Ricci solitons*, Recent advances in geometric analysis, Adv. Lect. Math. (ALM), vol. 11, Int. Press, Somerville, MA, 2010, pp. 1–38. MR2648937

[62] H.-D. Cao, *Geometry of complete gradient shrinking Ricci solitons*, Geometry and analysis. No. 1, Adv. Lect. Math. (ALM), vol. 17, Int. Press, Somerville, MA, 2011, pp. 227–246. MR2882424

[63] H.-D. Cao, B.-L. Chen, and X.-P. Zhu, *Recent developments on Hamilton's Ricci flow*, Surveys in differential geometry. Vol. XII. Geometric flows, Surv. Differ. Geom., vol. 12, Int. Press, Somerville, MA, 2008, pp. 47–112, DOI 10.4310/SDG.2007.v12.n1.a3. MR2488948

[64] H.-D. Cao and Q. Chen, *On locally conformally flat gradient steady Ricci solitons*, Trans. Amer. Math. Soc. **364** (2012), no. 5, 2377–2391, DOI 10.1090/S0002-9947-2011-05446-2. MR2888210

[65] H.-D. Cao and R. S. Hamilton, *Gradient Kähler-Ricci solitons and periodic orbits*, Comm. Anal. Geom. **8** (2000), no. 3, 517–529, DOI 10.4310/CAG.2000.v8.n3.a3. MR1775136

[66] H.-D. Cao and T. Liu, *Curvature estimates for 4-dimensional complete gradient expanding Ricci solitons*, J. Reine Angew. Math. **790** (2022), 115–135, DOI 10.1515/crelle-2022-0039. MR4472870

[67] H.-D. Cao, G. Tian, and X. Zhu, *Kähler-Ricci solitons on compact complex manifolds with $C_1(M) > 0$*, Geom. Funct. Anal. **15** (2005), no. 3, 697–719, DOI 10.1007/s00039-005-0522-y. MR2221147

[68] H.-D. Cao and N. Sesum, *A compactness result for Kähler Ricci solitons*, Adv. Math. **211** (2007), no. 2, 794–818, DOI 10.1016/j.aim.2006.09.011. MR2323545

[69] H.-D. Cao and D. Zhou, *On complete gradient shrinking Ricci solitons*, J. Differential Geom. **85** (2010), no. 2, 175–185. MR2732975

[70] H.-D. Cao and X.-P. Zhu, *A complete proof of the Poincaré and geometrization conjectures—application of the Hamilton-Perelman theory of the Ricci flow*, Asian J. Math. **10** (2006), no. 2, 165–492, DOI 10.4310/AJM.2006.v10.n2.a2. MR2233789

[71] X. Cao, B. Wang, and Z. Zhang, *On locally conformally flat gradient shrinking Ricci solitons*, Commun. Contemp. Math. **13** (2011), no. 2, 269–282, DOI 10.1142/S0219199711004191. MR2794486

[72] X. Cao and Q. S. Zhang, *The conjugate heat equation and ancient solutions of the Ricci flow*, Adv. Math. **228** (2011), no. 5, 2891–2919, DOI 10.1016/j.aim.2011.07.022. MR2838064

[73] J. A. Carrillo and L. Ni, *Sharp logarithmic Sobolev inequalities on gradient solitons and applications*, Comm. Anal. Geom. **17** (2009), no. 4, 721–753, DOI 10.4310/CAG.2009.v17.n4.a7. MR3010626

[74] G. Catino, P. Mastrolia, and D. D. Monticelli, *Classification of expanding and steady Ricci solitons with integral curvature decay*, Geom. Topol. **20** (2016), no. 5, 2665–2685, DOI 10.2140/gt.2016.20.2665. MR3556348

[75] P.-Y. Chan, *Curvature estimates for steady Ricci solitons*, Trans. Amer. Math. Soc. **372** (2019), no. 12, 8985–9008, DOI 10.1090/tran/7954. MR4029719

[76] P.-Y. Chan, *Gradient steady Kähler Ricci solitons with non-negative Ricci curvature and integrable scalar curvature*, Comm. Anal. Geom. **30** (2022), no. 2, 271–296. MR4516071

[77] P.-Y. Chan, *Curvature estimates and gap theorems for expanding Ricci solitons*, Int. Math. Res. Not. IMRN **1** (2023), 406–454, DOI 10.1093/imrn/rnab257. MR4530113

[78] P.-Y. Chan, Z. Ma, and Y. Zhang, *Hamilton-Ivey estimates for gradient Ricci solitons*, arXiv:2112.11025.

[79] P.-Y. Chan, Z. Ma, and Y. Zhang, *Volume growth estimates of gradient Ricci solitons*, J. Geom. Anal. **32** (2022), no. 12, Paper No. 291, 21, DOI 10.1007/s12220-022-01026-7. MR4487755

[80] P.-Y. Chan and B. Zhu, *On a dichotomy of the curvature decay of steady Ricci solitons. Part B*, Adv. Math. **404** (2022), Paper No. 108458, 40, DOI 10.1016/j.aim.2022.108458. MR4423811

[81] A. Chau and L.-F. Tam, *A note on the uniformization of gradient Kähler Ricci solitons*, Math. Res. Lett. **12** (2005), no. 1, 19–21, DOI 10.4310/MRL.2005.v12.n1.a2. MR2122726

[82] A. Chau and L.-F. Tam, *Gradient Kähler-Ricci solitons and complex dynamical systems*, Recent progress on some problems in several complex variables and partial differential equations, Contemp. Math., vol. 400, Amer. Math. Soc., Providence, RI, 2006, pp. 43–52, DOI 10.1090/conm/400/07529. MR2222464

[83] J. Cheeger, *Finiteness theorems for Riemannian manifolds*, Amer. J. Math. **92** (1970), 61–74, DOI 10.2307/2373498. MR263092

[84] J. Cheeger and D. G. Ebin, *Comparison theorems in Riemannian geometry*, revised reprint of the 1975 original, AMS Chelsea Publishing, Providence, RI, 2008, DOI 10.1090/chel/365. MR2394158

[85] J. Cheeger, M. Gromov, and M. Taylor, *Finite propagation speed, kernel estimates for functions of the Laplace operator, and the geometry of complete Riemannian manifolds*, J. Differential Geometry **17** (1982), no. 1, 15–53. MR658471

[86] B.-L. Chen, *Strong uniqueness of the Ricci flow*, J. Differential Geom. **82** (2009), no. 2, 363–382. MR2520796

[87] B.-L. Chen, S.-H. Tang, and X.-P. Zhu, *Complete classification of compact four-manifolds with positive isotropic curvature*, J. Differential Geom. **91** (2012), no. 1, 41–80. MR2944961

[88] B.-L. Chen, G. Xu, and Z. Zhang, *Local pinching estimates in 3-dim Ricci flow*, Math. Res. Lett. **20** (2013), no. 5, 845–855, DOI 10.4310/MRL.2013.v20.n5.a3. MR3207356

[89] B.-L. Chen and X.-P. Zhu, *A property of Kähler-Ricci solitons on complete complex surfaces*, Geometry and nonlinear partial differential equations (Hangzhou, 2001), AMS/IP Stud. Adv. Math., vol. 29, Amer. Math. Soc., Providence, RI, 2002, pp. 5–12, DOI 10.1090/amsip/029/02. MR1926430

[90] B.-L. Chen and X.-P. Zhu, *Ricci flow with surgery on four-manifolds with positive isotropic curvature*, J. Differential Geom. **74** (2006), no. 2, 177–264. MR2258799

[91] C.-W. Chen and A. Deruelle, *Structure at infinity of expanding gradient Ricci soliton*, Asian J. Math. **19** (2015), no. 5, 933–950, DOI 10.4310/AJM.2015.v19.n5.a6. MR3431684

[92] X. Chen and W. Ding, *Ricci flow on surfaces with degenerate initial metrics*, J. Partial Differential Equations **20** (2007), no. 3, 193–202. MR2348984

[93] X. Chen, P. Lu, and G. Tian, *A note on uniformization of Riemann surfaces by Ricci flow*, Proc. Amer. Math. Soc. **134** (2006), no. 11, 3391–3393, DOI 10.1090/S0002-9939-06-08360-2. MR2231924

[94] X. Chen, S. Sun, and G. Tian, *A note on Kähler-Ricci soliton*, Int. Math. Res. Not. IMRN **17** (2009), 3328–3336, DOI 10.1093/imrp/rnp056. MR2535001

[95] X. Chen and Y. Wang, *On four-dimensional anti-self-dual gradient Ricci solitons*, J. Geom. Anal. **25** (2015), no. 2, 1335–1343, DOI 10.1007/s12220-014-9471-8. MR3319974

[96] L. Cheng and Y. Zhang, *Perelman-type no breather theorem for noncompact Ricci flows*, Trans. Amer. Math. Soc. **374** (2021), no. 11, 7991–8012, DOI 10.1090/tran/8436. MR4328689

[97] O. Chodosh, *Expanding Ricci solitons asymptotic to cones*, Calc. Var. Partial Differential Equations **51** (2014), no. 1-2, 1–15, DOI 10.1007/s00526-013-0664-y. MR3247379

[98] O. Chodosh and F. T.-H. Fong, *Rotational symmetry of conical Kähler-Ricci solitons*, Math. Ann. **364** (2016), no. 3-4, 777–792, DOI 10.1007/s00208-015-1240-x. MR3466851

[99] B. Chow, *Small improvements for a trio of estimates for Ricci solitons*, Proc. Amer. Math. Soc., DOI: https://doi.org/10.1090/proc/16545, to appear.

[100] B. Chow, *The Ricci flow on the 2-sphere*, J. Differential Geom. **33** (1991), no. 2, 325–334. MR1094458

[101] B. Chow, S.-C. Chu, D. Glickenstein, C. Guenther, J. Isenberg, T. Ivey, D. Knopf, P. Lu, F. Luo, and L. Ni, *The Ricci flow: techniques and applications. Part I. Geometric aspects*, Mathematical Surveys and Monographs, vol. 135, American Mathematical Society, Providence, RI, 2007, DOI 10.1090/surv/135. MR2302600

[102] B. Chow, S.-C. Chu, D. Glickenstein, C. Guenther, J. Isenberg, T. Ivey, D. Knopf, P. Lu, F. Luo, and L. Ni, *The Ricci flow: techniques and applications. Part II. Analytic aspects*, Mathematical Surveys and Monographs, vol. 144, American Mathematical Society, Providence, RI, 2008, DOI 10.1090/surv/144. MR2365237

[103] B. Chow, S.-C. Chu, D. Glickenstein, C. Guenther, J. Isenberg, T. Ivey, D. Knopf, P. Lu, F. Luo, and L. Ni, *The Ricci flow: techniques and applications. Part III. Geometric-analytic aspects*, Mathematical Surveys and Monographs, vol. 163, American Mathematical Society, Providence, RI, 2010, DOI 10.1090/surv/163. MR2604955

[104] B. Chow, S.-C. Chu, D. Glickenstein, C. Guenther, J. Isenberg, T. Ivey, D. Knopf, P. Lu, F. Luo, and L. Ni, *The Ricci flow: techniques and applications. Part IV. Long-time solutions and related topics*, Mathematical Surveys and Monographs, vol. 206, American Mathematical Society, Providence, RI, 2015, DOI 10.1090/surv/206. MR3409114

[105] B. Chow, Y. Deng, and Z. Ma, *On four-dimensional steady gradient Ricci solitons that dimension reduce*, Adv. Math. **403** (2022), Paper No. 108367, 61, DOI 10.1016/j.aim.2022.108367. MR4405369

[106] B. Chow, M. Freedman, H. Shin, and Y. Zhang, *Curvature growth of some 4-dimensional gradient Ricci soliton singularity models*, Adv. Math. **372** (2020), 107303, 17, DOI 10.1016/j.aim.2020.107303. MR4128573

[107] B. Chow and R. Gulliver, *Aleksandrov reflection and nonlinear evolution equations. I. The n-sphere and n-ball*, Calc. Var. Partial Differential Equations **4** (1996), no. 3, 249–264, DOI 10.1007/BF01254346. MR1386736

[108] B. Chow and D. Knopf, *The Ricci flow: an introduction*, Mathematical Surveys and Monographs, vol. 110, American Mathematical Society, Providence, RI, 2004, DOI 10.1090/surv/110. MR2061425

[109] B. Chow and P. Lu, *Uniqueness of asymptotic cones of complete noncompact shrinking gradient Ricci solitons with Ricci curvature decay* (English, with English and French summaries), C. R. Math. Acad. Sci. Paris **353** (2015), no. 11, 1007–1009, DOI 10.1016/j.crma.2015.09.009. MR3419851

[110] B. Chow and P. Lu, *On κ-noncollapsed complete noncompact shrinking gradient Ricci solitons which split at infinity*, Math. Ann. **366** (2016), no. 3-4, 1195–1206, DOI 10.1007/s00208-016-1363-8. MR3563235

[111] B. Chow, P. Lu, and L. Ni, *Hamilton's Ricci flow*, Graduate Studies in Mathematics, vol. 77, American Mathematical Society, Providence, RI; Science Press Beijing, New York, 2006, DOI 10.1090/gsm/077. MR2274812

[112] B. Chow, P. Lu, and B. Yang, *Lower bounds for the scalar curvatures of noncompact gradient Ricci solitons* (English, with English and French summaries), C. R. Math. Acad. Sci. Paris **349** (2011), no. 23-24, 1265–1267, DOI 10.1016/j.crma.2011.11.004. MR2861997

[113] B. Chow, P. Lu, and B. Yang, *A necessary and sufficient condition for Ricci shrinkers to have positive AVR*, Proc. Amer. Math. Soc. **140** (2012), no. 6, 2179–2181, DOI 10.1090/S0002-9939-2011-11173-0. MR2888203

[114] B. Chow and L.-F. Wu, *The Ricci flow on compact 2-orbifolds with curvature negative somewhere*, Comm. Pure Appl. Math. **44** (1991), no. 3, 275–286, DOI 10.1002/cpa.3160440302. MR1090433

[115] S.-C. Chu, *Basic properties of gradient Ricci solitons*, Geometric evolution equations, Contemp. Math., vol. 367, Amer. Math. Soc., Providence, RI, 2005, pp. 79–102, DOI 10.1090/conm/367/06749. MR2112631

[116] S.-C. Chu, *Geometry of 3-dimensional gradient Ricci solitons with positive curvature*, Comm. Anal. Geom. **13** (2005), no. 1, 129–150. MR2154669

[117] S.-C. Chu, *Type II ancient solutions to the Ricci flow on surfaces*, Comm. Anal. Geom. **15** (2007), no. 1, 195–215. MR2301253

[118] T. H. Colding and W. P. Minicozzi, II, *Singularities of Ricci flow and diffeomorphisms*, arXiv:2109.06240.

[119] T. H. Colding and W. P. Minicozzi II, *Estimates for the extinction time for the Ricci flow on certain 3-manifolds and a question of Perelman*, J. Amer. Math. Soc. **18** (2005), no. 3, 561–569, DOI 10.1090/S0894-0347-05-00486-8. MR2138137

[120] T. H. Colding and W. P. Minicozzi II, *Singularities and diffeomorphisms*, ICCM Not. **10** (2022), no. 1, 112–116. MR4484469

[121] R. J. Conlon and A. Deruelle, *Expanding Kähler-Ricci solitons coming out of Kähler cones*, J. Differential Geom. **115** (2020), no. 2, 303–365, DOI 10.4310/jdg/1589853627. MR4100705

[122] R. J. Conlon, A. Deruelle, and S. Sun, *Classification results for expanding and shrinking gradient Kähler-Ricci solitons*, Geometry and Topology, to appear, arXiv:1904.00147.

[123] D. Cooper, C. D. Hodgson, and S. P. Kerckhoff, *Three-dimensional orbifolds and cone-manifolds*, with a postface by Sadayoshi Kojima, MSJ Memoirs, vol. 5, Mathematical Society of Japan, Tokyo, 2000. MR1778789

[124] A. S. Dancer and M. Y. Wang, *On Ricci solitons of cohomogeneity one*, Ann. Global Anal. Geom. **39** (2011), no. 3, 259–292, DOI 10.1007/s10455-010-9233-1. MR2769300

[125] P. Daskalopoulos and R. Hamilton, *Geometric estimates for the logarithmic fast diffusion equation*, Comm. Anal. Geom. **12** (2004), no. 1-2, 143–164. MR2074874

[126] P. Daskalopoulos, R. Hamilton, and N. Sesum, *Classification of ancient compact solutions to the Ricci flow on surfaces*, J. Differential Geom. **91** (2012), no. 2, 171–214. MR2971286

[127] P. Daskalopoulos and N. Sesum, *Eternal solutions to the Ricci flow on* $\mathbb{R}^2$, Int. Math. Res. Not., Art. ID 83610, 20, 2006, DOI 10.1155/IMRN/2006/83610. MR2264733

[128] Y. Deng and X. Zhu, *Three-dimensional steady gradient Ricci solitons with linear curvature decay*, Int. Math. Res. Not. IMRN **4** (2019), 1108–1124, DOI 10.1093/imrn/rnx155. MR3915297

[129] Y. Deng and X. Zhu, *Classification of gradient steady Ricci solitons with linear curvature decay*, Sci. China Math. **63** (2020), no. 1, 135–154, DOI 10.1007/s11425-019-1548-0. MR4050577

[130] Y. Deng and X. Zhu, *Higher dimensional steady Ricci solitons with linear curvature decay*, J. Eur. Math. Soc. (JEMS) **22** (2020), no. 12, 4097–4120, DOI 10.4171/jems/1003. MR4176787

[131] Y. Deng and X. Zhu, *Rigidity of* κ*-noncollapsed steady Kähler-Ricci solitons*, Math. Ann. **377** (2020), no. 1-2, 847–861, DOI 10.1007/s00208-019-01807-6. MR4099621

[132] A. Derdzinski, *A Myers-type theorem and compact Ricci solitons*, Proc. Amer. Math. Soc. **134** (2006), no. 12, 3645–3648, DOI 10.1090/S0002-9939-06-08422-X. MR2240678

[133] A. Deruelle, *Steady gradient Ricci soliton with curvature in* L^1, Comm. Anal. Geom. **20** (2012), no. 1, 31–53, DOI 10.4310/CAG.2012.v20.n1.a2. MR2903100

[134] A. Deruelle, *Smoothing out positively curved metric cones by Ricci expanders*, Geom. Funct. Anal. **26** (2016), no. 1, 188–249, DOI 10.1007/s00039-016-0360-0. MR3494489

[135] A. Deruelle, *Asymptotic estimates and compactness of expanding gradient Ricci solitons*, Ann. Sc. Norm. Super. Pisa Cl. Sci. (5) **17** (2017), no. 2, 485–530. MR3700376

[136] A. Deruelle, *Unique continuation at infinity for conical Ricci expanders*, Int. Math. Res. Not. IMRN **10** (2017), 3107–3147, DOI 10.1093/imrn/rnw110. MR3658133

[137] D. M. DeTurck, *Deforming metrics in the direction of their Ricci tensors*, J. Differential Geom. **18** (1983), no. 1, 157–162. MR697987

[138] D. M. DeTurck, *Deforming metrics in the direction of their Ricci tensors* (improved version), Collected Papers on the Ricci Flow, Series in Geometry and Topology, vol. 37, Int. Press, Somerville, MA, 2003, pp. 163–165.

[139] D. M. DeTurck and J. L. Kazdan, *Some regularity theorems in Riemannian geometry*, Ann. Sci. École Norm. Sup. (4) **14** (1981), no. 3, 249–260. MR644518

[140] S. K. Donaldson and P. B. Kronheimer, *The geometry of four-manifolds*, Oxford Mathematical Monographs, Oxford Science Publications, The Clarendon Press, Oxford University Press, New York, 1990. MR1079726

[141] G. Drugan, P. Lu, and Y. Yuan, *Rigidity of complete entire self-shrinking solutions to Kähler-Ricci flow*, Int. Math. Res. Not. IMRN **12** (2015), 3908–3916, DOI 10.1093/imrn/rnu051. MR3356743

[142] N. Dunford and J. T. Schwartz, *Linear operators. Part I, General theory*, with the assistance of William G. Bade and Robert G. Bartle, reprint of the 1958 original, Wiley Classics Library, A Wiley-Interscience Publication, John Wiley & Sons, Inc., New York, 1988. MR1009162

[143] L. P. Eisenhart, *Riemannian geometry*, 8th printing, Princeton Paperbacks, Princeton Landmarks in Mathematics, Princeton University Press, Princeton, NJ, 1997. MR1487892

[144] M. Eminenti, G. La Nave, and C. Mantegazza, *Ricci solitons: the equation point of view*, Manuscripta Math. **127** (2008), no. 3, 345–367, DOI 10.1007/s00229-008-0210-y. MR2448435

[145] L. C. Evans, *Partial differential equations*, 2nd ed., Graduate Studies in Mathematics, vol. 19, American Mathematical Society, Providence, RI, 2010, DOI 10.1090/gsm/019. MR2597943

[146] F.-Q. Fang, J.-W. Man, and Z.-L. Zhang, *Complete gradient shrinking Ricci solitons have finite topological type* (English, with English and French summaries), C. R. Math. Acad. Sci. Paris **346** (2008), no. 11-12, 653–656, DOI 10.1016/j.crma.2008.03.021. MR2423272

[147] M. Feldman, T. Ilmanen, and D. Knopf, *Rotationally symmetric shrinking and expanding gradient Kähler-Ricci solitons*, J. Differential Geom. **65** (2003), no. 2, 169–209. MR2058261

[148] M. Fernández-López and E. García-Río, *A remark on compact Ricci solitons*, Math. Ann. **340** (2008), no. 4, 893–896, DOI 10.1007/s00208-007-0173-4. MR2372742

[149] M. Fernández-López and E. García-Río, *A sharp lower bound for the scalar curvature of certain steady gradient Ricci solitons*, Proc. Amer. Math. Soc. **141** (2013), no. 6, 2145–2148, DOI 10.1090/S0002-9939-2013-11675-8. MR3034440

[150] M. H. Freedman, *The topology of four-dimensional manifolds*, J. Differential Geometry **17** (1982), no. 3, 357–453. MR679066

[151] D. Friedan, *Nonlinear models in $2 + \varepsilon$ dimensions*, Phys. Rev. Lett. **45** (1980), no. 13, 1057–1060, DOI 10.1103/PhysRevLett.45.1057. MR584365

[152] M. Furuta, *Monopole equation and the $\frac{11}{8}$-conjecture*, Math. Res. Lett. **8** (2001), no. 3, 279–291, DOI 10.4310/MRL.2001.v8.n3.a5. MR1839478

[153] D. Gabai, *The Smale conjecture for hyperbolic 3-manifolds:* Isom(M^3) $\simeq$ Diff(M^3), J. Differential Geom. **58** (2001), no. 1, 113–149. MR1895350

[154] H. Ge and W. Jiang, *ϵ-regularity for shrinking Ricci solitons and Ricci flows*, Geom. Funct. Anal. **27** (2017), no. 5, 1231–1256, DOI 10.1007/s00039-017-0420-0. MR3714720

[155] G. Giesen and P. M. Topping, *Ricci flow of negatively curved incomplete surfaces*, Calc. Var. Partial Differential Equations **38** (2010), no. 3-4, 357–367, DOI 10.1007/s00526-009-0290-x. MR2647124

[156] G. Giesen and P. M. Topping, *Existence of Ricci flows of incomplete surfaces*, Comm. Partial Differential Equations **36** (2011), no. 10, 1860–1880, DOI 10.1080/03605302.2011.558555. MR2832165

[157] D. Gilbarg and N. S. Trudinger, *Elliptic partial differential equations of second order*, reprint of the 1998 edition, Classics in Mathematics, Springer-Verlag, Berlin, 2001. MR1814364

[158] D. Glickenstein, *Riemannian groupoids and solitons for three-dimensional homogeneous Ricci and cross-curvature flows*, Int. Math. Res. Not. IMRN **12** (2008), Art. ID rnn034, 49, DOI 10.1093/imrn/rnn034. MR2426751

[159] H. Grauert, *On Levi's problem and the imbedding of real-analytic manifolds*, Ann. of Math. (2) **68** (1958), 460–472, DOI 10.2307/1970257. MR98847

[160] R. E. Greene and H. Wu, *Lipschitz convergence of Riemannian manifolds*, Pacific J. Math. **131** (1988), no. 1, 119–141. MR917868

[161] D. Gromoll and W. Meyer, *On complete open manifolds of positive curvature*, Ann. of Math. (2) **90** (1969), 75–90, DOI 10.2307/1970682. MR247590

[162] M. Gromov, *Metric structures for Riemannian and non-Riemannian spaces*, reprint of the 2001 English edition, based on the 1981 French original; with appendices by M. Katz, P. Pansu, and S. Semmes; translated from the French by Sean Michael Bates, Modern Birkhäuser Classics, Birkhäuser Boston, Inc., Boston, MA, 2007. MR2307192

[163] H.-L. Gu and X.-P. Zhu, *The existence of type II singularities for the Ricci flow on S^{n+1}*, Comm. Anal. Geom. **16** (2008), no. 3, 467–494. MR2429966

[164] P. Guan, P. Lu, and Y. Xu, *A rigidity theorem for codimension one shrinking gradient Ricci solitons in $\mathbb{R}^{n+1}$*, Calc. Var. Partial Differential Equations **54** (2015), no. 4, 4019–4036, DOI 10.1007/s00526-015-0929-8. MR3426102

[165] C. Guenther, J. Isenberg, and D. Knopf, *Linear stability of homogeneous Ricci solitons*, Int. Math. Res. Not., Art. ID 96253, 30, 2006, DOI 10.1155/IMRN/2006/96253. MR2264732

[166] B. Guo, D. H. Phong, J. Song, and J. Sturm, *Compactness of Kähler-Ricci solitons on Fano manifolds*, Pure Appl. Math. Q. **18** (2022), no. 1, 305–316, DOI 10.4310/PAMQ.2022.v18.n1.a9. MR4381854

[167] H. Guo, *Area growth rate of the level surface of the potential function on the 3-dimensional steady gradient Ricci soliton*, Proc. Amer. Math. Soc. **137** (2009), no. 6, 2093–2097, DOI 10.1090/S0002-9939-09-09792-5. MR2480291

[168] H. Guo, *Remarks on noncompact steady gradient Ricci solitons*, Math. Ann. **345** (2009), no. 4, 883–894, DOI 10.1007/s00208-009-0387-8. MR2545871

[169] M. Gutperle, M. Headrick, S. Minwalla, and V. Schomerus, *Spacetime energy decreases under world-sheet RG flow*, J. High Energy Phys. **1** (2003), 073, 20, DOI 10.1088/1126-6708/2003/01/073. MR1969853

[170] M. Guysinsky, B. Hasselblatt, and V. Rayskin, *Differentiability of the Hartman-Grobman linearization*, Discrete Contin. Dyn. Syst. **9** (2003), no. 4, 979–984, DOI 10.3934/dcds.2003.9.979. MR1975364

[171] S. J. Hall and T. Murphy, *On the linear stability of Kähler-Ricci solitons*, Proc. Amer. Math. Soc. **139** (2011), no. 9, 3327–3337, DOI 10.1090/S0002-9939-2011-10948-1. MR2811287

[172] R. S. Hamilton, *Harmonic maps of manifolds with boundary*, Lecture Notes in Mathematics, Vol. 471, Springer-Verlag, Berlin-New York, 1975. MR0482822

[173] R. S. Hamilton, *Three-manifolds with positive Ricci curvature*, J. Differential Geometry **17** (1982), no. 2, 255–306. MR664497

[174] R. S. Hamilton, *Four-manifolds with positive curvature operator*, J. Differential Geom. **24** (1986), no. 2, 153–179. MR862046

[175] R. S. Hamilton, *The Ricci flow on surfaces*, Mathematics and general relativity (Santa Cruz, CA, 1986), Contemp. Math., vol. 71, Amer. Math. Soc., Providence, RI, 1988, pp. 237–262, DOI 10.1090/conm/071/954419. MR954419

[176] R. S. Hamilton, *Eternal solutions to the Ricci flow*, J. Differential Geom. **38** (1993), no. 1, 1–11. MR1231700

[177] R. S. Hamilton, *The Harnack estimate for the Ricci flow*, J. Differential Geom. **37** (1993), no. 1, 225–243. MR1198607

[178] R. S. Hamilton, *A compactness property for solutions of the Ricci flow*, Amer. J. Math. **117** (1995), no. 3, 545–572, DOI 10.2307/2375080. MR1333936

[179] R. S. Hamilton, *The formation of singularities in the Ricci flow*, Surveys in differential geometry, Vol. II (Cambridge, MA, 1993), Int. Press, Cambridge, MA, 1995, pp. 7–136. MR1375255

[180] R. S. Hamilton, *An isoperimetric estimate for the Ricci flow on the two-sphere*, Modern methods in complex analysis (Princeton, NJ, 1992), Ann. of Math. Stud., vol. 137, Princeton Univ. Press, Princeton, NJ, 1995, pp. 191–200, DOI 10.1080/09502389500490321. MR1369139

[181] R. S. Hamilton, *Four-manifolds with positive isotropic curvature*, Comm. Anal. Geom. **5** (1997), no. 1, 1–92, DOI 10.4310/CAG.1997.v5.n1.a1. MR1456308

[182] R. S. Hamilton, *Non-singular solutions of the Ricci flow on three-manifolds*, Comm. Anal. Geom. **7** (1999), no. 4, 695–729, DOI 10.4310/CAG.1999.v7.n4.a2. MR1714939

[183] R. Haslhofer, *Uniqueness of the bowl soliton*, Geom. Topol. **19** (2015), no. 4, 2393–2406, DOI 10.2140/gt.2015.19.2393. MR3375531

[184] R. Haslhofer and O. Hershkovits, *Ancient solutions of the mean curvature flow*, Comm. Anal. Geom. **24** (2016), no. 3, 593–604, DOI 10.4310/CAG.2016.v24.n3.a6. MR3521319

[185] R. Haslhofer and R. Müller, *A compactness theorem for complete Ricci shrinkers*, Geom. Funct. Anal. **21** (2011), no. 5, 1091–1116, DOI 10.1007/s00039-011-0137-4. MR2846384

[186] A. E. Hatcher, *A proof of the Smale conjecture*, Diff(S^3) $\simeq$ O(4), Ann. of Math. (2) **117** (1983), no. 3, 553–607, DOI 10.2307/2007035. MR701256

[187] N. J. Hicks, *Notes on differential geometry*, Van Nostrand Mathematical Studies, No. 3, D. Van Nostrand Co., Inc., Princeton, N.J.-Toronto-London, 1965. MR0179691

[188] D. Hoffman, T. Ilmanen, F. Martín, and B. White, *Notes on translating solitons for mean curvature flow*, Minimal surfaces: integrable systems and visualisation, Springer Proc. Math. Stat., vol. 349, Springer, Cham, [2021] ©2021, pp. 147–168, DOI 10.1007/978-3-030-68541-6_9. MR4281668

[189] S. Hong, J. Kalliongis, D. McCullough, and J. H. Rubinstein, *Diffeomorphisms of elliptic 3-manifolds*, Lecture Notes in Mathematics, vol. 2055, Springer, Heidelberg, 2012, DOI 10.1007/978-3-642-31564-0. MR2976322

[190] G. Huisken, *Contracting convex hypersurfaces in Riemannian manifolds by their mean curvature*, Invent. Math. **84** (1986), no. 3, 463–480, DOI 10.1007/BF01388742. MR837523

[191] J. Isenberg, R. Mazzeo, and N. Sesum, *Ricci flow on asymptotically conical surfaces with nontrivial topology*, J. Reine Angew. Math. **676** (2013), 227–248, DOI 10.1515/crelle.2011.186. MR3028760

[192] T. Ivey, *Ricci solitons on compact three-manifolds*, Differential Geom. Appl. **3** (1993), no. 4, 301–307, DOI 10.1016/0926-2245(93)90008-O. MR1249376

[193] T. Ivey, *New examples of complete Ricci solitons*, Proc. Amer. Math. Soc. **122** (1994), no. 1, 241–245, DOI 10.2307/2160866. MR1207538

[194] T. Ivey, *Ricci solitons on compact Kähler surfaces*, Proc. Amer. Math. Soc. **125** (1997), no. 4, 1203–1208, DOI 10.1090/S0002-9939-97-03624-1. MR1353388

[195] T. A. Ivey, *On solitons for the Ricci flow*, ProQuest LLC, Ann Arbor, MI, 1992. Thesis (Ph.D.)–Duke University. MR2688811

[196] T. A. Ivey, *Local existence of Ricci solitons*, Manuscripta Math. **91** (1996), no. 2, 151–162, DOI 10.1007/BF02567946. MR1411650

[197] M. Jablonski, *Homogeneous Ricci solitons*, J. Reine Angew. Math. **699** (2015), 159–182, DOI 10.1515/crelle-2013-0044. MR3305924

[198] L. Ji, R. Mazzeo, and N. Sesum, *Ricci flow on surfaces with cusps*, Math. Ann. **345** (2009), no. 4, 819–834, DOI 10.1007/s00208-009-0377-x. MR2545867

[199] J. L. Kazdan and F. W. Warner, *Curvature functions for compact 2-manifolds*, Ann. of Math. (2) **99** (1974), 14–47, DOI 10.2307/1971012. MR343205

[200] J. R. King, *Exact polynomial solutions to some nonlinear diffusion equations*, Phys. D **64** (1993), no. 1-3, 35–65, DOI 10.1016/0167-2789(93)90248-Y. MR1214546

[201] B. Kleiner and J. Lott, *Notes on Perelman's papers*, Geom. Topol. **12** (2008), no. 5, 2587–2855, DOI 10.2140/gt.2008.12.2587. MR2460872

[202] S. Kobayashi and K. Nomizu, *Foundations of differential geometry. Vol. I*, reprint of the 1963 original, A Wiley-Interscience Publication, Wiley Classics Library, John Wiley & Sons, Inc., New York, 1996. MR1393940

[203] N. Koiso, *On rotationally symmetric Hamilton's equation for Kähler-Einstein metrics*, Recent topics in differential and analytic geometry, Adv. Stud. Pure Math., vol. 18, Academic Press, Boston, MA, 1990, pp. 327–337, DOI 10.2969/aspm/01810327. MR1145263

[204] B. Kotschwar, *On rotationally invariant shrinking Ricci solitons*, Pacific J. Math. **236** (2008), no. 1, 73–88, DOI 10.2140/pjm.2008.236.73. MR2398988

[205] B. L. Kotschwar, *A local version of Bando's theorem on the real-analyticity of solutions to the Ricci flow*, Bull. Lond. Math. Soc. **45** (2013), no. 1, 153–158, DOI 10.1112/blms/bds074. MR3033963

[206] B. Kotschwar, *Time-analyticity of solutions to the Ricci flow*, Amer. J. Math. **137** (2015), no. 2, 535–576, DOI 10.1353/ajm.2015.0012. MR3337803

[207] B. Kotschwar, *Kählerity of shrinking gradient Ricci solitons asymptotic to Kähler cones*, J. Geom. Anal. **28** (2018), no. 3, 2609–2623, DOI 10.1007/s12220-017-9922-0. MR3833809

[208] B. Kotschwar, *Identifying shrinking solitons by their asymptotic geometries*, Mean curvature flow, De Gruyter Proc. Math., De Gruyter, Berlin, [2020] ©2020, pp. 99–108, DOI 10.1515/9783110618365-010. MR4205016

[209] B. Kotschwar and L. Wang, *A uniqueness theorem for asymptotically cylindrical shrinking Ricci solitons*, arXiv:1712.03185.

[210] B. Kotschwar and L. Wang, *Rigidity of asymptotically conical shrinking gradient Ricci solitons*, J. Differential Geom. **100** (2015), no. 1, 55–108. MR3326574

[211] K. Kröncke, *Stability and instability of Ricci solitons*, Calc. Var. Partial Differential Equations **53** (2015), no. 1-2, 265–287, DOI 10.1007/s00526-014-0748-3. MR3336320

[212] Y. Lai, *3d flying wings for any asymptotic cones*, arXiv:2207.02714.

[213] Y. Lai, *O(2)-symmetry of 3d steady gradient Ricci solitons*, arXiv:2205.01146.

[214] Y. Lai, *A family of 3d steady gradient solitons that are flying wings*, J. Differential Geom., to appear, arXiv:2010.07272v2, 2022.

[215] J. Lauret, *Ricci soliton homogeneous nilmanifolds*, Math. Ann. **319** (2001), no. 4, 715–733, DOI 10.1007/PL00004456. MR1825405

[216] J. M. Lee, *Introduction to smooth manifolds*, 2nd ed., Graduate Texts in Mathematics, vol. 218, Springer, New York, 2013. MR2954043

[217] P. Li, *Geometric analysis*, Cambridge Studies in Advanced Mathematics, vol. 134, Cambridge University Press, Cambridge, 2012, DOI 10.1017/CBO9781139105798. MR2962229

[218] P. Li and S.-T. Yau, *On the parabolic kernel of the Schrödinger operator*, Acta Math. **156** (1986), no. 3-4, 153–201, DOI 10.1007/BF02399203. MR834612

[219] X. Li, L. Ni, and K. Wang, *Four-dimensional gradient shrinking solitons with positive isotropic curvature*, Int. Math. Res. Not. IMRN **3** (2018), 949–959, DOI 10.1093/imrn/rnw269. MR3801452

[220] Y. Li, *Ricci flow on asymptotically Euclidean manifolds*, Geom. Topol. **22** (2018), no. 3, 1837–1891, DOI 10.2140/gt.2018.22.1837. MR3780446

[221] Y. Li and B. Wang, *On Kähler Ricci shrinker surfaces*, arXiv:2301.09784.

[222] F. Lin and X. Yang, *Geometric measure theory—an introduction*, Advanced Mathematics (Beijing/Boston), vol. 1, Science Press Beijing, Beijing; International Press, Boston, MA, 2002. MR2030862

[223] J. Lott, *On the long-time behavior of type-III Ricci flow solutions*, Math. Ann. **339** (2007), no. 3, 627–666, DOI 10.1007/s00208-007-0127-x. MR2336062

[224] J. Lott and P. Wilson, *Note on asymptotically conical expanding Ricci solitons*, Proc. Amer. Math. Soc. **145** (2017), no. 8, 3525–3529, DOI 10.1090/proc/13611. MR3652804

[225] R. Mazzeo, Y. A. Rubinstein, and N. Sesum, *Ricci flow on surfaces with conic singularities*, Anal. PDE **8** (2015), no. 4, 839–882, DOI 10.2140/apde.2015.8.839. MR3366005

[226] J. D. Moore, *Lectures on Seiberg-Witten invariants*, 2nd ed., Lecture Notes in Mathematics, vol. 1629, Springer-Verlag, Berlin, 2001. MR1830497

[227] J. Morgan and G. Tian, *Ricci flow and the Poincaré conjecture*, Clay Mathematics Monographs, vol. 3, American Mathematical Society, Providence, RI; Clay Mathematics Institute, Cambridge, MA, 2007, DOI 10.1305/ndjfl/1193667709. MR2334563

[228] J. Morgan and G. Tian, *The geometrization conjecture*, Clay Mathematics Monographs, vol. 5, American Mathematical Society, Providence, RI; Clay Mathematics Institute, Cambridge, MA, 2014. MR3186136

[229] J. W. Morgan, *The Seiberg-Witten equations and applications to the topology of smooth four-manifolds*, Mathematical Notes, vol. 44, Princeton University Press, Princeton, NJ, 1996. MR1367507

[230] C. B. Morrey Jr., *The analytic embedding of abstract real-analytic manifolds*, Ann. of Math. (2) **68** (1958), 159–201, DOI 10.2307/1970048. MR99060

[231] J. Moser, *On Harnack's theorem for elliptic differential equations*, Comm. Pure Appl. Math. **14** (1961), 577–591, DOI 10.1002/cpa.3160140329. MR159138

[232] O. Munteanu, *The volume growth of complete gradient shrinking Ricci solitons*, arXiv:0904.0798.

[233] O. Munteanu, C.-J. Anna Sung, and J. Wang, *Poisson equation on complete manifolds*, Adv. Math. **348** (2019), 81–145, DOI 10.1016/j.aim.2019.03.019. MR3925015

[234] O. Munteanu and N. Sesum, *On gradient Ricci solitons*, J. Geom. Anal. **23** (2013), no. 2, 539–561, DOI 10.1007/s12220-011-9252-6. MR3023848

[235] O. Munteanu and J. Wang, *Smooth metric measure spaces with non-negative curvature*, Comm. Anal. Geom. **19** (2011), no. 3, 451–486, DOI 10.4310/CAG.2011.v19.n3.a1. MR2843238

[236] O. Munteanu and J. Wang, *Analysis of weighted Laplacian and applications to Ricci solitons*, Comm. Anal. Geom. **20** (2012), no. 1, 55–94, DOI 10.4310/CAG.2012.v20.n1.a3. MR2903101

[237] O. Munteanu and J. Wang, *Geometry of manifolds with densities*, Adv. Math. **259** (2014), 269–305, DOI 10.1016/j.aim.2014.03.023. MR3197658

[238] O. Munteanu and J. Wang, *Holomorphic functions on Kähler-Ricci solitons*, J. Lond. Math. Soc. (2) **89** (2014), no. 3, 817–831, DOI 10.1112/jlms/jdt078. MR3217651

[239] O. Munteanu and J. Wang, *Geometry of shrinking Ricci solitons*, Compos. Math. **151** (2015), no. 12, 2273–2300, DOI 10.1112/S0010437X15007496. MR3433887

[240] O. Munteanu and J. Wang, *Topology of Kähler Ricci solitons*, J. Differential Geom. **100** (2015), no. 1, 109–128. MR3326575

[241] O. Munteanu and J. Wang, *Conical structure for shrinking Ricci solitons*, J. Eur. Math. Soc. (JEMS) **19** (2017), no. 11, 3377–3390, DOI 10.4171/JEMS/741. MR3713043

[242] O. Munteanu and J. Wang, *Positively curved shrinking Ricci solitons are compact*, J. Differential Geom. **106** (2017), no. 3, 499–505, DOI 10.4310/jdg/1500084024. MR3680555

[243] O. Munteanu and J. Wang, *Structure at infinity for shrinking Ricci solitons* (English, with English and French summaries), Ann. Sci. Éc. Norm. Supér. (4) **52** (2019), no. 4, 891–925, DOI 10.11650/tjm/180806. MR4038455

[244] O. Munteanu and M.-T. Wang, *The curvature of gradient Ricci solitons*, Math. Res. Lett. **18** (2011), no. 6, 1051–1069, DOI 10.4310/MRL.2011.v18.n6.a2. MR2915467

[245] A. Naber, *Noncompact shrinking four solitons with nonnegative curvature*, J. Reine Angew. Math. **645** (2010), 125–153, DOI 10.1515/CRELLE.2010.062. MR2673425

[246] L. Ni, *Ancient solutions to Kähler-Ricci flow*, Math. Res. Lett. **12** (2005), no. 5-6, 633–653, DOI 10.4310/MRL.2005.v12.n5.a3. MR2189227

[247] L. Ni and N. Wallach, *On a classification of gradient shrinking solitons*, Math. Res. Lett. **15** (2008), no. 5, 941–955, DOI 10.4310/MRL.2008.v15.n5.a9. MR2443993

[248] L. Ni and N. Wallach, *On four-dimensional gradient shrinking solitons*, Int. Math. Res. Not. IMRN **4** (2008), Art. ID rnm152, 13, DOI 10.1093/imrn/rnm152. MR2424175

[249] M. Obata, *Certain conditions for a Riemannian manifold to be isometric with a sphere*, J. Math. Soc. Japan **14** (1962), 333–340, DOI 10.2969/jmsj/01430333. MR142086

[250] G. Perelman, *The entropy formula for the Ricci flow and its geometric applications*, arXiv:math.DG/0211159.

[251] G. Perelman, *Ricci flow with surgery on three-manifolds*, arXiv:math.DG/0303109.

[252] G. Perelman, *Finite extinction time for the solutions to the Ricci flow on certain three-manifolds*, arXiv:math.DG/0307245.

[253] S. Peters, *Convergence of Riemannian manifolds*, Compositio Math. **62** (1987), no. 1, 3–16. MR892147

[254] P. Petersen and W. Wylie, *On gradient Ricci solitons with symmetry*, Proc. Amer. Math. Soc. **137** (2009), no. 6, 2085–2092, DOI 10.1090/S0002-9939-09-09723-8. MR2480290

[255] P. Petersen and W. Wylie, *Rigidity of gradient Ricci solitons*, Pacific J. Math. **241** (2009), no. 2, 329–345, DOI 10.2140/pjm.2009.241.329. MR2507581

[256] P. Petersen and W. Wylie, *On the classification of gradient Ricci solitons*, Geom. Topol. **14** (2010), no. 4, 2277–2300, DOI 10.2140/gt.2010.14.2277. MR2740647

[257] D. H. Phong, J. Song, and J. Sturm, *Degeneration of Kähler-Ricci solitons on Fano manifolds*, Univ. Iagel. Acta Math. **52** (2015), 29–43, DOI 10.4467/20843828AM.15.004.3730. MR3438282

[258] S. Pigola, M. Rimoldi, and A. G. Setti, *Remarks on non-compact gradient Ricci solitons*, Math. Z. **268** (2011), no. 3-4, 777–790, DOI 10.1007/s00209-010-0695-4. MR2818729

[259] D. Ramos, *Gradient Ricci solitons on surfaces*, arXiv:1304.6391.

[260] D. Ramos, *Ricci flow on cone surfaces*, Port. Math. **75** (2018), no. 1, 11–65, DOI 10.4171/PM/2010. MR3854890

[261] T. Richard, *Canonical smoothing of compact Aleksandrov surfaces via Ricci flow* (English, with English and French summaries), Ann. Sci. Éc. Norm. Supér. (4) **51** (2018), no. 2, 263–279, DOI 10.24033/asens.2356. MR3798303

[262] P. Rosenau, *On fast and super-fast diffusion*, Phys. Rev. Lett. **74** (1995), 1056–1059.

[263] F. Schulze and M. Simon, *Expanding solitons with non-negative curvature operator coming out of cones*, Math. Z. **275** (2013), no. 1-2, 625–639, DOI 10.1007/s00209-013-1150-0. MR3101823

[264] W.-X. Shi, *Deforming the metric on complete Riemannian manifolds*, J. Differential Geom. **30** (1989), no. 1, 223–301. MR1001277

[265] M. Spivak, *A comprehensive introduction to differential geometry. Vol. III*, Publish or Perish, Inc., Boston, Mass., 1975. MR0372756

[266] J. Streets, *Ricci Yang-Mills flow on surfaces*, Adv. Math. **223** (2010), no. 2, 454–475, DOI 10.1016/j.aim.2009.08.014. MR2565538

[267] M. Struwe, *Curvature flows on surfaces*, Ann. Sc. Norm. Super. Pisa Cl. Sci. (5) **1** (2002), no. 2, 247–274. MR1991140

[268] R. Takahashi, *Asymptotic stability for Kähler-Ricci solitons*, Math. Z. **281** (2015), no. 3-4, 1021–1034, DOI 10.1007/s00209-015-1518-4. MR3421651

[269] G. Teschl, *Ordinary differential equations and dynamical systems*, Graduate Studies in Mathematics, vol. 140, American Mathematical Society, Providence, RI, 2012, DOI 10.1090/gsm/140. MR2961944

[270] W. P. Thurston, *Three-dimensional geometry and topology. Vol. 1*, edited by Silvio Levy, Princeton Mathematical Series, vol. 35, Princeton University Press, Princeton, NJ, 1997. MR1435975

[271] G. Tian and Z. Zhang, *Degeneration of Kähler-Ricci solitons*, Int. Math. Res. Not. IMRN **5** (2012), 957–985, DOI 10.1093/imrn/rnr036. MR2899957

[272] G. Tian and X. Zhu, *Uniqueness of Kähler-Ricci solitons on compact Kähler manifolds* (English, with English and French summaries), C. R. Acad. Sci. Paris Sér. I Math. **329** (1999), no. 11, 991–995, DOI 10.1016/S0764-4442(00)88625-5. MR1733907

[273] G. Tian and X. Zhu, *Uniqueness of Kähler-Ricci solitons*, Acta Math. **184** (2000), no. 2, 271–305, DOI 10.1007/BF02392630. MR1768112

[274] G. Tian and X. Zhu, *A new holomorphic invariant and uniqueness of Kähler-Ricci solitons*, Comment. Math. Helv. **77** (2002), no. 2, 297–325, DOI 10.1007/s00014-002-8341-3. MR1915043

[275] V. A. Toponogov, *Riemannian spaces containing straight lines* (Russian), Dokl. Akad. Nauk SSSR **127** (1959), 977–979. MR0108808

[276] P. Topping, *Lectures on the Ricci flow*, London Mathematical Society Lecture Note Series, vol. 325, Cambridge University Press, Cambridge, 2006, DOI 10.1017/CBO9780511721465. MR2265040

[277] P. M. Topping, *Uniqueness and nonuniqueness for Ricci flow on surfaces: reverse cusp singularities*, Int. Math. Res. Not. IMRN **10** (2012), 2356–2376, DOI 10.1093/imrn/rnr082. MR2923169

[278] P. M. Topping and H. Yin, *Smoothing a measure on a Riemann surface using Ricci flow*, arXiv:2107.14686.

[279] P. M. Topping and H. Yin, *Sharp decay estimates for the logarithmic fast diffusion equation and the Ricci flow on surfaces*, Ann. PDE **3** (2017), no. 1, Paper No. 6, 16, DOI 10.1007/s40818-017-0024-x. MR3625191

[280] K. K. Uhlenbeck, *Removable singularities in Yang-Mills fields*, Comm. Math. Phys. **83** (1982), no. 1, 11–29. MR648355

[281] N. Šešum, *Curvature tensor under the Ricci flow*, Amer. J. Math. **127** (2005), no. 6, 1315–1324. MR2183526

[282] N. Sesum, *Convergence of the Ricci flow toward a soliton*, Comm. Anal. Geom. **14** (2006), no. 2, 283–343. MR2255013

[283] X.-J. Wang and X. Zhu, *Kähler-Ricci solitons on toric manifolds with positive first Chern class*, Adv. Math. **188** (2004), no. 1, 87–103, DOI 10.1016/j.aim.2003.09.009. MR2084775

[284] B. Weber, *Convergence of compact Ricci solitons*, Int. Math. Res. Not. IMRN **1** (2011), 96–118, DOI 10.1093/imrn/rnq055. MR2755484

[285] G. Wei and W. Wylie, *Comparison geometry for the Bakry-Emery Ricci tensor*, J. Differential Geom. **83** (2009), no. 2, 377–405, DOI 10.4310/jdg/1261495336. MR2577473

[286] H. Whitney, *Differentiable manifolds*, Ann. of Math. (2) **37** (1936), no. 3, 645–680, DOI 10.2307/1968482. MR1503303

[287] E. Witten, *String theory and black holes*, Phys. Rev. D (3) **44** (1991), no. 2, 314–324, DOI 10.1103/PhysRevD.44.314. MR1117173

[288] J.-Y. Wu, *Comparison geometry for integral Bakry-Émery Ricci tensor bounds*, J. Geom. Anal. **29** (2019), no. 1, 828–867, DOI 10.1007/s12220-018-0020-8. MR3897035

[289] J.-Y. Wu, *Correction to: Comparison geometry for integral Bakry-Émery Ricci tensor bounds*, J. Geom. Anal. **30** (2020), no. 4, 4464–4465, DOI 10.1007/s12220-020-00450-x. MR4167291

[290] L.-F. Wu, *The Ricci flow on 2-orbifolds with positive curvature*, J. Differential Geom. **33** (1991), no. 2, 575–596. MR1094470

[291] L.-F. Wu, *The Ricci flow on complete* $\mathbf{R}^2$, Comm. Anal. Geom. **1** (1993), no. 3-4, 439–472, DOI 10.4310/CAG.1993.v1.n3.a4. MR1266475

[292] W. Wylie, *Complete shrinking Ricci solitons have finite fundamental group*, Proc. Amer. Math. Soc. **136** (2008), no. 5, 1803–1806, DOI 10.1090/S0002-9939-07-09174-5. MR2373611

[293] B. Yang, *A characterization of noncompact Koiso-type solitons*, Internat. J. Math. **23** (2012), no. 5, 1250054, 13, DOI 10.1142/S0129167X12500541. MR2914656

[294] N. Yang, *A note on nonnegative Bakry-Émery Ricci curvature*, Arch. Math. (Basel) **93** (2009), no. 5, 491–496, DOI 10.1007/s00013-009-0062-z. MR2563596

[295] H. Yin, *Normalized Ricci flow on nonparabolic surfaces*, Ann. Global Anal. Geom. **36** (2009), no. 1, 81–104, DOI 10.1007/s10455-008-9150-8. MR2520032

[296] H. Yin, *Ricci flow on surfaces with conical singularities*, J. Geom. Anal. **20** (2010), no. 4, 970–995, DOI 10.1007/s12220-010-9136-1. MR2683772

[297] T. Yokota, *Curvature integrals under the Ricci flow on surfaces*, Geom. Dedicata **133** (2008), 169–179, DOI 10.1007/s10711-008-9241-5. MR2390075

[298] T. Yokota, *Perelman's reduced volume and a gap theorem for the Ricci flow*, Comm. Anal. Geom. **17** (2009), no. 2, 227–263, DOI 10.4310/CAG.2009.v17.n2.a3. MR2520908

[299] T. Yokota, *Addendum to 'Perelman's reduced volume and a gap theorem for the Ricci flow' [MR2520908]*, Comm. Anal. Geom. **20** (2012), no. 5, 949–955, DOI 10.4310/CAG.2012.v20.n5.a2. MR3053617

[300] Q. S. Zhang, *Sobolev inequalities, heat kernels under Ricci flow, and the Poincaré conjecture*, CRC Press, Boca Raton, FL, 2011. MR2676347

[301] S. J. Zhang, *On a sharp volume estimate for gradient Ricci solitons with scalar curvature bounded below*, Acta Math. Sin. (Engl. Ser.) **27** (2011), no. 5, 871–882, DOI 10.1007/s10114-011-9527-7. MR2786449

[302] Y. Zhang, *Compactness theorems for 4-dimensional gradient Ricci solitons*, Pacific J. Math. **303** (2019), no. 1, 361–384, DOI 10.2140/pjm.2019.303.361. MR4045370

[303] Z.-H. Zhang, *On the completeness of gradient Ricci solitons*, Proc. Amer. Math. Soc. **137** (2009), no. 8, 2755–2759, DOI 10.1090/S0002-9939-09-09866-9. MR2497489

[304] Z.-L. Zhang, *Compact blow-up limits of finite time singularities of Ricci flow are shrinking Ricci solitons* (English, with English and French summaries), C. R. Math. Acad. Sci. Paris **345** (2007), no. 9, 503–506, DOI 10.1016/j.crma.2007.09.017. MR2375111

[305] Z. Zhang, *On the finiteness of the fundamental group of a compact shrinking Ricci soliton*, Colloq. Math. **107** (2007), no. 2, 297–299, DOI 10.4064/cm107-2-9. MR2284167

[306] Z. Zhang, *Degeneration of shrinking Ricci solitons*, Int. Math. Res. Not. IMRN **21** (2010), 4137–4158, DOI 10.1093/imrn/rnq020. MR2738353

[307] X. Zhu, *Kähler-Ricci soliton typed equations on compact complex manifolds with* $C_1(M) > 0$, J. Geom. Anal. **10** (2000), no. 4, 759–774, DOI 10.1007/BF02921996. MR1817785

Index

SELECTED PUBLISHED TITLES IN THIS SERIES

For a complete list of titles in this series, visit the
AMS Bookstore at **www.ams.org/bookstore/gsmseries/**.